MASONRY CONSTRUCTION IN ACTIVE SEISMIC REGIONS

Woodhead Publishing Series in Civil and Structural Engineering

MASONRY CONSTRUCTION IN ACTIVE SEISMIC REGIONS

Edited by

RAJESH RUPAKHETY

Earthquake Engineering Research Center, Faculty of Civil and Environmental Engineering, School of Engineering and Natural Sciences, University of Iceland, Selfoss, Iceland

DIPENDRA GAUTAM

Department of Architecture and Civil Engineering, City University of Hong Kong, Kowloon, Hong Kong

ELSEVIER

WP
WOODHEAD PUBLISHING
An imprint of Elsevier

Woodhead Publishing is an imprint of Elsevier
The Officers' Mess Business Centre, Royston Road, Duxford, CB22 4QH, United Kingdom
50 Hampshire Street, 5th Floor, Cambridge, MA 02139, United States
The Boulevard, Langford Lane, Kidlington, OX5 1GB, United Kingdom

Library of Congress Cataloging-in-Publication Data
A catalog record for this book is available from the Library of Congress

British Library Cataloguing-in-Publication Data
A catalogue record for this book is available from the British Library

ISBN: 978-0-12-821087-1 (print)
ISBN: 978-0-12-823196-8 (online)

For information on all Woodhead publications
visit our website at https://www.elsevier.com/books-and-journals

The front cover photo was taken by Yogi Kayastha at Rimi, Dolpa, Nepal

Publisher: Matthew Deans
Acquisitions Editor: Glyn Jones
Editorial Project Manager: Mariana C. Henriques
Production Project Manager: Vignesh Tamil
Cover Designer: Victoria Pearson

Typeset by SPi Global, India

Contents

11. Local out-of-plane failure modes in traditional block-masonry buildings

Claudia Casapulla

12. Mexican colonial churches: structural assessment and seismic behavior

Fernando Peña, Marcos M. Chávez, and Natalia García

13. Features and seismic response of large masonry structures: A case study of Singh Durbar main building, Nepal

Rishi Ram Parajuli

Contributors

Rabindra Adhikari Department of Civil Engineering, Institute of Engineering, Pulchowk Campus, Lalitpur, Nepal

Pilar Baquedano Department of Civil Engineering, ISISE, Institute of Science and Innovation for Bio-Sustainability (IB-S), University of Minho, Guimarães, Portugal

Rewati Baruwal Interdisciplinary Research Institute for Sustainability, Kathmandu, Nepal

Lalit Bhatt Interdisciplinary Research Institute for Sustainability, Kathmandu, Nepal

Claudia Casapulla Department of Structures for Engineering and Architecture, University of Naples Federico II, Naples, Italy

Marcos M. Chávez Instituto de Ingeniería, Universidad Nacional Autónoma de México, Mexico City, Mexico

Nimesh Chettri Civil Engineering and Architecture Department, College of Science and Technology, Royal University of Bhutan, Phuentsholing, Bhutan

Carlos Cuadra Akita Prefectural University, Yurihonjo, Japan

Andreea Duțu Technical University of Civil Engineering Bucharest, Bucharest, Romania

Rafael Ramírez Eudave Department of Civil Engineering, ISISE, Institute of Science and Innovation for Bio-Sustainability (IB-S), University of Minho, Guimarães, Portugal

Tiago Miguel Ferreira Department of Civil Engineering, ISISE, Institute of Science and Innovation for Bio-Sustainability (IB-S), University of Minho, Guimarães, Portugal

Stefano Galassi Department of Architecture, University of Florence, Florence, Italy

Natalia García Universidad Autónoma del Estado de Morelos, Cuernavaca, Mexico

Dipendra Gautam Department of Architecture and Civil Engineering, City University of Hong Kong, Kowloon, Hong Kong

Sandra Graus Department of Civil Engineering, ISISE, Institute of Science and Innovation for Bio-Sustainability (IB-S), University of Minho, Guimarães, Portugal

Polat Gülkan Başkent University, Ankara, Turkey

Jeevan Gwachha Lecturer Kathford International College of Engineering and Management, Lalitpur, Nepal

Yin Myo Min Htwe Department of Meteorology and Hydrology, Ministry of Transport and Communications, Nay Pyi Taw, Myanmar

Randolph Langenbach Conservationtech Consulting, Oakland, CA, United States

Paulo B. Lourenço Department of Civil Engineering, ISISE, Institute of Science and Innovation for Bio-Sustainability (IB-S), University of Minho, Guimarães, Portugal

Nuno Mendes Department of Civil Engineering, ISISE, Institute of Science and Innovation for Bio-Sustainability (IB-S), University of Minho, Guimarães, Portugal

Fabiana N. Miranda Department of Civil Engineering, ISISE, Institute of Science and Innovation for Bio-Sustainability (IB-S), University of Minho, Guimarães, Portugal

Teraphan Ornthammarath Department of Civil and Environmental Engineering, Faculty of Engineering, Mahidol University, Nakhon Pathom, Thailand

Javier Ortega Department of Civil Engineering, ISISE, Institute of Science and Innovation for Bio-Sustainability (IB-S), University of Minho, Guimarães, Portugal

Enrico Pagano Department of Architecture, Roma Tre University, Rome, Italy

Rishi Ram Parajuli University of Bristol, Bristol, United Kingdom

Fernando Peña Instituto de Ingeniería, Universidad Nacional Autónoma de México, Mexico City, Mexico

Nakhorn Poovarodom Department of Civil Engineering, Faculty of Engineering, Thammasat University, Bangkok, Thailand

Hugo Rodrigues RISCO, Department of Civil Engineering, University of Aveiro, Aveiro, Portugal

Nicola Ruggieri Superintendence of Archeology, Fine Arts and Landscape for the Province of Cosenza, Cosenza, Italy

Rajesh Rupakhety Earthquake Engineering Research Center, Faculty of Civil and Environmental Engineering, School of Engineering and Natural Sciences, University of Iceland, Selfoss, Iceland

Ginevra Salerno Department of Architecture, Roma Tre University, Rome, Italy

Manjip Shakya Khwopa Engineering College, Bhaktapur, Nepal

Binay Charan Shrestha Department of Urban Development and Building Construction, Government of Nepal, Kathmandu, Nepal

Rui Silva Department of Civil Engineering, ISISE, Institute of Science and Innovation for Bio-Sustainability (IB-S), University of Minho, Guimarães, Portugal

Giacomo Tempesta Department of Architecture, University of Florence, Florence, Italy

Miha Tomaževič Slovenian National Building and Civil Engineering Institute, Ljubljana, Slovenia

Saw Htwe Zaw Myanmar Earthquake Committee, Yangon, Myanmar

About the editors

Rajesh Rupakhety is a Professor in the Civil and Environmental Engineering Faculty and Director of Research at the Earthquake Engineering Research Centre at the University of Iceland. He obtained his PhD from the University of Iceland in earthquake engineering. He has a wide variety of research interests, mainly in solid and fluid mechanics and in engineering seismology. He is very active in the field and has worked on seismic design provisions for hydropower and wind plants in Iceland. He has published widely on construction and seismology.

Dipendra Gautam is a researcher in structural earthquake engineering. He has published more than 50 papers and refereed book chapters in top-tier journals and edited volumes. He works in the range of structural earthquake engineering, including structural dynamics, seismic vulnerability and reliability analysis, sustainable structural engineering, vernacular seismic resistant construction technologies, risk assessment, among others. He has experience in working after 2011 Eastern Nepal earthquake, 2015 Gorkha earthquake sequence, and 2016 Central Italy earthquake sequence.

CHAPTER

1

Testing and experimental simulation of seismic behavior for modeling and seismic resistance verification of masonry buildings

Miha Tomaževič

Slovenian National Building and Civil Engineering Institute, Ljubljana, Slovenia

1.1 Introduction

Masonry is a composite structural material, consisting of masonry units of various types and shapes, made of different materials and bound together in various ways with various types of mortar or glue. In the case of traditional masonry construction, stone and clay brick, either burned or sun-dried, prevail, with mortars of relatively poor quality, from mud mortar to lime, and sometimes, lime-cement mortar is used as a binding material. In the case of contemporary masonry, other materials, such as steel or polymer reinforcement and cement grout, are sometimes added to improve the load-bearing properties of masonry. Various traditional and modern technologies of laying the units, that is, types of masonry bond, can be found. Taking this into consideration, masonry can be considered as a typical representative of nonhomogeneous and anisotropic materials with limited elastic characteristics.

Although basic testing of masonry is required by design codes and standards (e.g., EN 1052-1, 1998; EN 1052-3, 2002) and a number of empirical equations have been proposed to calculate the mechanical properties of masonry on the basis of mechanical properties of constituent materials (e.g., EN 1996-1-1, 2005), additional specific testing of structural elements and assemblages is needed, not only to obtain actual values of mechanical

https://doi.org/10.1016/B978-0-12-821087-1.00012-0

properties of masonry but also to assess the values of parameters that determine seismic behavior of masonry structures, such as displacement and energy dissipation capacity, and others. The experience indicated that the correlation between the code-based calculations and experimental values of basic mechanical properties, such as compressive strength, is adequate for the type of masonry on the basis of which the equations have been derived but is not so good for the case of other masonry types (Tomaževič, 1999).

Testing is also needed to understand the behavior of masonry buildings and causes of earthquake damage, observed on the buildings subjected to strong earthquakes. On the basis of observations and quantified data measured during experimental simulation of the observed mechanisms, reliable mathematical models for structural verification can be developed. In case of the postearthquake recovery of existing masonry buildings and seismic rehabilitation projects, testing is the only means to verify the efficiency of methods for strengthening and repair prior to put them to examination by the real earthquake events, expected in the future. Last but not the least, experimental research is the basis for assessing the values of many parameters (coefficients/factors) given in the codes, which implicitly take into account the inelastic properties of masonry materials and structural systems. Determined by experiments, these parameters allow that equivalent static analysis methods for earthquake resistance verification be used instead of time-consuming and complex, scientific »exact« methods.

It is not the aim of this contribution to present state-of-the-art overview as regards the methods of seismic testing of masonry structural components and systems, as well as the methods for their seismic resistance verification. Only some basic concepts will be discussed based on the experience obtained at Slovenian National Building and Civil Engineering Institute, (ZAG) in Ljubljana, Slovenia, where experimental research in seismic behavior of masonry structures has been one of the major concerns in the past 60 years.

1.2 Seismic testing of masonry walls

1.2.1 General

Structural system of a masonry building consists of vertical elements, walls and piers, connected together at floor levels with horizontal elements, floors and roofs. In normal situations, walls and piers transfer vertical loads, carried by floors and roofs, into the foundation system. However, they should also resist in-plane and out-of-plane horizontal loads, induced by winds and earthquakes. Horizontal elements carry the vertical loads and distribute them to the walls. In the case of an earthquake, they distribute the induced loads to the walls and ensure integrity of the structure; namely, by vibrating as a box, masonry structure can utilize most of the available resistance of structural and cross-walls.

Masonry is a traditional building material, originally aimed at carrying vertical loads. It can be stressed relatively high in compression but is not a material suitable for carrying bending and shear, which develop in the structure when subjected to earthquakes. As a result, damage to traditional buildings in the case of moderate earthquakes was severe, and often, collapse on a large scale took place in the case of the strong ones. However, there were buildings observed which survived the earthquakes suffering only minor damage or even undamaged. On the basis of damage analysis and testing of materials, structural elements and/or buildings simulating the observed behavior, the reasons for poor or adequate behavior could be pointed out. The following conclusions can be made regarding the failure mechanisms of unreinforced masonry (URM) walls:

- When subjected to seismic loads acting in the plane of the wall, three main failure modes can be distinguished (Fig. 1.1): sliding shear mode, which is characterized by sliding of part of the wall on horizontal crack; diagonal tension shear mode, characterized by typical diagonally oriented cracks; and flexural mode, characterized by crushing of masonry at the compressed corner (toe). As the earthquake damage observations indicate, diagonal tension shear mode is predominant.
- When subjected to seismic loads orthogonal to its plane (out-of-plane), the failure mode depends on boundary conditions. Free standing walls are not stable. Due to small flexural resistance, overturning of such walls takes place along the bottom support (e.g., gable end walls).

A classic example of storey mechanism and shear failure of plain brick masonry walls without damage to spandrels is shown in Fig. 1.2. Following this mechanism, typical loads, acting on a masonry pier in seismic situation, are schematically presented in Fig. 1.3. This situation represents the basis for design of testing arrangements.

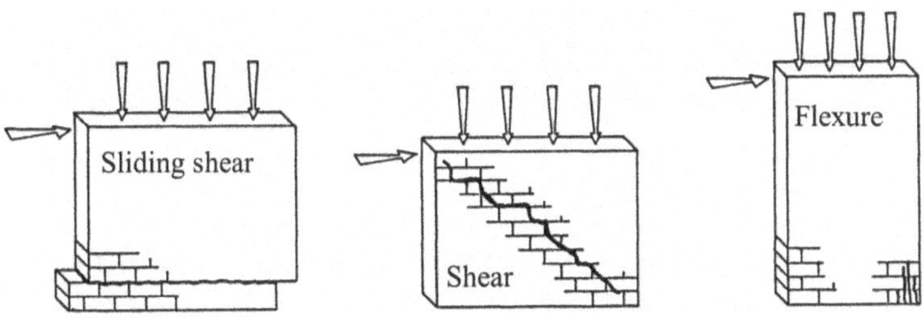

FIG. 1.1 Typical failure modes of unreinforced masonry walls (Tomaževič, 1999).

FIG. 1.2 Typical shear failure mechanism of an unreinforced brick masonry building (Montenegro earthquake, 1979).

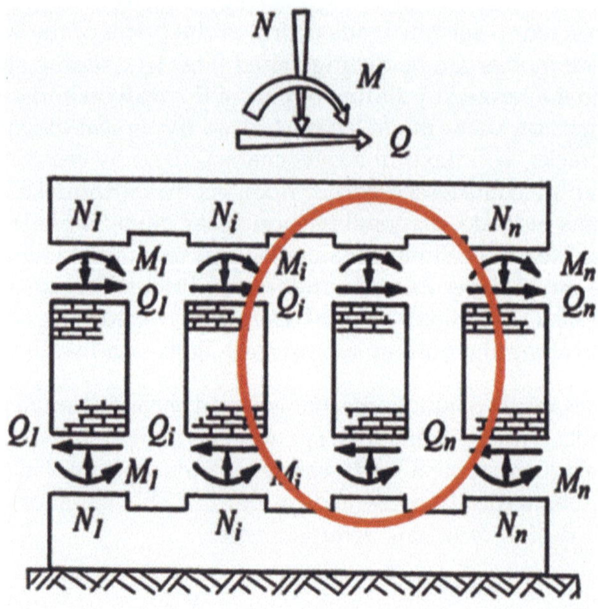

FIG. 1.3 Forces acting on a masonry pier during earthquake. *(After Tomaževič, M. (1999).* Earthquake resistant design of masonry buildings. *London: Imperial College Press/World Scientific.)*

To determine the parameters of seismic resistance, the observed failure mechanisms are simulated by testing the walls either in the laboratory or on site. Knowing the geometry of the walls, boundary restraints and loads acting on the walls during testing, as well as measuring the resulting response, equations that describe the observed damage patterns and failure mechanisms can be derived, and parameters to be used in these equations can be evaluated.

Whenever possible, the specimens are built and tested in the laboratory. In the case of existing masonry, however, it is not easy to fully reproduce the existing masonry walls in the laboratory, although thorough mechanical and chemical analyses of the properties of constituent materials, bricks, stone, and especially, mortar, had been previously carried out. Therefore original specimens are tested either in situ or cut from the existing buildings, transported to, and tested in the laboratory.

Typically, walls of height/length geometry aspect ratio of 1.0–1.5 are tested (Calvi, Kingsley, & Magenes, 1996; Mayes & Clough, 1975; Tomaževič, 2016), subjected to cyclic reversals of imposed displacements and constant vertical precompression. Cyclic shear tests are preferred because besides information regarding resistance capacity, they also provide information regarding the displacement and energy dissipation capacity of the structure. It should be borne in mind that, if the walls are tested under conditions that have little in common with seismic situation, the results of the tests cannot be directly applied in seismic resistance analysis.

1.2.2 Boundary restraints

In buildings, restraint conditions may change during the earthquake due to progressive damage to the wall and surrounding structural elements. Additional vertical stresses develop in the wall because of prevented rotation when subjected to lateral load. Because the real situation in the buildings is difficult to simulate and to avoid any uncertainties, which would prevent accurate evaluation of the measured parameters, tests are carried out under simplified but controlled boundary conditions. Usually, vertical load is kept constant during the test, whereas the in-plane acting lateral load is applied in the form of induced cyclically alternating displacements. Using displacements and not forces to control the tests makes possible the specimens be tested in the nonlinear range up until their final collapse.

Analyzing typical damage patterns occurred to URM walls subjected to earthquakes, it was believed that the actual situation will be best simulated if the upper and lower boundaries of the wall remain parallel during the test. Consequently, mechanical devices have been developed, usually in the form of pantographs, placed either on the top of the wall or laterally, which should keep the top and bottom boundaries of the specimen parallel during the tests (Fig. 1.4A). They transfer lateral

FIG. 1.4 Cyclic shear testing of masonry walls in structural laboratory of ZAG. (A) Testing of stone masonry using pantograph and (B) cyclic shear test of a block masonry wall, tested as vertical cantilever.

loads in the form of forces or displacements, induced by the actuators, to the tested specimens by means of horizontally moveable parts of the device, placed on the walls.

In the laboratory, test specimens are usually built on reinforced concrete (r.c.) foundation blocks, with r.c. bond beam at the top. Because the elements of the pantograph, placed on the walls, are not restrained to move vertically, rotation of the URM walls takes places in the usual case of the vertical load being kept constant during the test. Rocking of the specimen, which is usually not observed when testing low strength masonry walls, depends on the type and quality of masonry, the geometry of specimen (height/length ratio), and precompression ratio (ratio between the working compressive stresses and compressive strength of masonry). This means that, actually, the URM walls are not tested as fixed but as symmetrically restrained at both ends.

The walls can be tested also as simple vertical cantilevers (Fig. 1.4B). In this case, the foundation block on which the specimen is built is fixed to the testing floor. Lateral load is imposed by means of a programmable hydraulic actuator, connected to the bond beam on the top of the wall. To allow free motion of the upper part of the wall during cycling, rollers are placed between the steel elements placed on the top of the specimen for distribution of vertical load. However the arrangement for application, the vertical load restrains free rotation of the upper boundary of the walls. It has been found that in case where the vertical load is applied by two jacks, the position of moment inflection point is located in most cases somewhere in the upper half of the walls and sometimes near the mid-height of the specimen (Tomaževič & Gams, 2009). Actually, the walls are not tested as vertical cantilevers but as partly restrained, also at the top.

As a result of the testing method, phenomena typical for flexural mechanism can be observed in the initial phases of testing. In the case of very slender walls, however, where no diagonal cracks occur in the upper part of the wall and the resistance degrades after crushing of masonry at the compressed toes, this mechanism actually represents the flexural failure of the wall. In most other cases, however, the lateral resistance of the wall continues increasing until the principal tensile stresses in the mid-part of the wall cause diagonal cracks and splitting of the panel, that is, shear failure.

No such damage is observed when testing existing masonry walls in situ. In the case of the in situ test, the specimen is separated from the surrounding masonry at an appropriate place by vertically cutting the wall at both sides. To prevent the occurrence of damage to surrounding structural system, a strong enough part of the wall (reaction wall) should be found next to the test specimen to laterally support the actuator. In the vertical direction, the surrounding part of the floor structure is supported with wooden posts or otherwise to prevent the accidental collapse of the floor in the case of possible collapse of the specimen.

The specimen is instrumented with displacement- and strain-transducers and subjected to lateral load, simulating the effect of seismic action. Hydraulic actuator, placed at the mid-height and connected to the wall with a system of steel rods and supporting beams, is used to impose lateral displacements. Typical arrangement of testing a stone masonry wall is shown in Fig. 1.5, whereas details of damage to a brick masonry wall are shown in Fig. 1.6. In no case, horizontal tension cracks or crushing of units at supports have been observed (Sheppard & Tomaževič, 1985; Tomaževič, Klemenc, & Lutman, 2000).

Recent tests of a full-scale URM three-storey shear wall with openings, constructed with hollow clay units using general purpose mortar (Triller, Tomaževič, & Gams, 2016) confirmed the observation: observed crack pattern and digital optical measurements indicated pure shear behavior of the central masonry pier (Fig. 1.7).

Nowadays, mechanical devices are replaced by a system of computer-controlled hydraulic actuators (Fig. 1.8). A recent study regarding the influence of testing control on the test results has shown that a method where the resultant of axial forces acting at both vertical borders of walls is kept constant, the position of moment inflection point is fixed at the mid-height of the wall, but vertical deformations at the upper boundary

FIG. 1.5 Typical arrangement of an in situ test of a stone masonry wall (Tomaževič et al., 2000).

FIG. 1.6 Diagonal but no horizontal cracks or crushing of the bricks have been observed when testing a brick masonry wall at a very low compression in situ. Lower half of the tested wall is shown in the figure (Sheppard & Tomaževič, 1985).

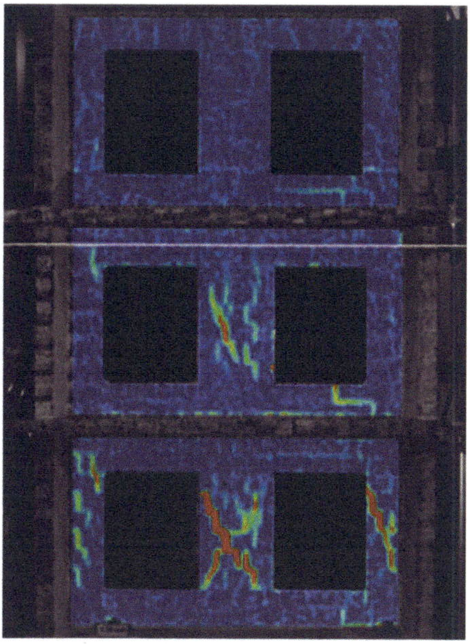

FIG. 1.7 Full-scale lateral resistance test of a three-storey block masonry building. Digital optical measurements did not indicate rotation of the middle pier at supports (Triller et al., 2016).

FIG. 1.8 Cyclic shear testing of masonry walls in structural laboratory of ZAG: testing of a block masonry wall where the upper boundary is restrained with a system of vertically acting actuators.

of the wall are allowed for the most reliable simulates the actual conditions of the wall in the masonry structural system (Triller, 2019).

Therefore when evaluating laboratory test results, one has to take into consideration the influence of testing arrangement and method on test results. Secondary damage, which precedes the final damage pattern and ultimate failure mode, does not have much impact on the values of attainable resistance; however, it could significantly influence the values of the measured deformability parameters. Unfortunately, it is not easy to filter out the effects of rocking and secondary damage from the measured total displacements by traditional measuring methods. It is expected that up-to-date digital optical measuring methods will be helpful in this regard.

1.2.3 Loading protocol

Because of the cyclic character of seismic loads, various cyclic lateral load patterns, applied statically or dynamically, are used to simulate the seismic loads in different laboratories. Usually, the walls are tested under constant vertical load, the magnitude of which is of the order of the expected working stresses in structural walls, which vary in the range of 5%–30% of the compressive strength of masonry (i.e., precompression ratio varies from 0.05 to 0.30). To make possible testing in the nonlinear range, horizontal displacements and not forces are imposed, simulating cyclic character of vibration of buildings during earthquake. There is no standardized displacement pattern; however, the most widely used one is of the shape of the three times repeated reversals of the same amplitude, which is gradually increased until the final collapse of the specimen. This way, strength and stiffness deterioration at repeated loading of the same amplitude can be studied.

The influence of frequency of application of cyclic lateral loads on the strength and ductility of URM walls has been already studied decades ago (Terčelj, Sheppard, & Turnšek, 1977; Williams & Scrivener, 1974). Static and dynamic responses of model masonry houses have been correlated (Paulson & Abrams, 1990), and a systematic study has been conducted also at the Slovenian National Building and Civil Engineering Institute (Tomaževič, Lutman, & Petković, 1996).

Owing to the nonlinearity and nonhomogeneity of masonry material, masonry walls are not perfectly elastic even in small deformation range. Changes in resistance and crack distribution in time are observed if the imposed lateral deformation is kept constant, resulting in strength and stiffness degradation at repeated displacement amplitudes at cycling. The phenomena are significant in the nonlinear range, which indicate the sensitivity of test results to the shape and frequency of time history

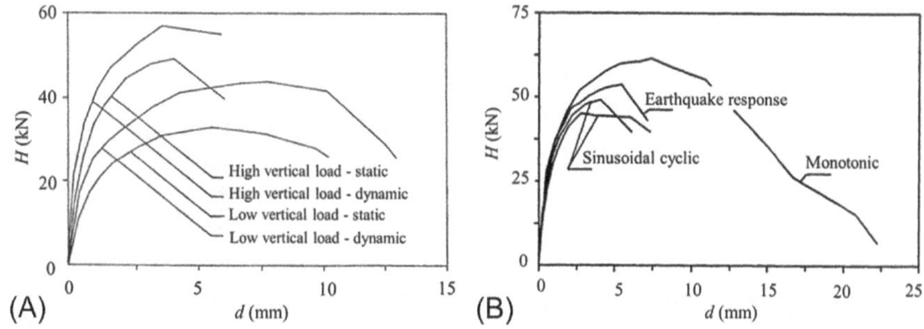

FIG. 1.9 Influence of testing procedures on the resistance and displacement capacity of masonry walls. (A) The influence of precompression and imposed strain rates at monotonic application of lateral loads and (B) the difference between various cycling patterns and monotonic application of lateral loads at the same level of precompression and imposed strain rates. *(After Tomaževič, M., Lutman, M., Petković, L. (1996). Seismic behavior of masonry walls: Experimental simulation.* Journal of Structural Engineering, ASCE, 122, *1040–1047.)*

of imposed lateral displacements (lateral load protocol) used for the simulation of seismic loads. The following main conclusions have been obtained regarding the influence of different loading procedures on the basic parameters of seismic resistance of walls (see Fig. 1.9):

- Higher values of lateral resistance and larger ultimate displacements have been obtained in the case of monotonic than in the case of cyclic loading procedures of any type.
- Higher values of lateral resistance have been obtained dynamically than statically; however, there was not much difference between static and dynamic protocols as regards the displacement capacity at the same level of vertical load.
- The level of vertical load was predominant parameter in all cases. When subjected to higher level of vertical load, lateral resistance of the wall was improved, but displacement capacity decreased at both static and dynamic types of loading, at all load patterns. The walls subjected to higher level of vertical load were more rigid than the walls tested at low vertical load, and larger stiffness values were determined by dynamic than by static tests. However, displacement and energy dissipation capacity were lower. The failure mechanism and propagation of damage were also affected by the level of the imposed vertical load.

The conclusions are in good agreement with previous and recent researches (Mergos & Beyer, 2014). It is, therefore, of relevant importance that the testing methods for the determination of parameters of seismic resistance of masonry walls to be used in the analysis are compatible with the testing methods, on the basis of which the design procedures have been developed.

1.3 Testing of small-scale models on simple earthquake simulators

To understand the seismic behavior of masonry structural systems and propose measures to ensure adequate earthquake resistance of buildings, testing of individual walls is not enough. For this purpose, testing of entire buildings in simulated earthquake conditions is necessary.

Nowadays, development of high technologies made possible the installation of large, multidegree of freedom shaking tables, which are capable of driving large masses of prototype-sized structures with a high degree of accuracy of simulation of the spatial, recorded, or artificial seismic ground motion. There is no better way to simulate seismic behavior of a structure in the laboratory. Unfortunately, costs of installation, maintenance, as well as experiments by using such devices are high. Therefore structures at reduced scale, that is, physical models of structures, are still tested on simple earthquake simulators in many laboratories.

There are advantages and disadvantages of model tests. Such tests are cheaper than full-scale tests. Testing facilities are simpler and, in many cases, hydraulic actuators, used to drive the platform when used as a shaking table, can be dismantled and used for other type of testing. The construction of models is cheaper, although many times not simpler than the construction of prototype; namely, the development of adequate model materials, the characteristics of which depend on the modeling scale, is a time-consuming, trial-and-error procedure. Of course, modeling techniques can be adjusted to possibilities. However, any modification requires caution when interpreting the test results.

Generally, only the overall behavior of structural system and its global failure mechanism can be determined by testing small-scale models and not the behavior of structural elements and details. In most cases, the possibility of modeling details limits the reduction of the size of the masonry building models. In the case where simple simulators are used, the excitation is usually one-directional, which provides better control of the observed mechanisms of structural behavior. However, one has to bear in mind that sometimes, seismic actions in the other directions (orthogonal to the plane of the elements, vertical direction) influence the observed mechanism.

Accuracy of simulation of seismic motion represents a technological problem of test control. As the experience shows, it is important that the induced shaking's spectral characteristics are similar to actual characteristics of earthquakes (Tomaževič & Velechovsky, 1992). Once the actual shaking table motion is accurately recorded, exact matching of the input and output motions is not of relevant importance because the actually recorded motion of the shaking table represents the earthquake to be considered as the imposed motion in the analysis and not the modeled earthquake, putting aside the fact that earthquakes with the same dynamic characteristics, duration, and acceleration time histories never occur twice.

The modeling techniques and similitude conditions are discussed elsewhere (e.g., Harris & Sabnis, 1999). Basic relationships between physical quantities measured on the prototype and model are determined on the basis of dimensional analysis (Langhaar, 1951). If a general quantity q_M has been measured on the model, the following relationship applies for the quantity q_P, which refers to the prototype:

$$q_P = q_M S_q, \tag{1.1}$$

where S_q is a scale factor. In the case where elastic behavior of the structure under static loads is studied, the relationships between the prototype and model structures are straightforward. However, if the dynamic behavior in the nonlinear range and failure mechanisms is investigated, they depend on the physical modeling technique. Depending on the mechanical characteristics of model materials with regard to the prototype ones, there are basically three possibilities:

- In the case where materials are used, whose strength properties are reduced at geometry scale, but other characteristics, such as strains, specific mass, and material damping, remain the same as in the case of the prototype, the models are called complete (true) models.
- In the case where the prototype materials are used, the models are called simple models.
- In the most frequent cases where not all mechanical properties of model material can be scaled according to the laws of the complete model similitude, such models are called distorted (general) models.

The theoretically obtained scale factors, referring to the characteristic physical quantities that determine the dynamic behavior, can be found in the literature (e.g., Harris & Sabnis, 1999; Langhaar, 1951; Tomaževič & Velechovsky, 1992). Generally speaking, however, to obtain the similitude in seismic behavior of models and prototypes, two basic requirements should be fulfilled:

- The similitude in dynamic behavior requires that the distribution of masses and stiffnesses along the height of the model and prototype are similar.
- The similitude in failure mechanisms, however, requires the same working stress level, that is, working stress/compressive strength of masonry ratio in masonry walls of the model and prototype structure.

In the case of the complete model, the requirements for similitude of mass distribution and failure mechanism are simultaneously fulfilled by fixing additional mass, which simulate the missing dead and live loads of floors to the floors of the model. Lead bricks, steel ingots, or concrete blocks, rigidly fixed to the floors, can be used. In the case of the simple models, however, by fulfilling the requirement for similitude of mass distribution, the level of working stresses in the load-bearing walls does not attain the level required for similitude of failure mechanism. Because the failure mode and resistance of masonry walls significantly depend on working stress level (e.g., Turnšek & Čačovič, 1970), the compressive stresses in the walls should be adjusted accordingly, however, without affecting the dynamic similitude, that is, required similitude in mass and stiffness distribution. Massless prestressing the walls is one of the possible solutions, using steel ropes fixed to the top slab and anchored into the foundation of the model. Soft springs at the top end of the ropes are used to keep the prestressing forces virtually constant during the shaking tests (Tomaževič & Velechovsky, 1992).

Although the mechanical properties of constituent materials, model masonry units, and mortar are modeled to a required degree of accuracy, the final correlation between the model and prototype masonry is assessed by comparing the results of tests of model and prototype structural elements, walls at both compression and cyclic shear loads. Prior to shaking table model tests, scaled masonry wallets are tested. Because the seismic behavior is concerned, besides the material parameters such as tensile strength, ductility, and energy dissipation capacity, the similitude of failure mechanism and damage propagation is also correlated.

As an example, the final collapse mechanisms of small-scale models of stone masonry houses, tested to study the efficiency of steel ties placed at floor levels, are shown in Fig. 1.10 (Tomaževič, Lutman, & Velechovsky, 1993). In addition, the correlation between prototype and model stone masonry, obtained by testing the prototype and model masonry walls, is shown in Fig. 1.11 (Tomaževič & Velechovsky, 1992).

1.4 Seismic resistance verification of masonry walls and buildings

1.4.1 Calculation of seismic resistance of masonry walls

As has been already mentioned, three main failure modes of URM walls can be observed when subjected to earthquake loads. In the case of low vertical load and poor-quality mortar, seismic loads may cause shearing of the wall into two parts and sliding of the upper part of the wall on one of the horizontal mortar joints. If the level of vertical load and axial

FIG. 1.10 Shaking table tests of models of stonemasonry houses on earthquake simulator at ZAG. (A) Separation of walls at low level of excitation in the case of model without wall ties and (B) model with tied walls resisted repeated strong shaking before collapse. The difference in failure mechanisms is obvious (Tomaževič et al., 1993).

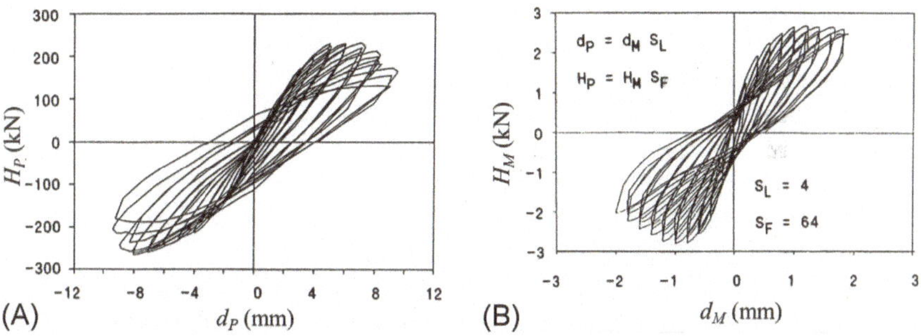

FIG. 1.11 Correlation between hysteretic behavior of prototype (A) and (B) 1:4 reduced-scale model stone masonry walls, indicating almost perfect matching between the required and actual properties of model masonry (Tomaževič & Velechovsky, 1992).

compression stresses in the wall are within the usual limits, the wall may fail either in shear or in bending. Shear failure, which is a typical mode of failure of masonry walls subjected to seismic loads, takes place where the principal tensile stresses, developed in the wall under a combination of vertical and horizontal loads, exceed the tensile strength of masonry. Characteristic diagonal cracks develop in the wall just before the attainment of lateral resistance. The cracks can either follow the mortar joints or pass through the masonry units or both. In the case of improved shear resistance and high moment/shear ratio, however, crushing of compressed zones at the ends of the wall usually takes place, indicating the flexural mode of failure.

To simplify the analysis and design, the values of sectional forces, stresses, and strains are usually determined based on the gross cross-sectional geometrical characteristics of the walls and assuming the elastic, homogeneous, and isotropic global properties of masonry as the structural material. These assumptions permit that equations, based on the simple theory of elasticity, can be used for seismic resistance verification, modified by taking into account the basic characteristics of the observed seismic behavior of masonry walls.

Although simplified, the equations reflect the actual failure mechanisms. Therefore the values of mechanical quantities, which determine the load-bearing capacity and deformability of masonry walls in the equations, such as the compressive strength of masonry, f, the shear, f_v, or tensile strength of masonry, f_t, the modulus of elasticity (secant modulus), E, and the shear modulus, G, should be determined experimentally by using testing procedures, compatible with the experiments on the basis of which the equations for calculations have been developed. One has to have in mind that these values do not represent actual physical values of relevant quantities but referential values, assigned to masonry assumed to be elastic, homogeneous, and isotropic materials. In accordance with assumptions for calculations, gross sectional areas of the tested specimens are taken into account when evaluating the values from test results. In a way, the equations used for calculation of resistance characteristic can be considered as empirical, statistical equations of the form known in the theory of elasticity.

To take into account full resistance and displacement capacity of walls and the complete structural systems, and to make the calculations simple, the actual hysteretic behavior, that is, hysteresis envelope, is idealized by a bilinear resistance envelope (Fig. 1.12). To idealize the experimental envelope, three limit states in the observed behavior of the tested walls are first defined:

- Crack (damage) limit, determined by displacement d_{cr} and resistance R_{cr} at the formation of the first significant cracks in the wall, which change the slope of the envelope.
- Maximum resistance, determined by maximum resistance R_{max}, attained during test, and corresponding displacement $d_{R_{max}}$.
- Ultimate state, determined by maximum displacement attained during test d_{max}, and corresponding resistance $R_{d_{max}}$.

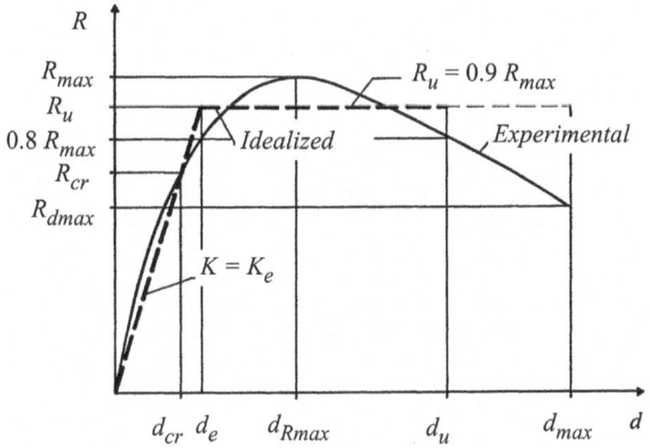

FIG. 1.12 Idealization of the experimentally obtained resistance curve of a masonry wall with indicated limit states (Tomaževič, 1999).

The effective stiffness of the wall is defined by a secant stiffness at the formation of cracks, which is called effective stiffness of the wall, $K_e = R_{cr}/d_{cr}$. The ultimate resistance of the idealized envelope R_u is evaluated by taking into account the equal energy dissipation capacity of an actual and idealized wall: the areas below the actual and idealized curves should be equal. As has been found by evaluating the results of tests of more than 60 walls, the average value of $R_u/R_{\max}$ ratio is 0.9 (Tomaževič, Lutman, Velechovsky, & Žarnić, 1986). Consequently, in the case of bilinear idealization of resistance envelope, the calculated values of maximum resistance should be multiplied by the factor of 0.9: $R_u = 0.9\ R_{\max}$. Arbitrarily, the ultimate idealized displacement d_u is defined as displacement where the resistance on the descending branch of experimental curve drops to 80% of the maximum.

In the case of the shear failure mode, the resistance capacity of the wall is calculated according to proposal of Turnšek and Čačovič (1970), who assumed that diagonal cracks at shear failure are caused by the principal tensile stresses that develop in the wall when subjected to a combination of vertical and lateral loads. Assuming that masonry is an elastic, a homogeneous and an isotropic material, the value of principal tensile stress can be easily evaluated knowing the geometry of wall and vertical and horizontal loads acting on the walls at that instant of time. Arbitrarily, the value of the principal tensile stress at the attainment of maximum resistance, $R_{\max}$, is called the tensile strength f_t (diagonal tension strength) of masonry. It has been found that the value of principal tensile stress at the formation of the first diagonally oriented cracks, which characterize shear failure mode, is of the order of 0.7–0.8 of the stress at the attained maximum resistance, that is, 0.7–0.8 f_t. The idea is simple and can be applied to all types of masonry, where diagonally oriented cracks determine the failure mechanisms. There is no need to develop models and equations based on the micromechanisms of interaction between units and mortar. However, the input values should be determined experimentally by testing specimens of each specific type of masonry, and the results of experiments should be evaluated on the basis of gross dimensions of the tested specimens.

Originally proposed equation has been modified to take into account the influence of geometry of the wall and distribution of actions (ratio between vertical and lateral load) at maximum resistance. The lateral resistance at shear failure, $R_{s,w}$, of a plain masonry wall panel failing in shear is evaluated by:

$$R_{s,w} = A_w \frac{f_t}{b} \sqrt{\frac{\sigma_o}{f_t} + 1}, \tag{1.2}$$

where A_w = cross-sectional area of the wall, $\sigma_o = N/A_w$ = the average compression stress due to vertical load N, and b = the shear stress distribution factor, depending upon the geometry of the wall and N/R_{max} ratio. In the case of the wall with geometrical aspect ratio $h/l = 1.5$, $b = 1.5$. In the case where the resistance envelope is idealized as a bilinear relationship, Eq. (1.1) is multiplied by 0.9.

Damage to structural walls, which could be attributed to predominant flexural behavior, is rarely observed in the case of simple masonry buildings. This is not only a result of specific properties of masonry materials, especially the low ratio between tensile and compressive strength, but also a result of structural characteristics, which cause predominant shear type behavior of the building when subjected to earthquake loads, that is, low moment to shear ratio.

Considering the fact that the behavior of masonry subjected to uniaxial compression is similar to that of concrete, a similar shape of an equivalent rectangular compressive stress block may be taken into account in the calculation of flexural resistance capacity of a masonry wall's section, as in the case of the concrete. The resulting ultimate bending moment, M_{Ru}, which determines the flexural capacity of the wall's section, is calculated by:

$$M_{Ru} = \frac{\sigma_0 t\, l^2}{2} \left(1 - \frac{\sigma_0}{f}\right), \tag{1.3}$$

where l = the length of the wall and t = the thickness of the wall.

1.4.2 Seismic resistance verification of simple masonry buildings

In the case of URM buildings that fulfill the basic requirements for rigid horizontal diaphragm floor action, concentration of damage at the bottommost storey has been observed by both postearthquake observations (see, e.g., Fig. 1.2) and experiments. Shear failure of walls at the ground floor and so-called storey mechanism indicate that the seismic resistance of regular low-rise masonry buildings generally depends on the resistance of the walls where the maximum seismic loads (base shear and bending moments) usually develop. Following the idea of seismic testing, where the structures are tested by imposing horizontal displacements and measuring the resistance to imposed displacements, the same idea has been also applied to calculation of seismic resistance. The calculation of storey resistance curve is based on the shear mechanism model and takes into account the actual, but idealized, resistance curves of all structural walls in the critical storey. The method of verifying the seismic resistance of simple masonry buildings by comparing the storey resistance curve to the design seismic action and global ductility requirements has been originally proposed and widely used more than four decades ago (Tomaževič, 1978). By today, significantly improved methods have been proposed by other researchers (e.g., Penna, Lagomarsino, & Galasco, 2014).

The following general assumptions are made in the calculation:

- When subjected to seismic loads, floors act as rigid horizontal diaphragms. Equal displacements of all structural walls in the critical storey are assumed.
- When subjected to earthquakes, the first mode of vibration is predominant. Therefore as good approximation, the triangular distribution of seismic forces and displacements along the height of the building can be assumed.
- The contribution of structural walls to the lateral resistance of the critical storey depends on the rigidity and deformation, as well as on lateral resistance of individual walls. Torsional effects are taken into account.
- The walls are carrying their part of seismic loads also in the nonlinear range. However, once the displacements of the storey exceed the assigned displacement capacity of the wall, the contribution of the wall in terms of rigidity and lateral resistance is no more taken into account. However, such walls still carry their part of the vertical loads.

In the calculation of the storey resistance envelope, the structure is displaced by a small value, assuming the chosen shape of distribution of displacements along the height of the building. The shear walls are deformed according to the assumed structural model, and the resisting forces in structural members are calculated. If torsional rotation takes place by displacing the mass centers of individual storeys, the displacements of individual walls are accordingly modified. The calculation is repeated step-by-step by increasing the imposed displacements. Once the walls enter into the nonlinear range, the structural system of the building and stiffness matrices are modified. The stiffness and resistance of individual walls in each step of calculation are determined

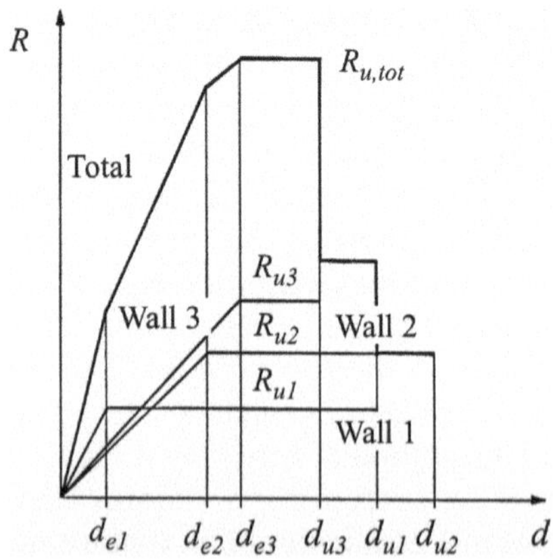

FIG. 1.13 Schematic presentation of construction of storey resistance envelope on the basis of idealized bilinear resistance-displacement relationships of individual walls (Tomaževič, 1999).

considering the calculated storey displacements and idealized resistance envelopes of structural walls in each storey. The procedure of calculation is schematically presented in Fig. 1.13.

As a result of calculation, the relationship between the resistance of the critical storey and interstorey drift, that is, the resistance envelope, is obtained. At the given imposed lateral displacement of the storey, the resisting storey shear is determined as a sum of resistances of structural walls in the storey under consideration. Experimental results obtained from shaking table tests of models of unreinforced, confined, and reinforced masonry buildings have been used to verify the validity of the proposed method (e.g., Tomaževič & Gams, 2012). In all cases, good correlation between experimental and theoretical results has been obtained (see Fig. 1.14), confirming the general usability of this relatively simple method for seismic resistance verification.

According to code requirements, the structural system should be verified for the required strength and ductility capacity. The strength and ductility assigned to the structure by the codes are compared with the capacity of the structure to dissipate the seismic energy and to withstand nonlinear deformations without losing the stability. In this regard, the extent of permissible damage occurring to the structure in the nonlinear range should be also carefully determined.

If the seismic resistance of the structure is verified using equivalent static analysis, as is the case of masonry buildings, the balance between the required strength and energy dissipation capacity is specified by the value of the behavior factor q (force reduction factor), which, according to traditional definition, represents the displacement and energy dissipation capacity of the structural system. Explained simply, factor q can be

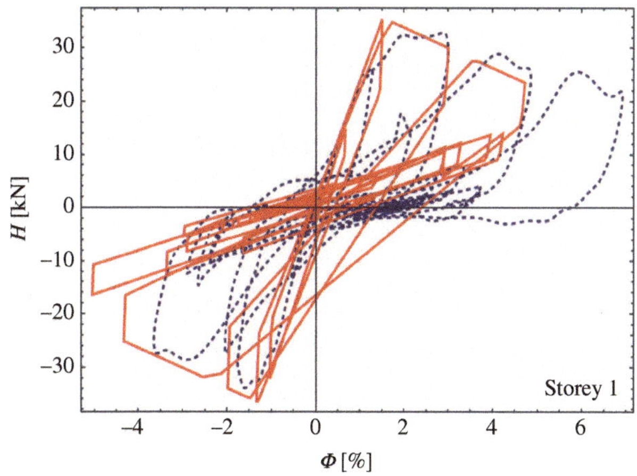

FIG. 1.14 Correlation between experimental (dashed line) and calculated (full line) hysteretic response of a masonry building model based on the calculated resistance curve and experimentally determined hysteretic rules (Tomaževič & Gams, 2012).

expressed in terms of global ductility of the structure by $q=\sqrt{2\mu-1}$, which gives an estimate for global ductility demand (see, e.g., Tomaževič, 1999):

$$\mu_u = \frac{q^2+1}{2}, \tag{1.4}$$

where $\mu_u = d_u/d_e$ = ultimate global ductility factor. In the assessment of behavior factor values, damage limitation requirements have been considered, which limit the available displacement capacity to be taken into account in the design. It has been found that, for the seismic resistance verification purposes, storey drift (relative storey displacement) up to three times the drift at the crack limit (storey rotation angle up to 1.2%) can be considered as acceptable when assessing ultimate global ductility without risking unrepairable damage to structure (Tomaževič, 2007). This means that for the case of the plain masonry buildings with walls, connected with bond beams and floors, behavior factor $q = 1.5$ or global ductility factor $\mu_u = 1.6$ in the case of resistance curve can be taken into account for seismic resistance verification.

Once the resistance curve is calculated, seismic resistance of the building can be verified. In case the calculated value of global ultimate ductility factor μ_u exceeds the value, attributed to behavior factor q, taken into account for the calculation of design seismic loads, and the resistance exceeds the code demand for resistance capacity at a given site, the resistance of the building under consideration fulfills the requirements for earthquake-resistant buildings.

Acknowledgments

The chapter summarizes some published results and unpublished experiences, obtained by the author and his colleagues within the framework of experimental part of research projects, financed in the past decades by public funds through what is today Slovenian Research Agency and industry. The material used in the preparation of this contribution has been referenced, so the reader may find the details in the papers listed in the list of references.

References

Calvi, G. M., Kingsley, G. R., & Magenes, G. (1996). Testing masonry structures for seismic assessment. *Earthquake Spectra, 12*, 145–162.

EN 1052-1 (1998). *Methods of tests for masonry—Part 1: Determination of compressive strength.* Brussels: CEN.

EN 1052-3 (2002). *Methods of tests for masonry—Part 3: Determination of initial shear strength.* Brussels: CEN.

EN 1996-1-1 (2005). *Eurocode 6: Design of masonry structures—Part 1-1: Common rules for reinforced and unreinforced masonry structures.* Brussels: CEN.

Harris, G. H., & Sabnis, G. M. (1999). *Structural modeling and experimental techniques.* Boca Raton: CRC Press.

Langhaar, H. L. (1951). *Dimensional analysis and theory of models.* New York: John Wiley & Sons.

Mayes, R. L., & Clough, R. W. (1975). *State-of-the-art in seismic shear strength of masonry—An evaluation and review.* Berkeley: Earthquake Engineering Research Center, University of California.

Mergos, P. E., & Beyer, K. (2014). Loading protocols for European regions of low to moderate seismicity. *Bulletin of Earthquake Engineering, 12*, 2507–2530.

Paulson, T. J., & Abrams, D. P. (1990). Correlation between static and dynamic response of model masonry structures. *Earthquake Spectra, 6*, 573–591.

Penna, A., Lagomarsino, S., & Galasco, A. (2014). A nonlinear macroelement model for the seismic analysis of masonry buildings. *Earthquake Engineering & Structural Dynamics, 43*, 159–179.

Sheppard, P., & Tomaževič, M. (1985). *Possibilities of revitalization of residential masonry buildings regarding the seismic safety, Research Report.* Ljubljana (in Slovene): ZRMK.

Terčelj, S., Sheppard, P., & Turnšek, V. (1977). The influence of frequency on the shear strength and ductility of masonry walls in dynamic loading tests. In: *Proceedings of the 6th world conference on earthquake engineering*Vol. 3, (pp. 2992–2999). , pp. 2992–2999. New Delhi.

Tomaževič, M. (1978). *Improvement of computer program POR.* Report ZRMK-IK, ZRMK, Ljubljana (in Slovene).

Tomaževič, M. (1999). *Earthquake resistant design of masonry buildings.* London: Imperial College Press/World Scientific.

Tomaževič, M. (2007). Damage as a measure for earthquake resistant design of masonry structures: Experience of Slovenia. *Canadian Journal of Civil Engineering, 34*, 1403–1412.

Tomaževič, M. (2016). Some considerations on testing and experimental simulation of seismic behaviour of masonry walls and buildings. In C. Modena, F. da Porto, & M. R. Valuzzi (Eds.), *Brick and block masonry: Trends, innovations and challenges, proceedings of the 16th international brick and block masonry conference, 26–30 June 2016* (pp. 37–53). Padova, Italy: CRC Press.

Tomaževič, M., & Gams, M. (2009). Shear resistance of unreinforced masonry walls. *Ingegneria Sismica, 26*(3), 5–18.

Tomaževič, M., & Gams, M. (2012). Shaking table study and modelling of seismic behaviour of confined AAC masonry buildings. *Bulletin of Earthquake Engineering, 10*, 863–893.

Tomaževič, M., Klemenc, I., & Lutman, M. (2000). In situ tests for the assessment of seismic resistance of old stone masonry houses. In *Proceedings of the 12th world conference on earthquake engineering*, New Zealand, CD ROM, paper 1975.

Tomaževič, M., Lutman, M., & Petković, L. (1996). Seismic behavior of masonry walls: Experimental simulation. *Journal of Structural Engineering, ASCE, 122*, 1040–1047.

Tomaževič, M., Lutman, M., & Velechovsky, T. (1993). Aseismic strengthening of old stone-masonry buildings: Is the replacement of wooden floors with R. C. slabs always necessary? *European Earthquake Engineering, 7*, 34–46.

Tomaževič, M., Lutman, M., Velechovsky, T., & Žarnić, R. (1986/1987). Seismic resistance of reinforced masonry walls—Test results, parts I and II. *Reports, ZRMK/IKPI-86/04/ZRMK/IKPI-87/06.,* .

Tomaževič, M., & Velechovsky, T. (1992). Some aspects of testing small scale masonry building models on simple earthquake simulators. *Earthquake Engineering & Structural Dynamics, 21*, 945–963.

Triller, P. (2019). *Seismic behaviour of multistorey masonry shear walls with openings (in Slovene) (Doctoral dissertation).* Ljubljana: University of Ljubljana.(2019). *https://repozitorij.uni-lj.si/IzpisGradiva.php?id=109565.*

Triller, P., Tomaževič, M., & Gams, M. (2016). Seismic behaviour of multistorey plain masonry shear walls with openings: An experimental study. In: C. Modena, F. da Porto, & M. R. Valuzzi (Eds.), *Brick and block masonry: Trends, innovations and challenges, proceedings of the 16th international brick and block masonry conference, 26–30 June 2016, Padova*, Italy: CRC Press, pp. 1949–1954.

Turnšek, V., & Čačovič, F. (1970). Some experimental results on the strength of brick masonry walls. In: *Proceedings of the 2nd international brick masonry conference*, pp. 149–156. Stoke-on-Trent.

Williams, D., & Scrivener, J. C. (1974). Response of reinforced masonry shear walls to static and dynamic cyclic loading. In: *Proceedings of the 5th world conference on earthquake engineering*, pp. 1491–1494. Rome.

CHAPTER

2

Seismic vulnerability of Himalayan stone masonry: Regional perspectives

Dipendra Gautam[a], Rajesh Rupakhety[b], Rabindra Adhikari[c], Binay Charan Shrestha[d], Rewati Baruwal[e], and Lalit Bhatt[e]

[a]Department of Architecture and Civil Engineering, City University of Hong Kong, Kowloon, Hong Kong, [b]Earthquake Engineering Research Center, Faculty of Civil and Environmental Engineering, School of Engineering and Natural Sciences, University of Iceland, Selfoss, Iceland, [c]Department of Civil Engineering, Institute of Engineering, Pulchowk Campus, Lalitpur, Nepal, [d]Department of Urban Development and Building Construction, Government of Nepal, Kathmandu, Nepal, [e]Interdisciplinary Research Institute for Sustainability, Kathmandu, Nepal

2.1 Background

Stone masonry is one of the oldest construction systems in Nepal. Stone masonry construction is especially dominant in the low-to-high mountains of Nepal from east to west (Gautam, Chaulagain, et al., 2018). Although more than 70% of Nepali building stocks is comprised of stone masonry (Gautam & Chaulagain, 2016; Gautam, Rodrigues, Bhetwal, Neupane, & Sanada, 2016; Parajuli & Kiyono, 2015), studies regarding their seismic vulnerability are limited. Gautam, Fabbrocino, and Santucci de Magistris (2018) and Gautam (2018) derived fragility functions for stone masonry buildings using the database of past earthquakes and concluded that stone masonry construction is highly vulnerable even in minor-to-moderate shaking. In their contribution, Gautam et al. (2018)

https://doi.org/10.1016/B978-0-12-821087-1.00014-4

classified stone masonry construction of Nepal as heritage stone masonry, rubble stone masonry (mud mortar), rubble stone masonry (cement mortar), and dressed stone masonry (cement mortar) with respective vulnerability class as C, A, A, and B per the EMS-98 vulnerability classification (Grunthal, 1998). Although stone masonry is gaining notable attention after the 2015 Gorkha earthquake, it is important to realize that seismic vulnerability of stone masonry constructions in Nepal, which varies widely from one region to another, is still not adequately quantified. Gautam, Chaulagain, et al. (2018) developed sets of fragility functions for several stone masonry building types in Nepal using heuristic approach. Furthermore, the fragility functions developed by Gautam (2018) using large number of damage data are insightful to understand the level of vulnerability in stone masonry construction. However, due to discrepancies in construction systems, fragility functions derived for a particular region may not be equally representative to other regions. Variation in type and shape of stone masonry units, type of mortar, presence/absence of mortar, size of wall and wythes, among others translate to variation in vulnerability of stone masonry in various parts of the country.

Uncertainties encountered in mechanical modeling and testing of stone masonry are well-recognized worldwide (Ahmad, Ali, Ashraf, Alam, & Naeem, 2012; António et al., 2013; Candela, Borri, Corradi, & Righetti, 2016; Costa, Arêde, Costa, & Oliveira, 2011; I. Senaldi, 2019; I.E. Senaldi, 2020, among others). However, experimental campaigns, forensic investigations, and numerical simulations are providing a strong ground to outline seismic behavior of stone masonry. Hitherto, seismic vulnerability assessment of stone masonry constructions has become popular and novel techniques for assessment, design, strengthening, and optimization are rapidly emerging (Abo-El-ezz, Nollet, & Nastev, 2013; Gautam, Prajapati, Paterno, Bhetwal, & Neupane, 2016; Maio, Estêvão, Ferreira, & Vicente, 2020; Ortega, Vasconcelos, Rodrigues, & Correia, 2018; Vintzileou, Mouzakis, Adami, & Karapitta, 2012, among others). To address the gap of exacting depiction of vulnerability of stone masonry in Nepal, this chapter presents a classified perspective of seismic vulnerability of stone masonry buildings considering the variation of construction systems in different regions of the country. Stone masonry from eastern, central, western, and far-western regions are considered as classes and damageability functions, which indicate the fraction of damage at a particular ground shaking, for stone masonry in mud mortar are derived. The main objective of this chapter is to provide a more scrutinized understanding of seismic damageability of stone masonry buildings in mud mortar. In addition, comparative damageability of stone masonry in regional scale is also presented.

2.2 Stone masonry in Nepal

Being the oldest and most economical construction system in Nepal, stone masonry can be found across the country except the southern Indo-Gangetic plain. From Siwalik to Higher Himalaya, stone masonry is constructed as mud mortar masonry, that is, stone in mud mortar, dry stone masonry (stacked stone pieces only), and stone masonry in cement mortar. Stone masonry in cement mortar is more recent practice in Nepal, and very few buildings use cement mortar. Most of stone in cement mortar construction is found in public buildings like the administrative offices, hospitals, and schools. Dry stone masonry construction is prevalent in many locations; however, they are not explicitly discernable as most of such constructions are mud plastered on both faces. Some stone masonry that makes use of semidressed stone units is not plastered. Mud plaster is also common in random rubble constructions. To improve binding characteristics, mud mortar used in masonry constructions is often mixed with cow dung. Stone masonry buildings in Nepal are one to three storied constructions. The first story is generally used as kitchen, the second story as living space, and the top story is generally used for storage. In some settlements, the first story is used as a cowshed as well.

Most of the stone masonry walls are multiwythed and the thickness of walls does not vary throughout the building. The outer and inner wythes generally have large stone units with filler stones in between. The filler stones are smaller stone units, either cut pieces left from the larger stones or naturally found smaller stones. In some cases, when larger stones are not readily available, the inner and outer wythes are made up of smaller stone units stacked together.

2.2.1 Eastern Nepal

Fig. 2.1 shows a typical stone masonry from eastern Nepal. Use of large stone pieces is common in this part of the country. In this region, stones used in the outer and inner wythes are usually large regular (rectangular) pieces as shown in Fig. 2.1. Rocks used in stone masonry are quite similar from the east to the west of the country, but workmanship and traditions of local construction differ. This creates regional variation in structural forms.

2.2.2 Central Nepal

As shown in Fig. 2.2, stone masonry construction in central Nepal uses more irregularly sized and shaped stones when compared with the eastern stone masonry construction. The stone units are not dressed and can have round/curved edges and surfaces. This leads to less binding

FIG. 2.1 A typical stone masonry construction in eastern Nepal.

FIG. 2.2 A typical stone masonry construction in central Nepal.

between the stone units. However, central region stone masonry construction system makes extensive use of massive and long cornerstones. The cornerstones are placed in alternate courses on orthogonal walls as shown in Fig. 2.3. In central and western Nepal, some masonry walls contain wooden wall belts called "*Nas*" in local language, as shown in Fig. 2.4. These are traditional equivalents of seismic bands. However, such bands are not always provided throughout the perimeter (see Fig. 2.4), which makes some parts of the building more vulnerable.

Stone masonry buildings in eastern and central Nepal are generally bigger in size than those found in the western and far western regions. In many locations where long stone masonry units are available, long and short stone masonry units are placed in alternate courses when possible. However, this is often not a mandatory practice, and depends to a large degree on availability and abundance of larger stone units.

2.2.3 Western Nepal

Stone masonry in western Nepal is different than those in the eastern and central regions of Nepal. Stone units used in western Nepal are more regular than those in central and eastern Nepal. A typical stone masonry building in western Nepal is shown in Fig. 2.5. Stone masonry buildings in western Nepal are typically smaller than those in eastern and central Nepal. This difference in size is more of a tradition than due to cultural or climatic differences.

In this part of the country, thick gable walls without opening are provided. The thickness of the gable wall is the same as that of the structural walls. Out of plane collapse of heavy gable walls which are not connected to the main structural system was commonly observed during the Gorkha earthquake. Such out plane collapse can be somewhat mitigated by providing timber bands.

2.2.4 Far western Nepal

The far western masonry construction system comprises relatively smaller buildings, but the stone units are loosely placed together making them highly vulnerable to ground shaking. Far western masonry constructions are generally two-storied (Fig. 2.6), and the vertical transportation system is sometimes placed outside the building as shown in Fig. 2.6.

The overall construction technique of stone masonry in Nepal and for that matter the Hind-Kush-Himalaya is similar. Regional variations are introduced by the quality of mud mortar, quality of stone units, and local traditions. The mud mortar used in eastern Nepal has higher proportions of organic clay than that in the rest of the country. Use of sedimentary and

FIG. 2.3 Cornerstones jutting on two orthogonal walls.

FIG. 2.4 Timber lacing in a stone masonry building.

FIG. 2.5 A typical stone masonry building in western Nepal.

FIG. 2.6 A typical stone masonry building from far west Nepal. *(With kind permission from Yogi Kayastha.)*

metamorphic rocks also discerns the quality of construction. Sedimentary rock stones are usually irregular in shape and difficult to form an interlocking pattern. This can result in disintegration of wythes as the mud mortar shrinks.

2.2.5 General characteristics of stone masonry

All stone masonry constructions in Nepal are rectangular in shape and 1–3 storied. The foundation is usually placed at 1–2 ft depth, with 1.5 ft being most common stronger stone units are generally placed in the foundation courses. The wall thickness of stone masonry constructions across Nepal is fairly uniform, with 18 in. being a typical value. The floor height is usually 6 ft but a significant variation in floor height can be also observed. The third story of stone masonry building will have the lowest floor height.

The flooring systems are flexible consisting of timber beams and joists. Long timber beams run from wall to wall in the longer direction of the building. Such beams are supported by walls and timber posts (Fig. 2.7). Flooring is laid over timber joists resting on the long timber beams. The flooring system comprises wooden planks or folded bamboo or small folded wooden logs supported by joists underneath a mud mortar masonry course that is generally 3–5 in. thick. Sometimes, wooden beams are provided to support joists; however, in most cases joists are directly laid on the wall. The wall thickness is similar in all stories. The stone

FIG. 2.7 Vertical timber post, beam, and joists supporting the flexible floor comprised by wooden planks and mud mortar masonry course.

FIG. 2.8 Top story with gable projection.

masonry wall in the third story is generally triangular in shape supporting double hip roofing as shown in Fig. 2.8. The vertical transportation system is either wooden ladder (Fig. 2.9) within the building in a small stair well or stepped stone wall placed outside (Fig. 2.6). The windows and doors in the western parts of the country are smaller than the ones in the central and eastern parts. The roofing system varies widely. The eastern region mainly uses thatched and corrugated galvanized iron sheets as the roofing materials. Three major roofing systems namely roof tiles, corrugated iron sheets, and thatched roof are common in other parts of the country. Most of the newer constructions use corrugated iron sheets as the roofing material.

2.3 Seismic performance of stone masonry construction: Historical perspectives

Since the known history of earthquakes in Nepal (for details see Gautam & Chaulagain, 2016), stone masonry has been the most vulnerable in the past earthquakes (Gautam, 2018; Gautam et al., 2018). The 1934 earthquake claimed 8519 lives in Nepal and most of these deaths were attributed to the collapse of stone masonry buildings in the mountains of central, eastern, and western Nepal (Rana, 1935). The areas with housing system dominated by stone masonry were the most affected by the 1833 earthquake (Bilham, 1995). Similarly, the 1966 and 1980 earthquakes in far-western Nepal caused extensive damage to stone masonry

FIG. 2.9 A typical wooden ladder used for vertical transportation in stone masonry buildings.

buildings. Although overall damage statistics of buildings have been reported since the 1833 earthquake in Nepal, details of damage mechanisms caused by earthquake before the 1988 Udaypur earthquake are not available. After the 1988 earthquake, the Asian Disaster Preparedness Center (ADPC) performed field reconnaissance in the affected areas and collected pictorial evidence of damage mechanisms, which was an important milestone to understand the behavior of stone masonry buildings during earthquakes (Gupta, 1988). The ADPC also suggested some improvements in construction techniques. The techniques suggested by the ADPC were later endorsed as the earthquake resistant construction practice in Nepal.

FIG. 2.10 Typical damage occurrence in stone masonry building in eastern Nepal during the 1988 earthquake *(With kind permission from Asian Disaster Preparedness Center, ADPC, Bangkok.)*

As shown in Fig. 2.10, gable collapse and vertical cracks were commonly observed in stone masonry building during the 1988 earthquake. The heavy gable and subsequent high lateral load exerted to the main structural system are the major causes of damage to the structural walls as shown in Fig. 2.10.

The 2011 earthquake in eastern Nepal mostly affected the eastern part of country. Different types of damages such as shear cracking, gable collapse, delamination of stone masonry wythes in the upper story, plaster cracks, total destruction of building, among others, were observed. Fig. 2.11 shows an example of in-plane shear damage on stone masonry wall. The propagation of shear cracks is usually facilitated by the joints between the random stone units. The shear cracks generally propagate around the weaker and smaller stone units. Stronger and bigger stone units occupy large area without mortar, and act as a barrier to crack propagation.

The 2011 earthquake highlighted many other weaknesses of stone masonry construction in Nepal. As shown in Fig. 2.12, a portion of the stone masonry wall is displaced due to lack of proper binding/anchorage in the wall. After some years of construction, mud mortar gets shrunk in stone masonry buildings, which causes loss of binding behavior between the stone units. This leads to separations or collapse of wall during earthquakes.

An example of separation of building segments and subsequent damage by the 2011 earthquake in eastern Nepal is shown in Fig. 2.13. It is common in rural Nepal to extend the house in the long wall direction to add

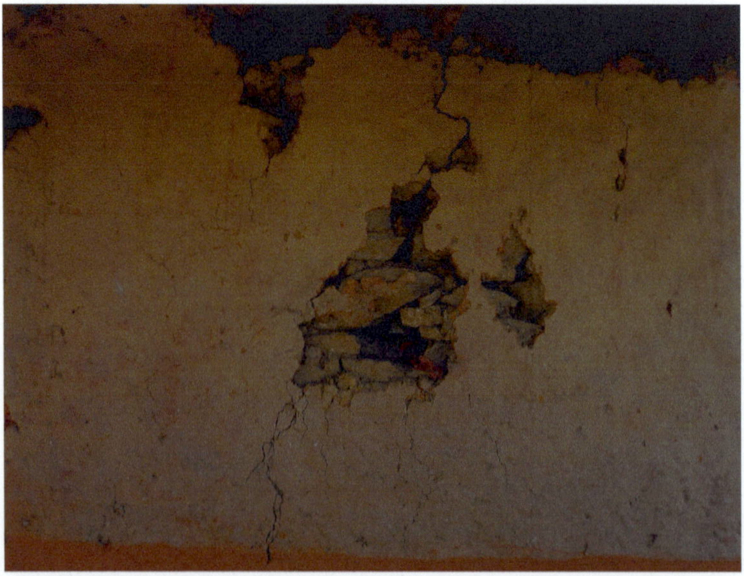

FIG. 2.11 In-plane shear damage on a stone masonry wall caused by the 2011 eastern Nepal earthquake.

FIG. 2.12 Corner separation in a stone masonry building caused by the 2011 eastern Nepal earthquake.

FIG. 2.13 Disintegration of extension of a stone masonry house caused by the 2011 earthquake.

some rooms either in the first or the second story. The extended sections are usually not anchored well to the main structure and get easily separated during lateral shaking.

An example of collapsed gable wall is shown in Fig. 2.14. Despite the light roof, the gable mass was not properly supported to resist out-of-plane shaking.

Load path continuity and uninterrupted/uniform distribution of loads from the upper stories to the ground is crucial for structural integrity. Fig. 2.15 highlights cracks caused by concentrated vertical load imposed on the wall by the floor beam. Such cracks under the floor beams were usually observed to propagate to nearby opening, creating a failure plane on the wall (Fig. 2.15).

As shown in Fig. 2.16, vertical cracks generally occur in the region where the cornerstones end. Furthermore, as shown in Fig. 2.16, long walls are more vulnerable than the short ones. Hence, damage is observed to be concentrated across the long wall. Local delamination is quite common in stone masonry due to lack of proper binding between stone units.

Fig. 2.17 shows an example of damage caused by inadequate wall plates. Wall plates formed by joining smaller pieces of timber that are not properly connected together is a weakness in seismic performance of masonry buildings. Damage or displacement of some parts of the wall plate leads to a discontinuity of load transfer path, and consequently

FIG. 2.14 Out-of-plane collapse of a stone masonry gable wall caused by the 2011 earthquake.

FIG. 2.15 In-plane crack extending from the floor beam to the door opening in a stone masonry house caused by the 2011 earthquake.

FIG. 2.16 Local delamination and vertical crack propagation in stone masonry building caused by the 2011 earthquake.

FIG. 2.17 Damage to wall plate and corner damage.

damage the structural walls. Lack of proper connection between wall plates at the corners of adjacent walls can lead to corner damage as shown in Fig. 2.17.

Fig. 2.18 shows an example of spandrel damage, which is generally not observed in stone masonry buildings but common in large brick masonry structures as reported by Adhikari, Jha, Gautam, and Fabbrocino (2019).

Interlocking and anchorage between the masonry units significantly restrict crack propagation. In lack of these features, cracks can grow and expand and form failure surfaces. An example of such failure is shown in Fig. 2.19, where the failure surface extends from the wall to the floor.

FIG. 2.18 Spandrel damage in a stone masonry building caused by the 2011 earthquake.

FIG. 2.19 Separation of wall and floor in a stone masonry building caused by the 2011 earthquake.

Lack of proper connections between the orthogonal walls makes the corners vulnerable to ground shaking. Local cracks at corners of masonry walls made from random units can grow and cause a wedge-shaped collapse as shown in Fig. 2.20.

An example of cracking and local bulging of stone masonry walls around large openings is shown in Fig. 2.21. In many cases, such cracks grow and cause local failure.

In many cases, interior partitioning walls are constructed after the main structure is completed. When such walls are not properly anchored to the main walls, they lack a mechanism to transfer out-of-plane inertia force to the main structure, and collapse, as is shown in Fig. 2.22. Such collapse can be local or, if the shaking is very strong, the whole partition wall may collapse.

Some stone masonry school buildings were constructed following the JICA (Japan International Co-operation Agency) model that introduced

FIG. 2.20 Wedge-shaped failure of corner and adjoining parts of a stone masonry building caused by the 2011 earthquake.

FIG. 2.21 Cracking and bulging of walls around large openings of a stone masonry house caused by the 2011 earthquake.

FIG. 2.22 Out-of-plane collapse of a partition wall in a stone masonry building caused by the 2011 earthquake.

tubular steel units as vertical reinforcement (Fig. 2.23). Although the overall performance of such buildings was better than those without such reinforcements, lack of binding and strength in the masonry units caused cracking and failure of parts of walls, an example of which is shown in Fig. 2.23. Although not extensive in this case, even partial out of plane collapse of heavy stone walls in such buildings poses a great threat to school children.

The 2015 Gorkha earthquake that struck eastern, central, and parts of western Nepal exposed several deficiencies in stone masonry constructions in these regions. Several damage mechanisms and performance scenarios are also reported by Gautam and Chaulagain (2016), Gautam, Rodrigues, et al. (2016), and Varum et al. (2017). The most dominant failure mechanism during the Gorkha earthquake was the out-of-plane collapse of structural walls and gable walls as shown in Figs. 2.24 and 2.25, respectively. As shown in Fig. 2.25, the heavy gable portion above the main structural system is highly vulnerable during lateral excitations. Fig. 2.25 also depicts that the wedge-shaped failure becomes prominent due to lack of continuity of cornerstones. The wedge-shaped failure at the junction of orthogonal walls is quite prominent when the connection between the orthogonal walls is compromised. Stone masonry buildings suffered damage even at low peak ground acceleration (PGA) as is reported in (Gautam, 2018).

FIG. 2.23 Local failure in a stone masonry school building caused by the 2011 earthquake.

FIG. 2.24 Out of collapse failure of a structural stone masonry wall.

FIG. 2.25 Wedge-shaped failure and cracking near the corners of a stone masonry wall.

Lack of integrity among and between structural and nonstructural elements is a prominent weakness of stone masonry walls in Nepal and other regions in the Himalaya. Separation of added wall from the main structural system is common if proper anchorage is not provided (Fig. 2.26). Lack of proper anchorage between walls of adjacent stories compromises a continuous load path and makes the buildings vulnerable to ground shaking. The flooring systems are flexible and not capable of safely transmitting inertia forces from higher stories to the lower ones. Brittle and complete collapse of many buildings constructed by piling up stone units was caused by the Gorkha Earthquake. An example of such collapse is shown in Fig. 2.27. Another common deficiency is the lack of interlocking between the inner and outer wythes, an example of which is shown in Fig. 2.28.

2.4 Seismic vulnerability functions for stone masonry buildings

As stone masonry buildings are more likely to sustain partial or complete collapse at relatively low shaking, they pose higher risk of injury and fatality to their occupants. Out of plane collapse of walls is critical in this regard. Furthermore, collapse of nonstructural components and other unanchored objects can be hazardous to the occupants.

FIG. 2.26 Separation of added portion from the main structural system.

FIG. 2.27 Piled up stone masonry units of damaged houses in the epicentral region (Barpak) of the 2015 Gorkha earthquake.

FIG. 2.28 A typical cross section of stone masonry wall showing the lack of connection between the inner and outer wythes.

Inspection of fatality and building collapse observed after the 1934, 1980, 1988, 2011, and 2015 earthquakes in Nepal show a strong correlation, as expected, between collapse and fatality as depicted in Eq. (2.1). Records of building collapse and fatality were obtained from archives and a linear function was fitted. As the function is constrained between 0 and 50,000 building collapse, a linear function was found to be the most significant. The correlation together with the data points is shown in Fig. 2.29.

$$f = 0.014^{*}c - 0.96 \quad \left(R^2 = 0.61\right) \tag{2.1}$$

where f indicates number of fatalities and c indicates number of building collapse.

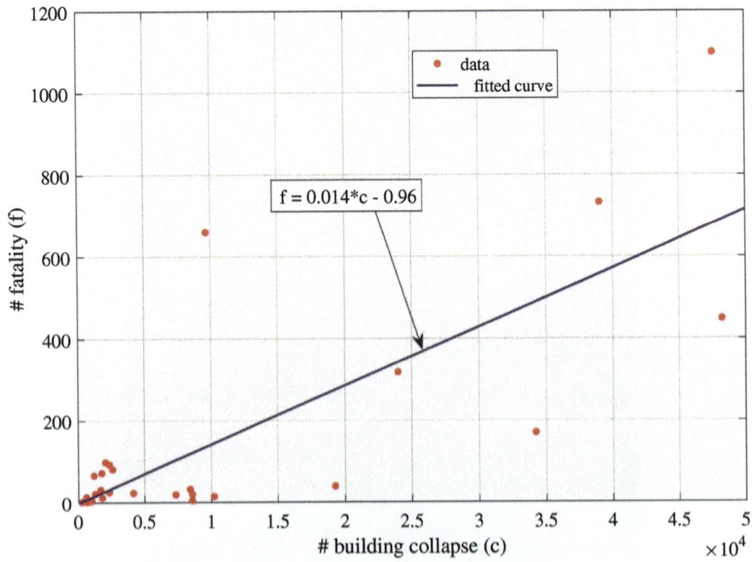

FIG. 2.29 Correlation between fatality and building collapse.

Number of fatalities seems to increase linearly with the number of collapsed buildings. The corresponding relation between the two is shown in the figure. The correlation between the two was found to be statistically significant. Although the proposed model shows a linear relation between the two variables, it needs to be used with caution. For example, the model prediction is lower than observed number of fatalities when the number of collapsed buildings is larger than 50,000. In general, such relations are difficult to quantify precisely. One of the important factors that govern such relationship is the state of occupancy of the buildings. In this context, the day and time of the earthquake plays a vital role. For example, fewer fatalities per collapsed buildings are expected if an earthquake happens during the day when people are working out in the fields. The opposite is expected if an earthquake happens during the night.

The earthquake damage data from 1934, 1980, 1988, 2011, and 2015 were reinterpreted to create seismic damageability functions for stone masonry buildings in Nepal. A representative damageability function for all stone masonry across Nepal is first developed classifying the building damage statistics into damage and nondamage classes. The PGA is used as intensity measure (IM). Each damage location, that is, affected district was assigned with a PGA value estimated using the ground motion prediction equation (GMPE) proposed by Jain, Roshan, Arlekar, and Basu (2000) for the Himalayan region as follows:

$$\ln(PGA) = b_1 + b_2 M + b_3 R + b_4 \ln(R) \quad (\sigma = 0.44) \tag{2.2}$$

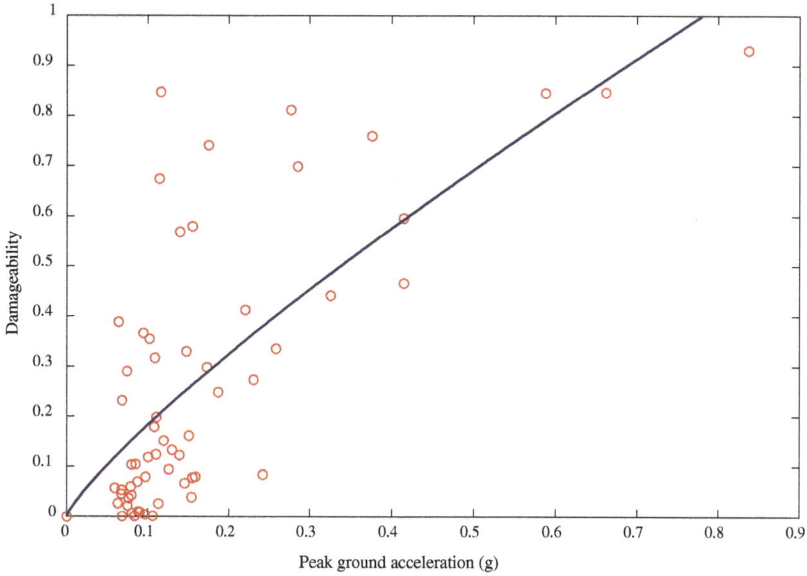

FIG. 2.30 Damageability function for Nepali stone masonry buildings.

where $b_1 = -3.443$; $b_2 = 0.706$; $b_3 = 0$; $b_4 = -0.828$; M is the magnitude of the earthquake; and R indicates the epicentral distance in kilometers.

The damageability, which is defined as the probability of sustaining any damage, of stone masonry buildings is modelled with a logit function as shown in Fig. 2.30. The total number of stone masonry building affected by all the earthquakes considered here is 2,015,646. Among them, 595,337 were damaged at various IM. As shown in Fig. 2.30, even at PGA of 0.15 g, damageability of stone masonry buildings is approximately 25%. Meanwhile, it reaches to 50% at 0.35 g and 75% a 0.5 g, which indicates that moderate to strong shaking would damage most of the stone masonry buildings. In the epicentral region of a strong earthquake, more than 95% of stone masonry buildings are expected to be damaged. It is noted that the damageability functions only indicate occurrence of damage and provide no information on the level of damage.

Damageability curves for stone masonry buildings in different regions of Nepal are presented in Figs. 2.31–2.34.

Fig. 2.31 shows the damageability function for the eastern stone masonry buildings. The damageability function shows that 20% probability of damage is expected at 0.1 g. Similarly, at 0.3, 0.4, and 0.5 g, probabilities of damage are 60%, 70%, and 85%, respectively. The damageability curve for stone masonry from central Nepal is shown in Fig. 2.32. As shown in Fig. 2.32, damageability of stone masonry buildings in central Nepal are 25%, 60%, and 80%, respectively, for 0.1, 0.3, and 0.5 g.

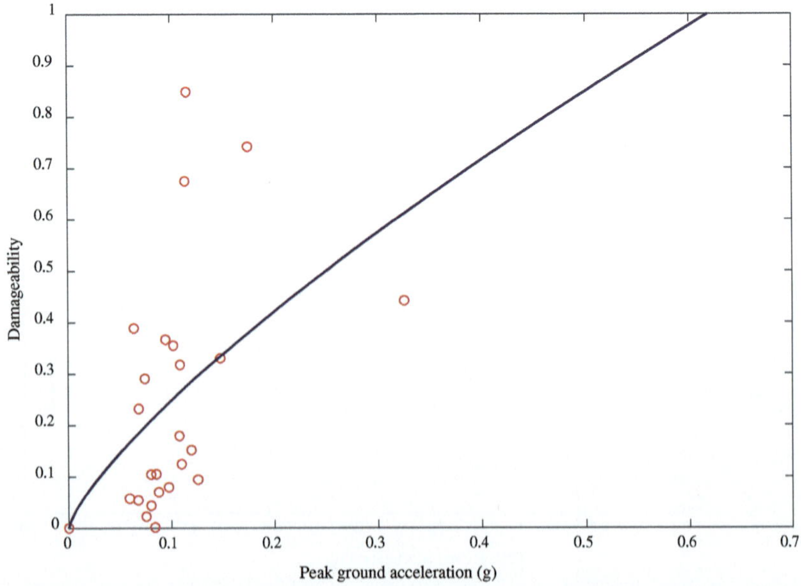

FIG. 2.31 Damageability function for stone masonry from eastern Nepal.

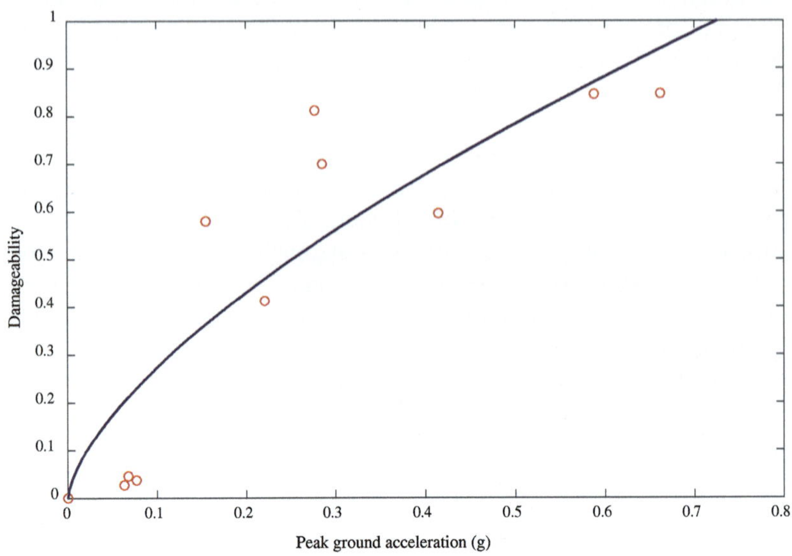

FIG. 2.32 Damageability function for stone masonry from central Nepal.

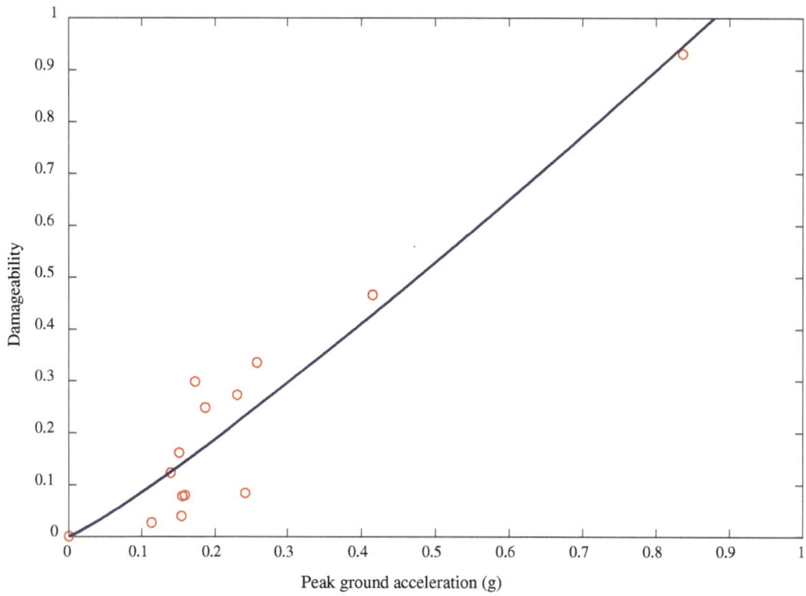

FIG. 2.33 Damageability function for stone masonry buildings from western Nepal.

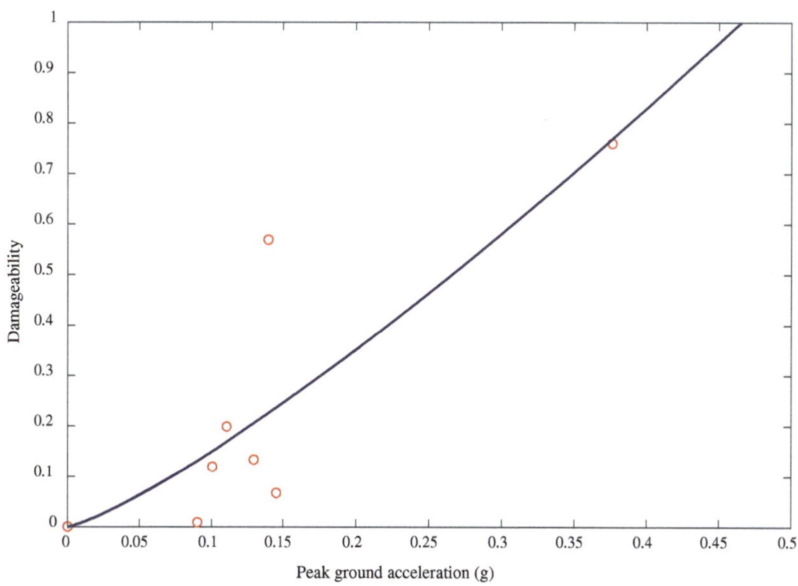

FIG. 2.34 Damageability function for stone masonry buildings from far western Nepal.

Comparing Figs. 2.31 and 2.32, damage to 100% of stone masonry buildings in eastern and central Nepal is expected at 0.9 and 0.7 g, respectively. This shows that stone masonry buildings in central Nepal are more vulnerable than those in eastern Nepal.

Fig. 2.33 shows the damageability curve for stone masonry buildings in western Nepal. The damageability curve shows that probabilities of damage at 0.1, 0.3, and 0.5 g are 10%, 30%, and 50%, respectively. The western stone masonry constructions are found to be less vulnerable when compared with the stone masonry buildings from other regions.

The damageability curve for the stone masonry buildings from far western Nepal is shown in Fig. 2.34. The damageability curve shows the probabilities of damage as 15%, 60%, and 90% for 0.1, 0.3, and 0.4 g. Thus, Figs. 2.31–2.34 depict that the central and far-western stone masonry buildings are more vulnerable than the eastern and western stone masonry buildings.

2.5 The changing landscape of stone masonry construction and strengthening in Nepal

Prior to the Gorkha earthquake, residential stone masonry construction was mostly carried out by untrained masons who decided the size, dimensions, and arrangement of all the structural and nonstructural components of the buildings. The traditional construction systems had some seismic features like wooden bands in some locations; however, they were not universal. After the Gorkha earthquake, most of the rural municipalities at least organized mason trainings to improve construction systems.

In 2015–16, the government of Nepal endorsed several construction techniques for improved seismic performance. Adherence to these endorsed techniques was made mandatory for receiving government grants for reconstruction. One of the most important features of the endorsed improvements is the provision of seismic bands made up of mortar. The endorsed models use reinforced concrete up to the damp proofing course. Three unreinforced concrete bands are then provided at different levels of the structural walls as shown in Fig. 2.35. Provision of such bands is expected to improve seismic performance. They help in integrating different structural elements thereby making them work together against seismic forces. They also provide integrity at corners. Being stronger in tension than the stone masonry units, they may also help by halting crack propagation, and reducing the changes of formation of a failure plane.

The vulnerability of heavy gable walls is not properly addressed in the endorsed models. An example of such a gable wall is shown in Fig. 2.36. In addition, other important aspects such as connections between the orthogonal walls, connection between structural and partition walls, and

FIG. 2.35 The government endorsed earthquake resistant stone masonry building construction practice after the Gorkha earthquake.

FIG. 2.36 Heavy gable portion in stone masonry above the lintel band.

connections between the main structural wall and the projections (Fig. 2.37) are not well-defined and implemented in the construction practice after the Gorkha earthquake. Although the projections as shown in Fig. 2.38 could help by providing buttressing action to the main wall, they may separate from the main wall, or collapse out of plane, the risks of which can be reduced by properly anchoring them to the main wall.

FIG. 2.37 An unconnected projection wall to support the extension of building.

FIG. 2.38 Splint and bandage retrofitting of stone masonry buildings.

A few stone masonry buildings in Nepal have been retrofitted after the 2015 Gorkha earthquake. One of the widely practiced retrofitting techniques that is being adopted in Nepal is splint and bandage (Fig. 2.38). As shown in Fig. 2.38, the corners are strengthened using splint and bandage and containment meshing is provided for other parts of the walls.

Several studies and experience from past earthquakes provide valuable insight on seismic performance of stone masonry buildings and ways to improve it. As discussed in this chapter, there are several features that contribute to the overall fragility of stone masonry buildings, and a holistic approach to address such weaknesses is necessary. These include, for example, load path regularity, connection between orthogonal walls, proper integrity at corners, reduction of gable mass or proper anchorage, interlocking between masonry units, proper flooring, and roofing systems. While seismic bands, on which the current construction practice partially addresses some of these weaknesses, they should not be taken as a cure-all remedy for improvement of seismic performance of stone masonry buildings.

2.6 Concluding remarks

Stone masonry buildings are abundant in many rural settlements around the Himalayan region. Given the high seismicity of the region and very fragile nature of stone masonry constructions, reducing impact of future earthquakes in the region relies heavily on improving seismic performance of stone masonry buildings. This chapter discusses different factors and mechanisms that contribute to seismic fragility of stone masonry buildings in different parts of Nepal. Strengths and weaknesses of masonry construction methods in different regions are discussed and put in the context of seismic performance. Several examples of damaged buildings are provided to highlight the weaknesses in stone masonry constructions and the resulting damage/failure mechanisms. An attempt is made to provide a preliminary idea on the correlation between building collapse and fatality. The results show that the number of fatalities grows exponentially with the number of collapsed buildings. The proposed relation is, however, not precise, as many other factors such as occupancy state of the buildings dictate the number of fatalities in a collapsed building. Damageability functions of stone masonry buildings in different regions of Nepal are presented using damage statistics from past earthquakes. These functions show that stone masonry buildings in Nepal are very fragile to seismic action, with those in central and far western regions of Nepal being more fragile than those in other regions of the country.

Some changes in construction practices of stone masonry building is ongoing after the government of Nepal endorsed, after the destruction caused by the 2015 Gorkha earthquake, some models of improved construction and made them mandatory for payment of reconstruction grants. The improved methods rely mostly on use of seismic bands. Furthermore, training of local masons on seismic resistant construction has been a positive development. It is, however, questionable, whether the government endorsed models are indeed seismic resistant as they proclaim to be. It is also important to realize that safety is a relative concept: it is not possible/feasible to construct buildings that can withstand all levels of possible ground shaking. In this context, it is imperative to relate safety to hazard. For example, a question that seems to remain unanswered is whether the government endorsed models are seismically safe for the expected hazard in the country. Although there is no doubt that they are marginally better than traditional constructions, it is too early to assume that they will resist future earthquakes safely. In this context, a holistic approach incorporating (1) understanding of seismic hazard in the different regions of the country, (2) levels of acceptable risk, and affordable limits of safety, (3) experimentally/scientifically tested methods of construction suitable for the local environment, and (4) continuing research and improvement of construction practices, needs to be focused on. Although improvement of seismic resistance of the built environment and thereby societal resilience to earthquakes is a slow and gradual process, major earthquakes provide an opportunity to engage the stakeholders in accelerating the process. Such opportunities are valuable also in the sense that they expose both the fragility of the built environment and lack of knowledge on how to tackle the problem. As the reconstruction cannot wait for scientific research to endorse improved construction methods, there is a danger of adoption of ad hoc solutions. The primary cause of this unfortunate scenario is the lack of interest among the stakeholders to treat increasing seismic resilience as a continuing process and not wait for the next earthquake to implement another set of marginal improvements. It can only be hoped that this changes with more emphasis placed on scientific research rather than relying on ad hoc solutions well before the next big one strikes.

References

Abo-El-ezz, A., Nollet, M. J., & Nastev, M. (2013). Analysis of uncertainties in seismic vulnerability modeling of stone masonry buildings. In: *Safety, reliability, risk and life-cycle performance of structures and infrastructures—Proceedings of the 11th international conference on structural safety and reliability, ICOSSAR 2013*.

Adhikari, R., Jha, P., Gautam, D., & Fabbrocino, G. (2019). Seismic strengthening of the Bagh durbar heritage building in Kathmandu following the Gorkha earthquake sequence. *Buildings*, *9*(5), 128.

Ahmad, N., Ali, Q., Ashraf, M., Alam, B., & Naeem, A. (2012). Seismic vulnerability of the Himalayan half-dressed rubble stone masonry structures, experimental and analytical studies. *Natural Hazards and Earth System Sciences*, *12*, 3441–3454.

António, A., Aníbal, C., Alexandre, C., Tiago, F., Ana, G., & Humberto, V. (2013). Experimental evaluation of the out-of-plane behaviour of traditional stone masonry buildings in Azores, Portugal. In: *ECCOMAS thematic conference—COMPDYN 2013: 4th international conference on computational methods in structural dynamics and earthquake engineering, proceedings—An IACM special interest conference.*

Bilham, R. (1995). Location and magnitude of the 1833 earthquake and its relation to the rupture zones of contiguous great Himalayan earthquakes. *Current Science, 69*(2), 155–187.

Candela, M., Borri, A., Corradi, M., & Righetti, L. (2016). Effect of transversal steel connectors on the behavior of rubble stone-masonry walls: Two case studies in Italy. In: *Brick and block masonry: Trends, innovations and challenges—Proceedings of the 16th international brick and block masonry conference, IBMAC 2016.*

Costa, A. A., Arêde, A., Costa, A., & Oliveira, C. S. (2011). In situ cyclic tests on existing stone masonry walls and strengthening solutions. *Earthquake Engineering and Structural Dynamics, 40*(4), 449–471.

Gautam, D. (2018). Observational fragility functions for residential stone masonry buildings in Nepal. *Bulletin of Earthquake Engineering, 16*, 4661–4673.

Gautam, D., & Chaulagain, H. (2016). Structural performance and associated lessons to be learned from world earthquakes in Nepal after 25 April 2015 (M_W 7.8) Gorkha earthquake. *Engineering Failure Analysis, 68*, 222–243.

Gautam, D., Chaulagain, H., Rupakhety, R., Adhikari, R., Neupane, P., & Rodrigues, H. (2018). Vernacular masonry construction in Nepal: History, dynamics, vulnerability and sustainability. In H. Rodrigues (Ed.), *Masonry: Design, materials and techniques.* New York: Nova Science Publishers Inc..

Gautam, D., Fabbrocino, G., & Santucci de Magistris, F. (2018). Derive empirical fragility functions for Nepali residential buildings. *Engineering Structures, 171*, 617–628.

Gautam, D., Prajapati, J., Paterno, K. V., Bhetwal, K. K., & Neupane, P. (2016). Disaster resilient vernacular housing technology in Nepal. *Geoenvironmental Disasters, 3*(1), 1–14.

Gautam, D., Rodrigues, H., Bhetwal, K. K., Neupane, P., & Sanada, Y. (2016). Common structural and construction deficiencies of Nepalese buildings. *Innovative Infrastructure Solutions, 1*(1), 1.

Grunthal, G. (1998). *European macroseismic scale 1998.* Luxemburg: European Seismological Commission.

Gupta, S. P. (1988). *Eastern Nepal earthquake 21 August 1988.* Bangkok: Asian Disaster Preparedness Center.

Jain, S. K., Roshan, A. D., Arlekar, J. N., & Basu, P. C. (2000). Empirical attenuation relationships for the Himalayan earthquakes based on Indian strong motion data. In: *Proceedings of the sixth international conference on seismic Zonatio.*

Maio, R., Estêvão, J. M. C., Ferreira, T. M., & Vicente, R. (2020). Cost-benefit analysis of traditional seismic retrofitting strategies integrated in the renovation of stone masonry buildings. *Engineering Structures, 206*, 110050.

Ortega, J., Vasconcelos, G., Rodrigues, H., & Correia, M. (2018). Assessment of the efficiency of traditional earthquake resistant techniques for vernacular architecture. *Engineering Structures, 173*, 1–27.

Parajuli, R. R., & Kiyono, J. (2015). Ground motion characteristics of the 2015 Gorkha earthquake, survey of damage to stone masonry structures and structural field tests. *Frontiers in Built Environment. 1*, 23. https://doi.org/10.3389/fbuil.2015.00023.

Rana, B. J. B. (1935). *Nepal ko Maha Bhukampa (The great earthquake of Nepal).* Kathmandu, (in Nepali) Kathmandu: Jorganesh Press.

Senaldi, I. , et al. (2019). Experimental seismic response of a half-scale stone masonry building aggregate: Effects of retrofit strategies. In R. Aguilar et al. (Ed.) *RILEM Bookseries.* Switzerland: Springer International Publishing.

Senaldi, I. E. , et al. (2020). Experimental seismic performance of a half-scale stone masonry building aggregate. *Bulletin of Earthquake Engineering, 18*, 609–643.

Varum, H., Dumaru, R., Furtado, A., Barbosa, A. R., Gautam, D., & Rodrigues, H. (2017). Seismic performance of buildings in Nepal after the Gorkha earthquake. In D. Gautam, & H. Rodrigues (Eds.), *Impacts and Insights of the Gorkha Earthquake* (pp. 47–63). Elsevier.
Vintzileou, E., Mouzakis, H., Adami, C. E., & Karapitta, L. (2012). Strengthening of historical stone masonry buildings: Experimental testing and modeling of a 2-storey plain masonry building. In: *Life-cycle and sustainability of civil infrastructure systems—Proceedings of the 3rd international symposium on life-cycle civil engineering, IALCCE 2012.*

CHAPTER

3

Traditional timber-laced masonry construction in Turkey known as *himiş*

Polat Gülkan[a] *and Randolph Langenbach*[b]

[a]Başkent University, Ankara, Turkey, [b]Conservationtech Consulting, Oakland, CA, United States

3.1 Introduction

A feature of traditional Ottoman construction practice that is echoed widely in the northern parts of the Mediterranean basin is the use of timber lacing in masonry walls. This timber-laced masonry construction can be divided into two broad types: (1) the use of continuous horizontal timbers (*hatıl*) embedded into bearing wall masonry, and (2) the insertion of masonry in between the columns, beams, and studs of a complete timber frame, which is referred to in Turkish as *hımış* (upper stories in Fig. 3.1 (left); and (Fig. 3.2 (left))). Both construction types frequently exist in the same building, with the bearing wall forming a strong, sometimes fortified, first story base to the structure, and the *hımış* used for the upper floors. The timber framework of the *hımış* has studs rarely more than 0.60-m apart. The studs are themselves tied at mid-story height by other timbers. Because the masonry is only one wythe (leaf) in thickness, the walls are light enough to be supported on the cantilevered timbers. The infill masonry is either brick or rubble stone. The rubble stone type usually consists of small stones set in a thick lime or mud mortar. The local variations in the actual composition of the *hımış* structural system are substantial.

Masonry buildings with or without timber lacing continue to provide shelter to people in many parts of the world. However, even when laced with timber beams, posts and diagonals, as in *hımış*, or timber ring beams, as in *hatil* construction, the impression of most engineers is that the seismic performance of such buildings is poor, and that these forms of construction

https://doi.org/10.1016/B978-0-12-821087-1.00015-6

FIG. 3.1 Safranbolu: (left) Two story of *hımış* resting on one story of *hatil* construction. (right) A fragment of a *hatil*-reinforced stone wall viewed from what had been the interior side, showing the horizontal timber lacing, which is made up with horizontal timbers on the interior and exterior sides of the wall, held together with short cross-pieces. *(Photos by Randolph Langenbach).*

FIG. 3.2 (left) This house developed only interior plaster cracks in 1999 earthquake in the heart of Gölcük, although it was surrounded by collapsed RC buildings. The father of the man in the photo built the house in the middle of the 20th century. (right) View in Adapazari with *hımış* house on left beside several blocks of pancake collapsed RC buildings. *(Photos by Randolph Langenbach.)*

must be discarded as if they are just another form of "unreinforced masonry." They do not lend themselves easily to rational engineering analysis as engineering is now practiced. In addition, the drift toward seemingly sophisticated computer-based analysis is at odds with older forms of masonry construction that are often shaped by tradition.

FIG. 3.3 View of *hımış* house with very little damage next to collapsed four-story RC building after the 1999 Düzce earthquake (Dogangün, 2006). *(Photo by Adem Doğangün.)*

Indeed, most guidelines for traditional construction are a distillation of centuries old practices, supported in only a few exceptional cases by lab experience. The notoriously poor earthquake resistance of adobe and of unreinforced masonry dwellings creates a false urge to dismiss the entire category of traditional forms of masonry construction, even those described here. Yet, this is a misleading impression because there is ample merit in properly conceived and built traditional buildings, even if unsophisticated, whether held together by a timber skeleton, or with timber bands, or even sometimes with little more than good quality floor and roof diaphragms well bolted or framed into the walls (Langenbach, 2007). In fact, it is important to acknowledge that many traditional timber-laced buildings actually meet or exceed the modern definition of reinforced masonry when the timbers are installed correctly and can serve as the tensile and bending reinforcement (Fig. 3.3).

With rare exceptions, buildings in many nonmetropolitan settlements continue to be hand-built by itinerant builders, often violating some of the provisions of current codes. Because of their small size, there is not much in-plant industrialization available in Turkey for most of these buildings. The partial introduction of modern materials and systems into the building process, such as the light weight steel beams used in houses in Iran, or brittle hollow clay tile block in the Turkish-reinforced concrete frame structures, sometimes can result in a poorer earthquake performance than has been demonstrated by traditional construction in recent earthquakes (Langenbach, 2004).

Complicating matters is the fact that popular perceptions of risk may diverge from actual risk. This has frequently been the case for reinforced concrete frame construction, where poor quality construction in this material has so often produced buildings that are more dangerous than the traditional unreinforced masonry buildings they replaced, contrary to the hollow promises made about the merits of concrete buildings. We have little information to be able to determine whether the contemporary nonengineered houses in areas that have undergone strong earthquakes would have performed better or worse if they had been built using entirely premodern materials and systems instead of reinforced concrete or concrete blocks.

To explore if this is true, we focus on Turkey, where traditional forms of construction dating from before the Ottoman Empire have continued into the present. For this chapter, we will focus first on the construction type that in Turkish is called *hımış*, a word that has an imprecise translation into English as "filled." The timber framework is filled with suitably graded brick or rubble stone, or adobe, bound together with lime or mud mortar of widely varying ingredients and quality.

3.2 Traditional timber-laced masonry in Turkey

Earthquakes have been common in many parts of Turkey. This provides an opportunity to study the influence of this risk on local building traditions over the centuries. In addition, the traditional construction practices of the Ottoman Period were not limited to Turkey because the Ottoman Empire, which lasted some six and one-half centuries, had a broad cultural influence over the Middle East and southeastern Europe. With the related Moghul Empire to the east, this cultural influence extended across a sizable swath of the globe, from the Balkans to the Middle East and into Afghanistan and Pakistan and large parts of northern and western India, and particularly the two parts of Kashmir now controlled by Pakistan and India.

Turkey has a moderate climate and, historically, an abundance of wood as well as stone and clay. The Turkish Ottoman-style house, with its tiled roof and overhanging timber-and-brick bays above a heavy ground floor stone wall, has become an identifiable icon recognized worldwide. Where they survive, the overhanging upper stories, or jetties, contribute to the visual vitality and delight of historic Turkish towns. The protrusions also serve a structural purpose because the joists that support the bays cantilever over the lower-story masonry walls and, with their infill masonry, hold these thick masonry walls firmly in place. This compressive force of the weight gives the heavy masonry walls (which are usually reinforced with *hatil* embedments) added strength against lateral forces, as shown in Fig. 3.4 (Gülkan & Langenbach, 2004).

Over the past two centuries, many of the more common houses throughout the northern part of Anatolia have been constructed entirely in *hımış*, usually with brick infilling. Examples of this later type of *hımış*

FIG. 3.4 Houses in Safranbolu, a World Heritage Site, showing *hımış* upper stories with projecting bays, resting on ground floor walls of *hatil* construction. *(Photos by Randolph Langenbach.)*

construction can be found in the areas affected by the 1999 Kocaeli and Düzce earthquakes, which thus provided a unique opportunity to examine their performance in earthquakes that were severe enough to cause widespread collapses of more recent buildings.

Structurally, the masonry in *hımış* is very different from bearing wall masonry. The timber frame remains an important element by providing tensile reinforcement and confinement for the thin layer of masonry infill. The masonry is usually only one wythe of brick, or a thin membrane of partially dressed rubble stone set in a thick mud or lime mortar. If bricks are used, they are sometimes placed at angles so as to form a decorative pattern on the facade as in Figs. 3.5 and 3.6, or one that simply provides a tighter protection against the elements on the exterior facades.

The light timbers and brick infill form walls that are only 100–120 mm thick, a seemingly fragile sheet of thin masonry reinforced with small timbers only lightly nailed together. This combination of a single wythe of masonry held in a framework of light timbers would seem to be vulnerable for safety in earthquakes, yet the widespread reconnaissance of the damage districts after both of the 1999 and subsequent earthquakes has proven otherwise (Fig. 3.6).

Hımış construction is a variation on a shared construction tradition that has existed through history in many parts of the world from ancient Rome almost to the present. In Britain, where it became one of the identity markers of the Elizabethan Age, it was called "half-timber." In Germany it was called "*Fachwerk*," in France (and Haiti), "*colombage*," in Indian and Pakistani Kashmir, as "*dhajji dewari*" (Langenbach, 2010a). In parts of Central and South America a variant was called "*bahareque*" and "*taquezal*" (Correia, Lourenco, & Varum, 2015; Langenbach, 2010b). Ancient Roman examples have been unearthed in Herculaneum, several involving interior partitions,

FIG. 3.5 (left) A two-story *hımış* house in the heart of the damage district of Gölcük. This shows the braces from floor plate to post. Although this building had been abandoned for years, and its brick back wall had been entirely removed for building materials, the earthquake only did the damage that is visible where the bricks have been dislodged at the bottom story and plinth. (right) This is a house in Yuva, Çankırı District, after the 2001 Çankırı earthquake. It shows clearly the "working" (in the engineering sense of the term) of the *hımış* framework with the infill masonry. Because there is no plaster coating on the Gölcük photo, such working of the *hımış* along its joints is not visible. *(Photos by Randolph Langenbach.)*

FIG. 3.6 (left) A *hımış* house in Düzce that has more artistically laid brickwork than is common. It survived the 1999 earthquakes with no damage visible on the exterior. (right) A long abandoned house in Gölcük showing the single wythe, 100-mm thick *hımış* wall after the rear facade had been dismantled for salvage. This is the same house as in Fig. 3.5. This house was derelict and lacking its rear facade from before the 1999 earthquake, but it survived with little more damage than is visible here where some brick panels are missing from the lower part of the wall. *(Photos by Randolph Langenbach.)*

FIG. 3.7 House on the left and interior wall on right are in Herculaneum. They were restored after being found by archeologists after having been buried by the pyroclastic flow from Mt. Vesuvius eruption in AD 79. The original construction from about two millennia before the present, is similar to *hımış* and *dhajji dewari*. *(Photos by Randolph Langenbach.)*

but one involving the construction of an entire two-story row house (Fig. 3.7). The palaces at Knossos have been identified as having possessed timber lacing of both the horizontal and the infill frame variety (Kienzle, 1998). This sets the date of what can be reasonably described as timber-laced masonry construction back to as early as 1500–2000 BCE.

Hımış continued in common use for new house construction in Turkey up until it was rapidly displaced by reinforced concrete frame construction beginning in the middle of the 20th century. This is a relatively late date for the survival of a construction method that is not much different from what was common throughout Europe in the Middle Ages and even the Ancient World, but which has long since gone out of use in most of those other countries. Thus one must ask whether its relatively recent continued use resulted to any extent from the perceptions of earthquake risk. Should it be recognized as an example of a "local seismic culture," as other traditional elements have been?

The answer to this question is complicated by other contributing factors, such as the efficiency and economy of the system compared with using thicker dressed bearing-wall masonry or timber alone. However, partial evidence that seismicity influenced its development and longevity can be found by turning to 18th-century Portugal, where after the great

1755 Lisbon earthquake, the government mandated that reconstruction of the Baixa area include the construction of interior and exterior timber-frame-with-masonry-infill-walls, a system of construction that was named "*gaiola*" (cage) that has also become known as Pombalino construction from its invention under the oversight of Sebastião José de Carvalho e Melo who had become appointed as Marquês de Pombal by King Joseph I (Fig. 3.8) (Langenbach, 1998).

Only 11 years after the Lisbon earthquake, several large earthquakes occurred in the Marmara basin in Turkey. The two largest of these occurred in May and August of 1766, an eventful year for earthquakes in Turkey. These caused widespread destruction in a 300-km wide region that enclosed Istanbul (Ambraseys, 2009; Ergin, Güçlü, & Uz, 1967; Pınar & Lahn, 1952). Most reports from the period focus on the damage sustained by religious, military, and institutional buildings—but all refer to damage in residential buildings caused by what was arguably the most destructive sequence of earthquakes that the imperial capital had witnessed. Contemporary etchings show large fires consuming Istanbul on top of collapsed edifices, confirming both the prevalence of timber houses and explaining why the earthquakes entered the annals of history as the "major apocalypse.".

In Italy, only 28 years following the 1755 earthquake in Lisbon, an earthquake struck Calabria that is believed to have killed approximately 50,000 people. After that tragedy, the *Casa Baraccata*, which is another variation of timber frame with infill construction, was developed and even patented and then used in southern Italy. There was a strong influence from the Portuguese *gaiola* in its development, and like the *gaiola*, the timbers were usually *hidden behind a layer of masonry on the building's exterior, and on the interior* they were hidden by a layer of plaster (Ruggieri, 2015) (Fig. 3.9).

FIG. 3.8 *Gaiola*, also known as *Pombalina* construction, Baixia, Lisbon Portugal. *(Photos by Randolph Langenbach.)*

FIG. 3.9 *Casa Baracatta* in an abandoned building, Mileto, Calabria, Italy. *(Photos by Randolph Langenbach.)*

3.2.1 Does *himiş* exhibit ductile behavior?

Before we embark upon a close examination of analytical and experimental studies of traditional timber construction similar to the Turkish *hımış* under seismic conditions, it will be useful to conjecture what the flow path for internal forces under gravity and lateral force conditions in the system might be. The construction sequence of a timber frame house is that the timber skeleton is erected first, and then the infill is laid in according to some pattern. If the structure is taller than a single story, this sequential construction process is repeated.

At the floor level, there exists some form of a horizontal transfer beam connected to the top ends of the vertical posts or studs. The floor load then flows into those vertical elements, and eventually to the foundation. During this stage the infill masonry is subjected only to gravity loads, that is, its own weight. Neither the roof nor the floor load is a part of what they carry in the vertical direction.

This simplified structural performance is completely changed when different levels of the structure undergo drift. With the lateral force, the timber system will rack and change its geometry, but this is impeded by the masonry component. The degree of interaction is a function of the infill masonry material, its mortar, and the timber lacing geometry including the nailing of the joints and the workmanship quality, so it is not easy to generalize for any particular conclusions. These facts have continued to bedevil engineers in their pursuit of rational design procedures.

How does *hımış* resist earthquakes? Inspections of the interiors of some of the *hımış* houses in 1999 following two earthquakes in the Sea of Marmara region, where such construction is widespread, provided a more complete understanding of the behavior of *hımış* as a structural system. There were clusters of *hımış* buildings in the heart of these districts. These houses, mostly dating from the early part of the 20th century predated the reinforced-concrete apartment blocks that lay in ruins nearby. Most of these older *hımış* houses remained intact. Only a few had significant damage, and of those that were inspected, the damage was often related to structural changes made subsequent to their construction or to extensive decay of the timbers. Structural changes included removal of portions of the exterior and/or interior *hımıs* walls in order to enlarge windows or door openings, or to join rooms together, etc like that seen in Fig. 3.10 (left).

In one example, a second story bathroom had been encased in walls of concrete, making it a heavy and stiff element. This caused the end of the house with that bathroom to collapse, leaving the rest of the structure intact.

The otherwise good performance of the *hımış* houses was confirmed by earthquake reconnaissance teams that conducted detailed statistical studies in several areas of the damage district, finding a wide difference in the percentage of modern reinforced concrete buildings that collapsed when compared with those of traditional construction. In one district in the hills

FIG. 3.10 (left) Example of damage to a *hımış* house from enlarging the windows from 1999 earthquake in Turkey. (right) Building in central Baramulla in Kashmir, India, after 2005 earthquake showing collapse of URM wall with the *dhajji dewari* top floor resilient enough to have bridged over the collapse. Notice that the *dhajji dewari* panels do not have diagonal braces, except at the corners. *(Photos by Randolph Langenbach.)*

above Gölcük, where 60 of the 814 reinforced concrete, four-to-seven-story structures collapsed or were heavily damaged, only 4 of the 789 two-to-three-story traditional structures collapsed or had been heavily damaged. The reinforced-concrete buildings accounted for 287 deaths against only 3 in the traditional structures (Gülhan & Özyörük, 2000).

In the heart of the damage district in Adapazari, where the soil was poorer, records by reconnaissance teams showed that 257 of the 930 reinforced concrete structures collapsed or were heavily damaged and 558 were moderately damaged. By comparison, *none* of the 400 traditional structures of all types, including some of unreinforced masonry without timber lacing, collapsed or were heavily damaged, and only 95 were moderately damaged (Langenbach, 2010b).

It was evident that the infill masonry walls responded to the stress of the earthquake by "working" along the joints between the infilling and the timber frame; the straining and sliding of the masonry confined by the timbers dissipated a significant amount of the energy of the earthquake. The only visible manifestation of this internal movement was the presence of cracks in the interior plaster on the walls of the rooms, revealing the pattern of the timbers embedded in the masonry underneath (Fig. 3.11). This level of damage was evident in every house.

On the exterior, unless the masonry was covered with stucco, damage was mostly not visible because it was confined to hidden parts of the mortar joints. The bricks themselves infrequently were displaced sufficiently for a separation to be visible, except where in some cases small sections of the infill masonry were shaken out of the frame. The movement was primarily along the interface between the timbers and the brick panels where a construction joint already exists. Thanks to the timber studs that subdivided the infill, the loss of portions or all of several masonry panels did not lead progressively to the destruction of the rest of the wall or put the building at risk of collapse. The closely spaced studs prevented the propagation

FIG. 3.11 (left) Düzce showing the house where the photo of the *hımış* wall (center) was taken after the earthquake that collapsed the RC buildings behind it. The "working" of the infill masonry in the *hımış* is shown by the debonding of plaster over the timber framing of the panels. (right) Collapse of a typical brittle interior hollow clay tile infill wall in a partially collapsed RC building. *(Photo (left) by Roger Bilham, NOAA/NCEI (middle and right) by Randolph Langenbach.)*

of "X" cracks within any single panel, and thus reduced the possibility of having all of the masonry falling out of an entire framed wall.

An important additional positive factor in the performance of the walls was the common use of weak rather than strong mortar, as counterintuitive as that may seem. The traditional use of mud or weak lime mortar encouraged sliding along the bed joints instead of cracking through the masonry units when the masonry panels deformed. This served to dissipate energy and reduced the likelihood of incompatibility between rigid masonry panels and the flexible timber frame. The basic principle in this weak, flexible-frame-with-masonry-infill construction is that there are no strong and stiff elements to "attract" the full lateral force of the earthquake. The buildings thus survive the earthquake ground shaking by not fully engaging with it, in the same way that a palm tree can survive a hurricane which can topple nearby oak trees.

This "working" during an earthquake can continue for a long period before the degradation advances to a destructive level. The comparison of the postearthquake view of the *hımış* wall in the center of Fig. 3.11 with the destroyed hollow clay tile wall in a partially collapsed RC building in Gölcük shown on the right in Fig. 3.11 are in support of this assertion (Langenbach, 2000).

Thus we can say that, while *hımış* structures do not have much lateral strength, they do have substantial lateral drift capacity. These buildings respond to seismic forces by swaying with them rather than by attempting to resist them with rigid materials and strong, but brittle, connections. This is not an elastic response, but a postelastic one that manifests resilience rather than strength. When these structures lean in an earthquake, they do so with incremental low-level cracking that is distributed throughout

FIG. 3.12 (left) The modest two-story *hımış* house in Gölcük not only coped with extreme ground motions during the M7.4 Kocaeli earthquake in 1999 when thousands of RC buildings experienced abject collapse, but also saved its inhabitants from harm to life and limb, and provided shelter afterwards. (right) Exposed base of a *hımış* house following the 1998 earthquake in the Adana region. Note masonry crushing at ends and the working of the base causing the de-bonding of the stucco at the bottom of the middle wall panel. *(Photos by Polat Gülkan,)*

the wall because of the confinement of the masonry by the timber framework. The effect of a severe working along the base of a *hımış* house in the 1998 Adana earthquake is given in Fig. 3.12 where crushing at ends of the base caused by overturning and the sliding arrested by the now exposed timber lacing are evident.

In other words, although the masonry itself is brittle, a masonry wall with mortar that is weaker than the masonry units behaves as if it were ductile. By confining the masonry and maintaining the structural integrity of the building, the timber components permit larger drifts than would be the case if the wall were unreinforced masonry.

3.3 The search for scientific proof

Experimental research on typical Turkish or other traditional timber frame with masonry infill buildings is scant, while that for modern steel and concrete frame or reinforced concrete block structural systems is plentiful. This has left a gap within the structural engineering community in terms of acceptance of such structural systems for new construction, or even for changes of occupancy such as the re-use of such buildings for schools and community buildings. The most likely response one might hear from a local building official is *"That may be fine, but show me your calculations."*

Structural calculations are usually not available or are based on assumptions that do not fully reflect the structural system at hand. Owner-built buildings with wide differences from one building to another do not lend themselves to structural calculations, and this is particularly true for those built largely of traditional forms of masonry.

Adding to this problem is the fact that, while earthquakes are a natural structural threat for which the codes contain structural requirements, unlike tornados or tsunamis, but it is the only one of these natural threats for which the codes allow a design for force levels that are less than that which would prevent structural damage from a "design level earthquake." In other words, the codes allow for design calculations based on linear elastic analysis with reduced force levels. The objective of these codes is thus not to prevent structural damage, but to reduce the likelihood of collapse. To require a structural strength that would avoid structural damage for anything other than nuclear power plants and "immediate occupancy" buildings would be so costly as to make ordinary construction impossible. Thus ordinary code conforming buildings may even have to be replaced after a design level earthquake, but the code is aimed toward reducing the likelihood of collapses that would be lethal to the occupants.

Then one can ask: *"what is an engineer to do with a premodern structural system that deliberately has very little elastic strength, and for which the capacity and resilience in an earthquake is almost entirely a product of its postelastic behavior, as is the case with* hımış?" If calculations were to be done from a code analysis standpoint, a conventionally trained engineer would most likely simply treat the infill masonry as dead weight and analyze the structure as a timber frame using linear-elastic structural analysis equations.

However, structurally these buildings are not frames but solid walls, and the masonry has a major lateral force-resisting role, and thus it is incorrect to treat it simply as dead load. The engineering term that can probably best represent it from a structural standpoint is "confined masonry," but that term now has a very specific and different meaning in the field of reinforced concrete infill-frame construction because the construction of the masonry walls precedes the placement of the concrete for the confining frame. For example, conventional "confined masonry" most often involves infill masonry panels laid with strong but brittle cement mortar, and the size of such panels is that of the entire opening between the beams and columns of the moment-frame or shear wall structure that confines them.

In a *hımış* or *dhajji dewari* building, the postelastic resilience comes from the early onset of friction between the masonry infill and the timbers, rather than the nonlinear deformations of the steel or reinforced concrete beams and columns—as would be necessary in modern reinforced concrete or steel construction. The sustained interest of the structural engineering community in unraveling the postelastic properties of reinforced concrete frames with masonry infill to more fully understand the effects (1) of different mortar and masonry unit characteristics, and (2) different layouts of door and window openings, bears testimony to the difficulties that must be overcome. For seismic risks, the difficulties are compounded by the fact that the infill components are under biaxial loads, where the in-plane behavior is driven by drift, but the out-of-plane response is inertia force controlled, with a strong interaction between them.

It is now useful to digress from the principal narrative of this chapter to examine a number of experimental and theoretical investigations that have focused on the seismic behavior of *hımış*-like load carrying systems in different regions. It is important to gloss over the local variations between these different systems to focus on the common mechanical similarities that will enable later generalizations.

First, even when the timber frame is an imperfect truss with some diagonals connecting only onto the vertical posts, the gravity loads are transmitted by these posts to a sill resting on the foundation system, or onto a bearing wall reinforced with *hatillar*. The strength and stiffness of the infill masonry is only mobilized when the ground motion during an earthquake causes racking of the timber framework. The degree to which the infill and the timbers form a composite structure when racked is not open to straightforward formulation. Connection details, nailing of the joints, strength of the infill, the strength and brittleness of the mortar material strength and brittleness and workmanship all play mechanical roles that are difficult to formulate.

The roof or floor diaphragm in timber frame buildings are not rigid, so the seismic inertia effects caused by gravity loads transmitted to the vertical panels are proportional to the tributary areas of the panels. Then, there is no need to distribute the lateral load among the vertical panel elements in proportion with their rigidities because a flexible diaphragm does not transmit in-plane forces to the resisting components in the same way as a rigid diaphragm.

Another trait that sets *hımış* apart from the more customary forms of rural home construction is the degree of informality that goes into its creation. The lateral load capacity for a house that is entirely made of brick masonry might be assessed on the basis of the cumulative areas of the solid wall segments in a particular direction, but this device will not work when the wall segments are nonhomogenous, with wide variations in their subdivision and even the workmanship.

3.3.1 Construction of *himiş homes* in Turkey after the 1999 earthquakes

Interestingly, while one might have thought that surviving buildings could serve as models for a government program for reconstructions after an earthquake, in Turkey one can only find individual examples of self-built *hımış* houses in the large 1999 damage districts, such as that shown in Fig. 3.13C. Years before, the father of the man in Fig. 3.13B had constructed the *hımış* house in Fig. 3.13A. Prior to the 1999 earthquake, this man had started construction of a reinforced concrete home for himself and his family. When he saw the extent of damage and collapse of concrete buildings around him in 1999, he switched to *hımış*, as seen under construction in Fig. 3.13C.

A total of some 60,000 new homes were built by the government in Turkey following the devastation caused by the two 1999 earthquakes, yet not a single one of the new buildings was of the type described in this chapter (Gülkan, 2000). The one completed postearthquake example shown in Figs. 3.14 was constructed after the earthquake as a model in Düzce by a charitable organization.

3.3.2 *Hımış* in comparison with *dhajji dewari* in Pakistan after 2005 Kashmir earthquake

For a more extensive program of reconstruction using traditional timber with infill masonry, one may turn to Pakistan after the 2005 Kashmir

FIG. 3.13 *Hımış* timber frame house on the left was constructed by the father of the man in the center who was constructing a house of RC construction when the 1999 earthquakes convinced him to switch to himis, which is shown on the right under construction in Düzce. *(Photos by Randolph Langenbach.)*

earthquake. Unlike 1999 earthquakes in Turkey, where the damage was much more focused on urban areas that were mainly filled with reinforced concrete multistory construction (with scattered examples of *hımış* construction), the epicenter of the earthquake in Pakistan was in the foothills of the Himalaya Mountains, where there were many buildings of rubble stone with mud mortar. Many of these were completely leveled by the earthquake, as seen in Fig. 3.15.

The traditional construction similar to *hımış* that had once been common in the area is known as *dhajji dewari* (which means "patchwork quilt wall" in ancient Persian), whereas the more recent owner-built houses were simply built of rubble stone with mud mortar. In the damage district, most of these collapsed, as seen in Kamsar, near Muzaffarabad, close to the epicenter of the quake, shown in Fig. 3.15.

FIG. 3.14 A *hımış* house constructed by UMCOR, a charitable organization for relief.

FIG. 3.15 Collapse of unreinforced concrete block masonry houses in Kamsar near Muzaffarabad, Pakistan. *(Photo credit: Naeem, A., Qaisar A., Javed, M., Hussain, Z., Naseer, A., Ali, S. M., Irshad A., & Ashraf. M. (2005).* A summary report on Muzaffarabad earthquake, Pakistan. *Peshawar: Earthquake Engineering Center at the Department of Civil Engineering, N-W.F.P. University of Engineering and Technology. Available on http://www.Reliefweb.int.)*

At first, the Pakistan government provided financial assistance only if an affected family reconstructed their house with reinforced concrete or concrete block as specified by the government's Earthquake Reconstruction and Rehabilitation Authority (ERRA), but many rural families set about rebuilding for themselves with the only materials at hand from the ruins of their houses, and the trees from their woods and farmland. Only by so doing could they avoid the strenuous carrying of the heavy materials for reinforced cement block construction on their backs to remote sites often only accessible on foot. For this rebuilding, they often resurrected the *dhajji dewari* technology for the simple reason that, when they climbed out from under their collapsed houses, the survivors could see that the only houses still standing in their village after the earthquake were of *dhajji* construction. Even in the remote rural village of Thub, near Bagh, the reinforced concrete general store collapsed, but the nearby two and a half story block of shops of *dhajji* construction remained standing Fig. 3.20A.

About a year and a half after the earthquake, ERRA came to understand that *dhajji dewari* was the only proven safe construction system that was practical in the mountain areas of Pakistani Kashmir, and thus approving its use for financial assistance. Approximately a year later they approved *bhatar*, the bearing-wall masonry equivalent of *hatillar* in Turkey.

What makes this experience relevant to the search for engineering science verification of *hımış* and *dhajji* is that following this government sponsored reconstruction program, the London-based international structural engineering firm, Ove Arup & Partners, undertook a highly detailed computational nonlinear finite element analysis of *dhajji dewari* construction The author, Kubilây Hiçyılmaz, reported that in Pakistan he came across *dhajji dewari* in 2005 and 2006 when on secondment to an NGO to teach earthquake engineering to the affected population (Hiçyılmaz, 2011).

This project coincided with a physical testing program undertaken in the Earthquake Engineering Center (EEC) at the University of Engineering and Technology, Peshawar (UET Peshawar), Pakistan. The conclusion was that the level of correlation was good, as was illustrated in Hiçyılmaz (2011) with Fig. 3.16. Unfortunately, the UET Peshawar testing did not include anything similar to the Arup "lack of fit." In their tests, the diagonals ran from corner-to-corner. As can be seen in Fig. 3.16B and C, the diagonal bracing tended to jack the frame perimeter members apart, as could be seen in the Arup FE model in Fig. 3.19, a detail of which is shown in Fig. 3.16C.

Looking at Fig. 3.16, one can see that the most prominent feature of both the physical model and the analytical model is the "X" bracing in the *dhajji* wall. The story that unfolds from this is focused on this particular feature, which is also relevant to *hımış* construction in Turkey. The Arup report

FIG. 3.16 (A) The ERRA sponsored manual on *dhajji* construction. (B) A detail view of the UET model of a *dhajji* wall with corner-to-corner X braces, showing the side post having been jacked off of the sill plate by the braces. (C) View of the Arup model showing the same results. This is before the implementation of the "lack of fit." *((A) Credit: Tom Schacher, Swiss Agency for Development and Cooperation. (B) & (C) Images from Hiçyılmaz, K. (2011). Dhajji Dewari, affordable seismically resistant and sustainable housing: Non linear response history analysis Revision C, job # 215631. Arup Gulf Ltd.)*

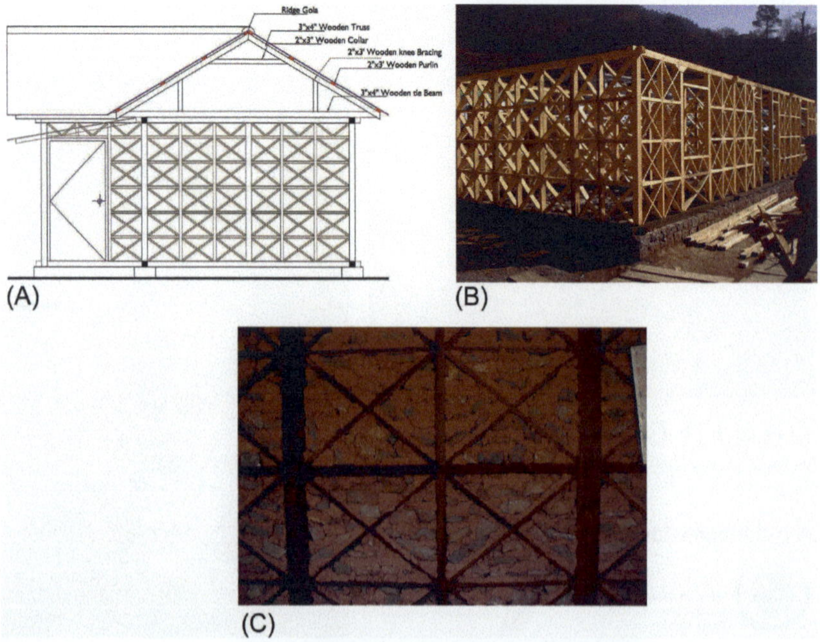

FIG. 3.17 (A) Typical side elevation of an "engineered" *dhajji dewari* building after the 2005 Pakistan earthquake, (B) "engineered" *dhajji* building frame under construction, (C) post-2005 earthquake house with X bracing as shown in (A) and (B) with rubble stone with mud mortar infill. *(Images from World Housing Encyclopedia (WHE), EERI and IAEE (2020). Available at: www.world-housing.net/.)*

states that their choice of their model for the FE analysis was the following: *"In collaboration with UN-Habitat Pakistan, it was decided to base the engineering analysis as closely as possible on the building layout shown in* Fig. 3.17." (Hiçyılmaz, 2011).

Indeed, this configuration with X braces has now been embraced as an accepted structural system for the reconstruction of houses throughout the damage district in Pakistani Kashmir (Fig. 3.17B). This can be compared with Fig. 3.18 of a house constructed by the rural carpenter without any formal training. In this case, the diagonals do not follow any particular pattern, nor are they meant to be artistic. Rather, the one thing they do *not* do is extend from corner-to-corner as they would if this were in a structural truss.

Traditional *dhajji dewari* often has many diagonals in more random patterns but making an FE model demands repetition which further reinforces the use of the repeated X bracing. This X-braced model is

FIG. 3.18 Ad hoc version of a *dhajji dewari* house, with its owner-builder, located in the rural mountain area between Bagh and Muzaffarabad, in Pakistan Kashmir. *(Photos by Randolph Langenbach.)*

described in Fig. 3.17A as "engineered" (Hiçyılmaz, 2011). The new construction using X bracing seen in Fig. 3.17B is also described as "engineered" in the Earthquake Engineering Research Institute (EERI) World Housing Encyclopedia entry for *dhajji dewari* (WHE, 2020).

What is particularly interesting is that first runs of the FE model in Fig. 3.19, which took almost 3 days of continuous computer time even after enlarging the masonry units to reduce their quantity, produced a video that showed the model appearing to burst apart after only four simulated earthquake shakes. From the very first reversal, the vertical groups of X-braced panels rocked back and forth, breaking free from the timber that forms the sill, while the masonry, held in place by the X braces, did not distort until all of it fell into a heap. This process began with the vertical posts being jacked up off of the sill plate by the corner-to-corner X braces. Then, after only four shakes, they simply spread apart, as can be seen in Fig. 3.19C. Thus by being locked into the corners of each panel, the diagonal braces prevented the postelastic mobilization of the infill masonry from sliding against the timber framework and then against itself by forming sliding joints within the mud mortar. It is this sliding back and forth without falling apart that is needed to produce the large amount of hysteretic damping from the friction that is required to maintain the integrity

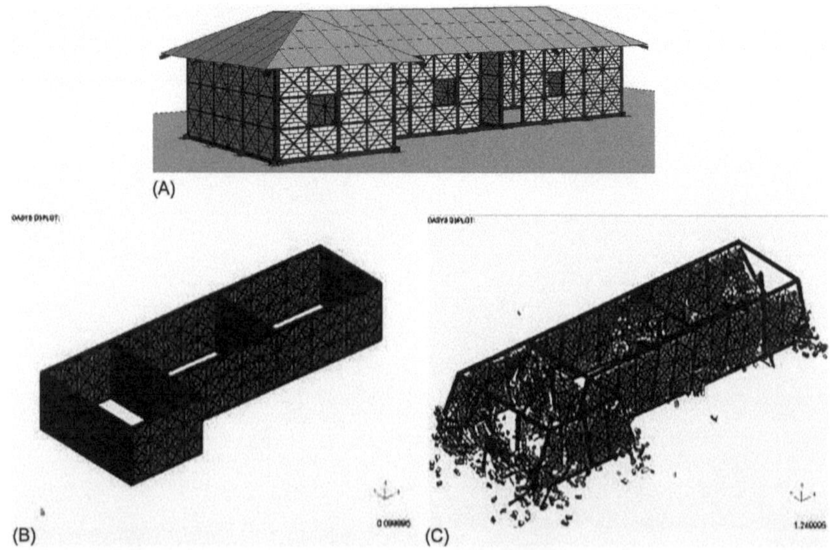

FIG. 3.19 (A) The FE model in Hiçyılmaz (2011), (B) intact walls, and (C) the walls after only four cycles of virtual loading. *(Images from Hiçyılmaz, K. (2011). Dhajji Dewari, affordable seismically resistant and sustainable housing: Non linear response history analysis Revision C, job # 215631. Arup Gulf Ltd.)*

of the structure through the earthquake, but this is only available if the panels can distort as they rock back and forth, without breaking apart.

It is certainly plausible that having a coherent diagonally braced frame with a single wythe of masonry infill would be stronger and more resilient in an earthquake than a random and fully nonengineered assembly of diagonal timbers within the masonry infill. However, further research findings disproved this assumption, and in the 2005 earthquake, there was no evidence in the reconnaissance of the damage district that observed this kind of collapse with any *dhajji* structure. In fact it was quite the opposite, which explains why the rural farmers had resurrected the *dhajji* system to rebuild their collapsed rubble stone houses (Fig. 3.20).

The suggestion was then made to remove the diagonals from the Arup FE model, but the complexity and cost of "building" such a mathematically rigorous model on a computer made this impossible. The solution that the analyzing team arrived at was to shorten the diagonals, which is described as "lack of fit." Hiçyılmaz also experimented with the addition of overburden weight, which in real life would be the weight of the roof and attic floor. Once the overburden weight was sufficient to cause the frame with its infill masonry to distort, the hysteresis loops grew larger with each increase in overburden weight as shown in Fig. 3.21.

FIG. 3.20 (A) This roadside two story dhajji block of shops on the road to Thub survived with no more damage than the visible hole in the wall. (B) The large house of *dhajji* construction at the top of the slope above the village of Thub in Pakistani Kashmir survived the earthquake intact, while the rubble stone with mud mortar houses in the village below crumbled as seen on the right in (D). (C and D) Views of the reconstruction of a house in the village of Thub reviving a style of *dhajji dewari* that had been characteristic in the mountain areas of northern Pakistan. The rubble stone is laid in between timber studs on short diagonal timbers. The ruin of the rubble stone house being replaced is on the right, and the home owner is seen standing with his carpenter next to his new *dhajji* home. *(Photo (C) by Maggie Stephenson. Other photos by Randolph Langenbach.)*

In brief, as can be seen in Fig. 3.21, the Arup findings revealed that there was a 9%–21% increase in energy absorption when the braces were reduced in length, which allowed the masonry units to slide across each other (Hiçyılmaz, 2011).

The analytical features, which mimic "reduction of hysteresis pinching" and "energy absorption," are among the most significant findings of this FE analysis of *dhajji* construction. An increase in gravity load is necessary to activate this feature, but in real life at least some overburden

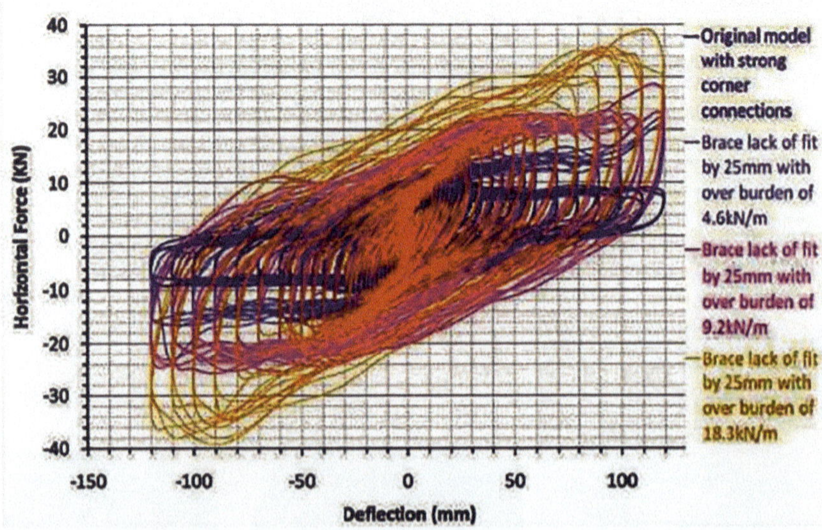

FIG. 3.21 Arup model: Hysteresis loop comparison. *(From Hiçyılmaz, K. (2011). Dhajji Dewari, affordable seismically resistant and sustainable housing: Non linear response history analysis Revision C, job # 215631. Arup Gulf Ltd.)*

is usually inherent in the urban buildings in Srinagar, some of which are as tall as four stories with *dhajji* construction for every story with very few or no diagonal braces (Fig. 3.24). Moreover, the single story buildings seen in the hillside villages were of a different configuration of studs, braces, and mud-mortared stonework, as in Fig. 3.20.

3.4 *Hımış* in comparison with *perete de paiantă* construction in Romania

Further confirmation is found in general findings in relation to the lateral drift resisting mechanism of similar studies in Romania by Dutu, Niste, Spatarelu, Dima, and Kishiki (2017), as confirmed by Parisse (2019). These works focus on timber-frame-with-infill masonry construction of a similar type as *hımış* and *dhajji dewari*, which in Romania is called "*perete de paiantă*" or *paiantă* wall construction.

Dutu and colleagues constructed two timber and masonry models shown in Fig. 3.22, with the only difference being the angle and length of the braces. What makes this relevant to the discussion about Turkish *hımış* is that, when comparing it with the X bracing that has been adopted

FIG. 3.22 Specimen S1 (left) and S2 (center) with masonry infill and diagonal struts in the last test cycle. These photos on the right show how much the diagonal braces in S1 were able to slide along the post in response to the lateral force, giving greater ductility and energy dissipation than corner-to-corner diagonals did in (S2). *(Photos by Andreea Dutu (from Parisse, F. (2019).* Numerical modelling of the seismic performance of Romanian traditional timber-framed buildings. *Master's Thesis in Structural Analysis of Monuments and Historical Constructions, University of Minho).)*

for most of the new *dhajji* construction in Pakistan after the 2005 earthquake, the common form of braces did not form concentric bracing. Instead, it may be that the braces were used as a temporary support for the columns during the construction of the house.

Despite a lack of present-day knowledge about any connection to earthquake risks, the lab testing reveals a significant difference in terms of seismic engineering. In short, the Dutu and Parisse research confirms a significant difference in performance between the model S1 with the corner-to-side diagonal braces, and that of S2 with the corner-to-corner diagonals. The four photos in Fig. 3.22 show the friction sliding of the diagonals. The tests revealed that the energy dissipation of S1 was higher than S2, with a value of 3.22 kN m compared with 2.68 kN m. The damping in S1 was higher as well, and S2 did show heavier damage than S1 (Dutu et al., 2017; Dutu, Sakata, Yamazaki, & Shindo, 2015; Parisse, 2019).

Perhaps most comparable with the Arup numerical tests is the finding that S1 showed only 1 mm of uplift in the bottom connection, while that in S2 was up to 25 mm, a difference that is large enough to break the structural connection that would hold the frame together (Parisse, 2019, p. 45). This is an important difference because what one sees in the numerical model, without the settings for "lack of fit," shows a breaking apart at the base and top of each set of panels. Thus these tests reveal an impressive difference in uplift in the bottom connection for the lower ends of the vertical posts, together with the sliding at the upper ends. Dutu et al. (2017) and Parisse (2019) both concluded that the brace configuration in S1, seen also in the houses shown in Fig. 3.23, is "considered representative" of the houses of *Perete de Paiantă* construction in Romania.

FIG. 3.23 Historic houses in Romania. Both show braces which do not go corner-to-corner, but instead connect to the posts just below the plate, as Dutu et al. (2017) has reported as typical. This is similar to *hımış* construction in Turkey. Note: The top of the brick wall in the photo on the right is slightly obscured by the timber siding covering the gable. *(Photos by Andreea Dutu.)*

FIG. 3.24 (left) A three-story *hımış* building in Gölcük, Turkey after 1999 earthquake which has no visible damage. (right) A three-story *dhajji dewari* building in Srinagar, Kashmir, with no diagonals. *(Photos by Randolph Langenbach.)*

One can also see in Fig. 3.24 an example of a three story building in the heart of the damage district of the 1999 earthquake in Gölcük, with diagonals that only go from bottom plate to the side posts. Next to it is a three story building *dhajji dewari* building in Srinagar with no diagonal braces. From those observations we now return to *hımış* in Turkey, where we find a strong parallel with brace configurations in Romania documented by Dutu and Parisse.

3.4.1 *Hımış* in Turkey

Returning to Turkey, it is interesting to compare the Romanian study with the principal characteristics of Turkish *hımış* construction. The lateral force testing of the two frames with the differing configuration of the diagonal braces brings into focus a very interesting and important characteristic of almost all *hımış* construction in Turkey. It turns out that hardly any of the lateral bracing to be found in Turkey extends from corner-to-corner. The braces almost always stop short of the corner of the panels. They then connect to the columns of the frame rather than the top beams, as can be seen in Figs. 3.1–3.6 at the beginning of this chapter.

While the Hiçyılmaz (2011) investigation and the studies by Dutu et al. (2017) and Parisse (2019) provide a direct comparison of the difference in terms of seismic resilience and ductility of the two systems of bracing, this observation about the braces in Turkish *hımış* finds that they are almost always connected below the top of the panel. The practical explanations for this detail (such as ease of construction and the bracing of the structure before the masonry has finished being laid to serve that purpose) are all plausible. However, it is likely, historically, that the earthquake risk may have played some part. In fact, earthquakes may have contributed to the earlier adoption of this method, which may have initiated a tradition even if the reasons for this practice have been forgotten over time. For example, in Northern Europe and Great Britain, where earthquakes are not a factor, X-braces and other corner-to-corner diagonals are common, but not in Turkey or the surrounding countries, at least until this recent widespread use of cross braces in Pakistan for government-related reconstructions (described earlier) following the 2005 earthquake.

The next question inevitably must be: why did this not lead to a revival of this feature when the Government of Pakistan's accepted *dhajji dewari* construction for financial assistance, and why it is not in the government-issued design guidelines and instructions to carpenters and masons? This is not an easy question for us to answer. In fact, when they resurrected the *dhajji* construction system for their own homes to replace their collapsed rubble stone houses, the itinerant rural farmer-builders used the more flexibly braced construction tradition. Only rarely did they use corner-to-corner X-braced infill walls (Fig. 3.20).

The answer to the question about the design of diagonal braces in post-earthquake Pakistani *dhajji* (*above*) may be as profound as it is simple. Structural engineering has gone through a revolution over the past century. In the past decades of the 19th century, buildings of increasing height began to exceed the practical height limits for bearing wall masonry structures, and with the invention of the Bessemer converter for the making of industrial quantities of steel, the building of multistory steel frame structures became practical and competitive. This achievement required a shift in engineering practice from a largely empirical process of the design of solid walls, vaults, and domes of masonry, to one of rigorous analysis. To meet this new demand, the teaching and practice of the structural engineering of buildings of all sizes moved away from the design of solid wall structures with post and beam interiors to the analysis and design of frames.

Until the 19th century, frame structures and the internal framework of buildings were most often made of timber, with the lateral forces resisted by masonry walls acting in shear and/or by braces within the heavy timber framework. These changes also had an impact on remote rural mountain environments in a postdisaster environment when central governments become involved, as it did in Pakistan after 2005 Kashmir earthquake, and in Turkey after the 2001 Çankırı earthquake affected villages around Orta in a sparsely populated rural area in the province of Çankırı.

In the villages of Yuva and Elden, the government condemned the original *hımış* houses when their inspectors incorrectly concluded that they had been damaged beyond repair. The government then proceeded in these two villages to rebuild them on new sites. Sadly, most of these new houses have remained unoccupied because the government had completely failed to understand the sociology of the situation, focusing instead on their assessment of some very remote technical risks. The new housing was remote from the barns and fields for the farm animals, and also from the mosque and general store that remained in the middle of their old villages, which, in the case of Elden, was many kilometers distant (Fig. 3.25).

Even more shocking is that many of the new houses were constructed of unreinforced hollow clay tile masonry with concrete roofs, a system that is far more dangerous than that of any *hımış* house, even in its damaged state. None of the *hımış* houses collapsed, and the level of damage was such that they could have easily been repaired as was proved in a nearby farm village of Asagi-Kay. There the population was younger and still actively farming, more than were the aging families of Yuva and Elden. The problem was, and continues to be, that the government damage inspectors were untrained and without knowledge of the seismic resilience of the traditional forms of masonry construction found in all of these villages (Figs. 3.26). In all of these villages, the masonry infill did the "working" (in the engineering sense of the word) that kept these seemingly weak and flimsy buildings standing. Cracks were most often apparent only where the wall had been plastered, but in the eyes of the inspectors, trained only to understand how frames are analyzed, and to look only for cracks, they concluded the houses were unsafe. Often, they then persuaded the occupants that they were as well.

In fact, the sad fate of these traditional *hımış* houses after the earthquake is emblematic of the revolutionary changes to the field of structural engineering of the transition from its focus on solid walls to frames, because to these inspectors taught only about frames, it seemed inconceivable that, with the abundance of visible cracks and some gaps in the masonry, these houses could still be safe. In truth, the cracks are a product of exactly the kind of energy dissipation that gave these buildings the resilience they demonstrated, even in the much larger Kocaeli and Düzce earthquakes in 1999 (Aschheim, Gülkan, & Sezen, 2000).

An interesting fact about the historical development of the modern reinforced concrete and steel frame construction, and of the portal frame analysis that evolved for engineers in the late 19th and early 20th centuries, is that thick masonry infill and cladding was at that time an often-accepted element of the early steel and reinforced concrete buildings. This was true even though then, as now, it was not included in the engineering

FIG. 3.25 (A) This *hımış* house in Yuva, Turkey, shows the couple who lived in it until the 2001 Çankırı earthquake earthquake standing in front of it. The earthquake damage was limited to an opening of cracks in the plaster and exterior stucco only, as can be seen in (B). A view of Yuva with its traditional houses is in (C), and the government constructed new houses are visible in (D), and in (E) one of those houses under construction before being stuccoed reveals the use of the unreinforced hollow clay tile blocks with a concrete attic floor slab—a far more dangerous construction was used for these new houses, than is found in the older homes like that seen in (A). The couple in (A) have not wanted to move to the new houses, because the old village seen in (C), down the road on the left side of (D) is where their gardens, barns and mosque are are located, all of which is alongside a tree-lined river. *(Photos by Randolph Langenbach.)*

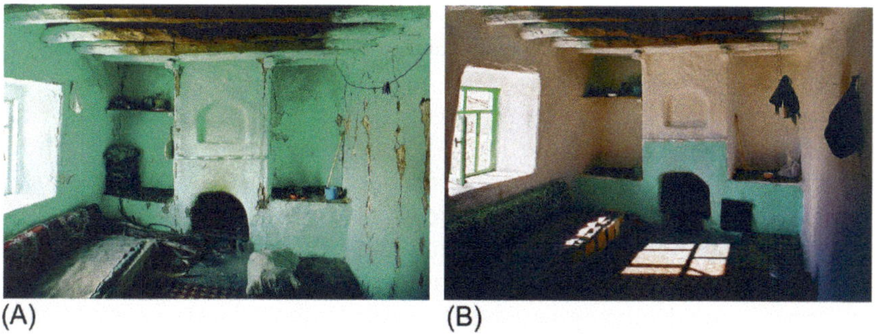

FIG. 3.26 (A) A *hımış* farmhouse in Asagi-Kay after quake, (B) after repairs done by the owners without government assistance. *(Photos by Randolph Langenbach.)*

calculations for lateral resistance, except as dead weight. In fact, the eminent author in 1901 of the first textbook on the first generation of steel frame multistory construction, Joseph Kendall Freitag, states frame construction is *"dependent largely for its efficiency upon the exterior and interior [masonry] walls and partitions which serve to brace the structure…* [but there is] *still no definite or even approximate values [that] can be given to such items, except their purely static resistance and weight* (Freitag, 1901).

In terms of engineering practice, the linear-elastic portal frame analysis of such structures has come to define most of the day-to-day professional engineering work for multistory buildings for many years. Moment frames provide lateral resistance in both shear and flexure of the framing members. The enclosure and partition walls that turn this open framework into a useable building are routinely ignored in the structural calculations except as dead weight. The advantage of this approach is that it has allowed for a coherent mathematics-based engineering approach to building design by separating the infinite complexity of a finished building with all of its parts from that of the primary structural system, the frame.

3.4.2 Tests on Turkish *hımış* in Turkey

Now we turn to a seismic engineering study focused directly on *hımış* in Turkey. It also includes physical testing of full scale models that reproduced *hımış* walls identical to those in actual houses. All of the sources for these walls were from houses in Safranbolu, a World Heritage Site in northern Anatolia with an acclaimed collection of historic houses of *hatillar* and *hımış* traditional construction. What makes this particularly interesting is that the tested models reproduced not only several different configurations of timber framing but also the infilling or cladding historically used on these frames. This included both fired brick and adobe infill, and *bagdadi* and *samdolma* cladding, instead of infill (Aktaş, Akyüz, Türer, Erdil, & Sahin-Güçhan, 2014).

This project was begun by a project team of professors, together with two PhD students, at Middle East Technical University (METU), under a grant from TÜBİTAK. The grant funded the construction of full size frames which reproduced wall sections from houses in the World Heritage site of Safranbolu in northern Anatolia. This consisted of eight frames with the timber arrangements shown in Fig. 3.27. After the frames were constructed, the testing and analysis came to be the subject of a PhD Thesis in Restoration by Yasemin Aktaş.

The first reversed cyclic loading tests were without infill and then with infill. All specimens with no infill had a specific pattern of nailing at the joints. After the nails had "worked" as the frames went through imposed drifts, they were replaced with the same pattern of nailing before the laying in of the infill masonry. The frames were then forced to go through a similar drift history (Aktaş, 2011).

The infill material (adobe or brick) and the cladding (either *bagdadi where* 40-mm wide wood strips shown at right in Fig. 3.28) or *samdolma* (100-mm wide laths) were the variables considered. Fig. 3.27 shows the measured cyclic response of these frames. It shows that both cladding and infill increased hysteresis loops when compared with the bare frames (Aktaş et al., 2014).

Aktaş et al. (2014) lead off the discussion of the findings by explaining that the purpose of their study was specifically to find an engineering science explanation for the frequent empirical observations of the good performance of *hımış* construction in earthquakes: "*There have been numerous studies reporting that the traditional Ottoman hımış houses with timber-frame construction are likely to behave more seismically resistant under earthquake loading than the more recent reinforced concrete construction type. All of these studies remained qualitative in nature and were based on visual observation.... For this reason, the seismic resistance...*[including] *energy dissipation capacity and secant stiffness... of traditional Ottoman hımış houses with timber-frame construction was evaluated experimentally in this study.*"

When comparing the bare timber frames with the different infilled frames, the principal conclusions of the investigators can be summarized as follows. The infill or cladding increased the strength compared with the bare frames in the order of 1.74–2.25 times, and the increase in stiffness caused by infill and cladding ranged from 2.2 to 4 times the initial values. The study is useful in confirming that the capacity and stiffness of the timber-alone model would approximately double when infilled or clad in similar fashion to the types of construction described.

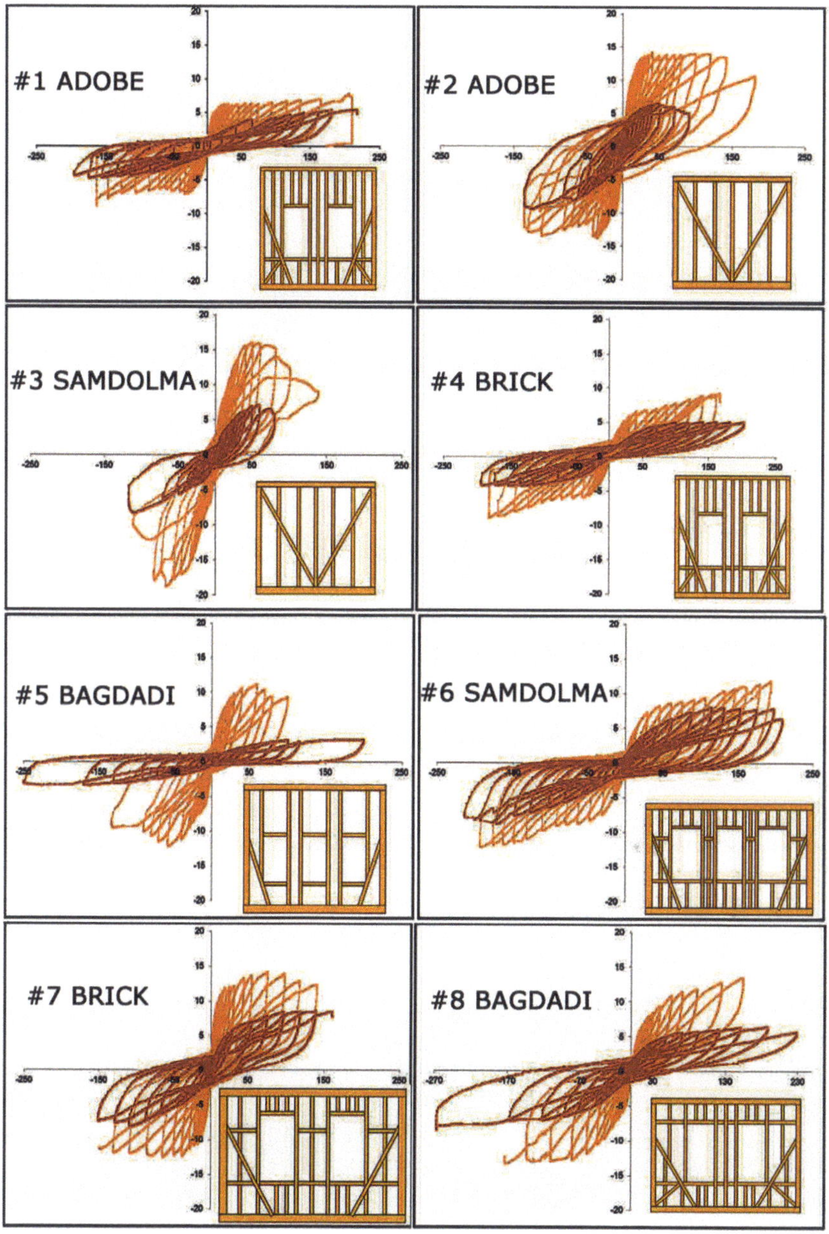

FIG. 3.27 Lateral load (kN) graphs obtained at the end of each test with *dark and light lines* indicating tests without and with infill/cladding, respectively (Aktaş et al., 2014).

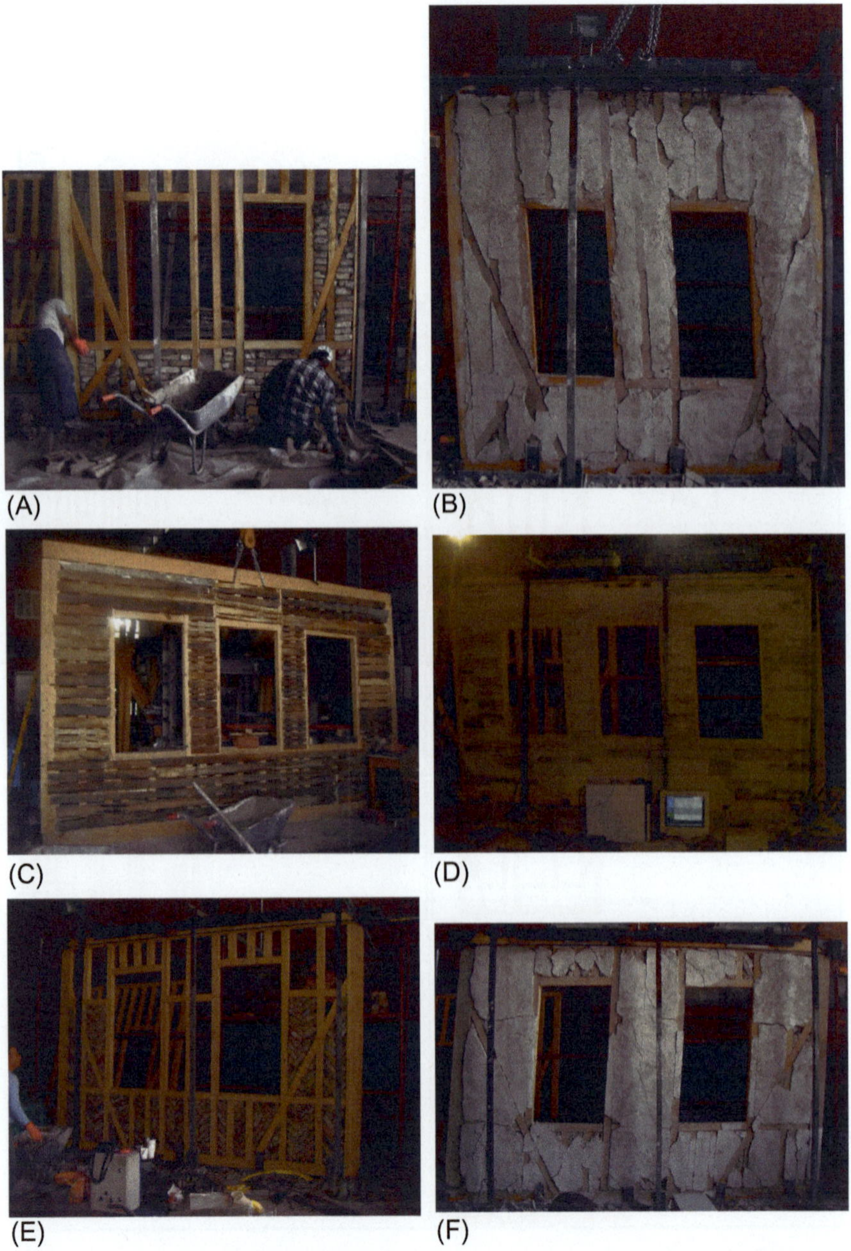

FIG. 3.28 (A) Frame #4 with brick infill being laid into it, (B) Frame #4 after test, (C) Frame #6 with *Şamdolma*, (D) Frame #6 after test, (E) Frame #7 with Brick infill, (F) Frame #7 after test. These views are selected from Aktaş's 2011 Thesis because they are the only photographs which are illustrative of the testing of *hımış* with masonry infill, and also which show for comparison, the testing of a timber frame with *şamdolma* cladding instead of infill masonry. *(Photos by Aktaş, Y. A. (2011).* Evaluation of seismic resistance of traditional ottoman timber frame houses. *A Thesis Submitted to the Graduate School of Natural and Applied Sciences of Middle East Technical University (METU).)*

3.5 Conclusion

As interesting and extensive as this project was, it has so many variables in terms of the timber layout of the walls and the multiple different infill and cladding materials, that there are limits to its broader scientific lessons. Perhaps the most important information to emerge is that, despite all of these extreme variations in timber framing geometries and variations in the infill, plus the use of different cladding systems instead of infill, they all are, in the end, more similar with regard to lateral force resisting performance than they are different.

From the perspective of how structural engineering is practiced today, all types of traditional construction can be very difficult to analyze rigorously simply because they are so changeable and adaptable for use by largely untrained carpenters and masons, and therefore do not fit a single standard. Their greatest asset is their accessibility for safe home construction by untrained people, including the homeowners themselves, who often lack any more than a basic training or experience in building construction.

Perhaps the best proof of the efficacy of these systems for houses that are safe in earthquakes is that this summary shows that the hysteresis loops from so many very different frames with different infills or cladding prove that this wide range of systems in or on these different frames all provide significant amounts of energy dissipation. The best explanation for this may be that in houses of only one to three stories, there is an inherent level of redundancy that makes the buildings with their solid interior and exterior walls of any kind of highly damped energy dissipating construction inherently resilient in earthquakes, not so much because of their strength but because of their inherent flexibility.

All one needs to do is to compare them with the hundreds of reinforced concrete buildings that collapsed in Turkey in 1999. Consider that the reinforced concrete moment frame buildings depend on their column/beam intersections for both strength and ductility, but the fact that access to that ductility is only when the bending of these joints goes beyond their elastic limit. Such inelastic bending is by definition damage—exactly the kind of damage to such moment frame structures that can rapidly lead to their collapse. This is true unless the *engineering* of frames with their connections goes beyond the minimum requirements of the codes, and that the *construction* of the joints and the rest of the building is done to a very high standard. Also, consider that if a reinforced concrete building is of a shear wall design, likewise the shear walls in both directions have to be designed and constructed well and located correctly to be fully effective.

In *hımış* structures, all of the walls are expected to participate in its resilience. It is precisely because the elastic strength of the masonry walls with mud or lime mortar is low that the friction and energy dissipation begins almost at the onset of any earthquake shaking. Counter-intuitively, this has proven to be of great benefit because it is the energy dissipation that helps prevent the structure from resonating with the earthquake vibrations. With the redundancy provided by having so many walls in play, and with the early onset of postelastic deformation of the *hımış* walls, the *hımış* buildings as a whole are not likely to be overwhelmed. The recent earthquakes in Turkey have proven this to be the case.

References

Aktaş, Y. A. (2011). *Evaluation of seismic resistance of traditional ottoman timber frame houses.* A Thesis Submitted to the Graduate School of Natural and Applied Sciences of Middle East Technical University (METU).

Aktaş, Y. A., Akyüz, U., Türer, A., Erdil, B., & Sahin-Güçhan, N. (2014). Seismic resistance evaluation of traditional Ottoman timber-frame hımış houses: Frame loadings and material tests. *Earthquake Spectra, 30*(4), 1711–1732.

Ambraseys, N. (2009). *Earthquakes in the eastern Mediterranean and the Middle East: A multidisciplinary study of seismicity up to 1900.* Cambridge University Press. ISBN 9780521872928.

Aschheim, M., Gülkan, P., & Sezen, H. (2000). Chap. 11: Performance of buildings of 1999 Kocaeli, Turkey, earthquake reconnaissance report. *Earthquake Spectra, 16*, . Supplement A.

Dogangün, A. (2006). Traditional wooden buildings and their damages during earthquakes in Turkey. *Engineering Failure Analysis, 13*(6), 981–996.

Correia, M. R., Lourenco, P. B., & Varum, H. (2015). *Seismic retrofitting: Learning from vernacular architecture* (1st ed.). CRC Press. ISBN 9781138028920.

Dutu, A., Sakata, H., Yamazaki, Y., & Shindo, T. (2015). In-plane behavior of timber frames with masonry infills under static cyclic loading. *Journal of Structural Engineering, 142*, 04015140. https://doi.org/10.1061/(ASCE)ST.1943-541X.0001405.

Dutu, A., Niste, M., Spatarelu, I., Dima, D. I., & Kishiki, S. (2017). Seismic evaluation of Romanian traditional buildings with timber frame and mud masonry infills by in-plane static cyclic tests. *Engineering Structures, 167*, 655–670. Elsevier.

Ergin, K., Güçlü, U., & Uz, Z. (1967). *A catalog of earthquakes for Turkey and surrounding area, 11 A.D. to 1964 A.D.* Institute of Earth Physics, Istanbul Technical University.

Freitag, J. K. (1901). *Architectural engineering with especial reference to high building construction, including many examples of Chicago office buildings* (2nd ed.). NYC: John Wiley & Sons.

Gülhan, D., & Özyörük, G. İ. (2000). The behaviour of traditional building systems against earthquake and its comparison to reinforced concrete frame systems: Experiences of the Marmara earthquake damage assessment studies in Kocaeli and Sakarya. In *Earthquake-safe: Lessons to be learned from traditional buildings*. Istanbul: ICOMOS.

Gülkan, P. (2000). *What emerged from the rubble.* Washington, DC: ProVention, World Bank.

Gülkan, P., & Langenbach, R. (2004). *The earthquake resistance of traditional timber and masonry dwellings in Turkey. In: 13th World Conference on Earthquake Engineering. Vancouver, B.C., Canada: Paper 2297* Available at(2004). www.conservationtech.com.

Hiçyılmaz, K. (2011). *Dhajji Dewari, affordable seismically resistant and sustainable housing: Non linear response history analysis.* Revision C, job # 215631Arup Gulf Ltd..

Kienzle, P. (1998). *Conservation and reconstruction of the palace of Minos at Knossos:* (pp. 122–134). (DPhil thesis)York: IAAS.

Langenbach, R. (1998). *Parades Pombalinas and the 'invention' of earthquake-resistant construction. In: International seminar on seismic risk and rehabilitation of stone masonry housing. Azores.* Available at www.conservationtech.com.

Langenbach, R., 2000. Keynote address: "Intuition From The Past: What Can We Learn From Traditional Construction," In: Proceedings of the International Conference: Earthquake-Safe: Lessons to be Learned from Traditional Construction, Istanbul, Turkey, November 15, 2000. Available at www.conservationtech.com.

Langenbach, R. (2004). Soil dynamics and the earthquake destruction of the Arg-e Bam. *Iranian Journal of Seismology and Earthquake Engineering, 5(4) and 6(1).*

Langenbach, R. (2007). *Preventing pancake collapses: Lessons from earthquake-resistant traditional construction for modern buildings of reinforced concrete. Keynote Address paper, India Chapter of American Concrete Institute, Mumbai.* Iranian Journal of Seismology and Earthquake Engineering. Available at *www.conservationtech.com.*

Langenbach, R. (2010a). *Rescuing the baby from the bathwater: Traditional masonry as earthquake-resistant construction. In: Keynote address paper, 8th international masonry conference, Dresden Germany, July.* Iranian Journal of Seismology and Earthquake Engineering. Available at www.conservationtech.com.

Langenbach, R. (2010b). *Earthquake resistant traditional construction is not an oxymoron: The resilience of timber and masonry structures in the Himalayan Region and beyond and its relevance to heritage preservation in Bhutan. In: The Royal Government of Bhutan international conference on disaster management and cultural heritage "living in harmony with the four elements" 12-14 December 2010.* Iranian Journal of Seismology and Earthquake Engineering. Available at www.conservationtech.com.

Parisse, F. (2019). *Numerical modelling of the seismic performance of Romanian traditional timber-framed buildings.* Master's Thesis in Structural Analysis of Monuments and Historical Constructions, University of Minho.

Pınar, N., & Lahn, E. (1952). *An annotated catalog of earthquakes in Turkey.* (in Turkish)Ministry of Public Works, Directorate of Construction and Development. Publication No. 6(36).

Ruggieri, N. (2015). Seismic performance evaluation of timber-framed masonry walls: Experimental tests and numerical modelling. In N. Ruggieriet al. (Ed.)*Historical earthquake-resistant timber frames in the Mediterranean area.* Switzerland: Springer International Publishinghttps://doi.org/10.1007/978-3-319-16187-7_8. (Chapter 8).

World Housing Encyclopedia (WHE), EERI and IAEE (2020). Available at:(2020). www.world-housing.net/.

Further reading

Bankoff, G., Langenbach, R., & Stephenson, M. (2014). *Culture risk and the built environment. In World disasters report, 2014.* International Federation of Red Cross and Red Crescent Societies Available online at(2014). *https://www.ifrc.org/world-disasters-report-2014.* (chapter 5).

Dutu, A. , et al. (2012). Technical note: Timbered masonry for earthquake resistance in Europe. *Materiales de Construcción, 62*(308), 615–628.

CHAPTER

4

Traditional earth construction in Latin America: A review on the construction systems and reinforcement strategies

Pilar Baquedano, Rafael Ramírez Eudave, Fabiana N. Miranda, Sandra Graus, and Tiago Miguel Ferreira

Department of Civil Engineering, ISISE, Institute of Science and Innovation for Bio-Sustainability (IB-S), University of Minho, Guimarães, Portugal

4.1 Introduction

The traditional use of earthen-based construction techniques in Latin America is the result of a long process of learning, continuously developed. As in other regions of the world, the knowledge related with earth constructions has been a selective process in which successful actions are transcendent, and failures are collectively reminded. Especially in territories exposed to earthquakes, in which these events have caused significant human, economic, and cultural losses.

Despite this, the use of earthen-based techniques has continued, trending to the optimal use of materials, geometries, and spaces. As a construction material, raw earth has indisputable advantages, such as excellent acoustic and thermal insulation. It is, furthermore, an economical and sustainable solution due to the possibility of reuse of the material.

Specific techniques such as adobe, rammed earth, and wattle and daub have been successful strategies, in a regional level, for using local soils as the most important material for vernacular housing and even for monumental architecture (Fig. 4.1).

https://doi.org/10.1016/B978-0-12-821087-1.00011-9

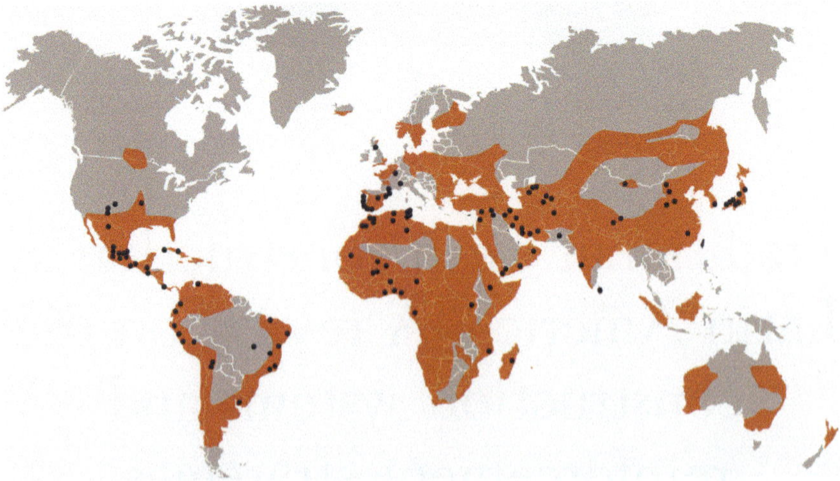

FIG. 4.1 Areas of earth architecture and earth architecture (Gandreau & Delboy, 2010).

As a result of centuries of continual use, several improvements and retrofitting strategies have also been developed (Guerrero Baca, 2007). To mitigate seismic risks, elements such as wires, wood, or natural fibers have been traditionally incorporated. Nowadays, in many Latin American countries, it is common to use geogrid, steel meshes, or other technologies, some of them still in test, like the LC-TRM strengthening system.

Most of the earthen-based constructions can be classified into three different and well-differenced techniques: adobe, rammed earth, and wattle and daub. This chapter discusses the evolution, similarities, and differences between these three traditional construction techniques and some seismic retrofitting solutions addressed to Mexico-, Peru-, Chile-, and Bolivia-specific circumstances.

4.2 A review of traditional systems in Latin America

4.2.1 Adobe

In many parts of the world, adobe is the specific term that defines a mud-brick. The use of these sundried blocks dates back to 8000 BCE (Houben & Guillaud, 1994). As a technique, make adobe involves the understanding of site location, natural resources, climate, and traditional building practices where the earth is moulded in regular pieces to be used as masonry units in combination with earth mortars. This configuration permits different scales and organisations for the work, as well as other advantages resulting from serial production.

However, adobe structures are vulnerable to the effects of natural phenomena such as earthquakes, rain, and floods. Traditional adobe constructions are especially sensitive to earthquake ground shaking. The seismic deficiencies of adobe buildings are caused by its heavyweight, low strength, and brittleness. During earthquakes, these structures develop high levels of seismic forces that they are unable to resist, often failing abruptly (Blondet, Villa Garcia, & Brzev, 2003).

Although the existence of numerous local varieties, adobe blocks are used to be fabricated with some general procedures. Blocks are usually composed of approximately 20% of clay and 80% of sand (Borreguero Sánchez, 2015). The soil and sand mixture is generously hydrated and receives fibrous materials as straw before being mixed. The resulting mixture is poured into parallelepiped moulds. Once they have a proper consistency, they are left to dry in ambient conditions. Sometimes, this process can take as long as 5 years to have perfect dried pieces. One of the most representative properties of clay is the volumetric changes before and after the drying process, known as shrinkage. Natural fibers are added to control the shrinkage and to reduce inducted damage and cracking. Due to the relatively small scale of the blocks, the adobe technique is an adequate system for soils with an elevated proportion of clay.

The use of individual blocks allows creating multilayered walls and, consequently, complex and robust structures that can even replicate geometric solutions for masonry, as vaults (Catalán Quiroz, Moreno-Marínez, Galván, & Arroyo Matus, 2019). The use of thick walls is a common strategy to prevent damages in a seismic context (Guerrero Baca, 2019) (Fig. 4.2).

The thermic, energetic, and economic benefits of adobe aimed to develop contemporary variations of this technique, such as superadobe or "earthbag," which use the earth mixture confined into a bag. This

FIG. 4.2 Housing production in El Obispo, Malinaltepec, Guerrero, Mexico (Hastings García & Huerta García, 2015).

proposal has been studied as an alternative for new constructions with improved seismic performance (Canadell Ruiz, 2014).

Adobe is the most commonly used earth construction technique in Mexico, Chile, Bolivia, and Peru, being still widely adopted today. The size of adobe blocks varies according to the place, type of use, and historical time of the construction, being the old adobes larger than those used today. An example of these differences is the adobe pieces found in old churches and haciendas, in which the adobes had about 0.35 × 0.70 × 1.0 m, differently from the current blocks, which measure about 0.30 × 0.60 × 0.10 m (Ferrada & Segovia, 2007).

The use of adobe in the Bolivian region dates back to Tiwanaku civilisation, one of the most important civilisations of the Andean region, which used adobe since approx. 500 CE. Archaeological excavations on the site evidence the use of adobe in dwellings and residential areas (Janusek, 2009), see Fig. 4.3.

In the Bolivian highlands, adobe mixture includes pieces of straw with a length of approximately 10 cm as an additive to control the microcracking during the drying process. It is possible to find also the use of natural plasticizer additive from the gel of cactuses (Cactaceae family), a common specie in the Andean region, to add resistance, as the straw additive to the adobe brick. The cactuses gel additive is mostly used in restorations of archaeological adobe structures.

In Peru, construction with adobe is a traditional feature mostly associated with archaeological building remains. It can be found in "huacas," they were important administrative and religious complexes from the pre-Hispanic period, commonly located on the coast of Peru. These massive constructions present a variety of shapes, sizes, and styles of masonry such as the conic adobes of Sechín, Casma Valley (2000–1250 BCE) and the

FIG. 4.3 Cobble and adobe structural foundations in Chi'iji Jawira, Tiwanaku, Bolivia (Janusek, 2009).

rectangular adobes from the Mochica civilization (100–800 CE). Two massive adobe buildings, known as Huaca del Sol and Huaca de la Luna (Fig. 4.4), are part of this complex located in La Libertad, Peru.

Another example is Chan Chan, one of the world's largest mud citadels (1200–1440 CE) located in the north coast of Peru. This citadel extends over an area of about 28 km^2. At the peak of the empire, it housed more than 50,000 people. The Chan Chan archaeological site (see Fig. 4.5) has survived many severe earthquakes over the past 600 years (Blondet, Vargas, & Tarque, 2008), as well as to El Niño-Southern Oscillation (ENSO), which represents a severe threat to adobe-based structures. In its most intense manifestation, ENSO is responsible for heavy torrential rains during the summer period (Takahashi, 2014).

Nowadays, adobe is still used in the traditional construction of houses and small-scale public equipment all over Latin America, especially in rural communities. The preference for this material is not only related to its constructive and economic advantages but also with the traditional sense that involves the preparation of the bricks and the community construction of the building. See Fig. 4.6A.

Most of these earthen houses are built without any professional intervention or structural reinforcement, tending to mimic the architectural

(A) (B) (C)

FIG. 4.4 (A, B) Huaca del Sol (Photos from S. Graus); (C) Group of bricks from Huaca de la Luna (Tsai, 2012).

FIG. 4.5 Chan Chan citadel, La Libertad, Perú (Proyecto Especial Complejo Arqueológico Chan Chan: chanchan.gob.pe).

FIG. 4.6 (A) Construction of adobe wall in Tutucucho, La Paz, Bolivia (Photo from R. Bello). (B) A contemporary adobe house in Cusco, Peru.

features of contemporary clay masonry houses with several stories, irregular plan, and large windows and door openings (Fig. 4.6B), which make them extremely vulnerable to the seismic action (Blondet, 2010).

4.2.2 Rammed earth

Though some soils may be inadequate for adobe production because of the lack of clays, they can still be used for construction with the rammed-earth system. Because this construction system also needs less water than adobe production, it is a suitable alternative in places with limited access to water sources.

The most representative aspect of rammed-earth walls is that the whole structural element is configured by a mould, which is sequentially filled with layers of earth that are compressed using an appropriate ram. Formworks can be displaced just a few minutes after the ramming process, and then the operation can be repeated in the next section. This process allows having homogeneous structures by using a low amount of wood (Fig. 4.7). Due to the low rates of clay in the used soil, there are low shrinkage rates. The considerable thickness of rammed-earth walls provides excellent isolation performance and long durability when compared with regular masonry and adobe. It is also an economic strategy because the only specialized works are the ones related to the formwork (Viñuales, 2007).

Contrary to adobe constructions, widely spread in almost all Mexican territory, rammed earth is geographically concentrated in the north of the country. This is because of the limited access to timber and water that limit the possibilities of adobe production. Another important reason lies in the hard weather conditions in the desert, where thermal isolation by mass is the most effective low-tech system. The low energy and material resources needed in the rammed-earth constructions are attractive for contemporary designers. In Mexico, there are several examples of the new use of rammed earth, displaying a broad set of spatial possibilities as well as climatic

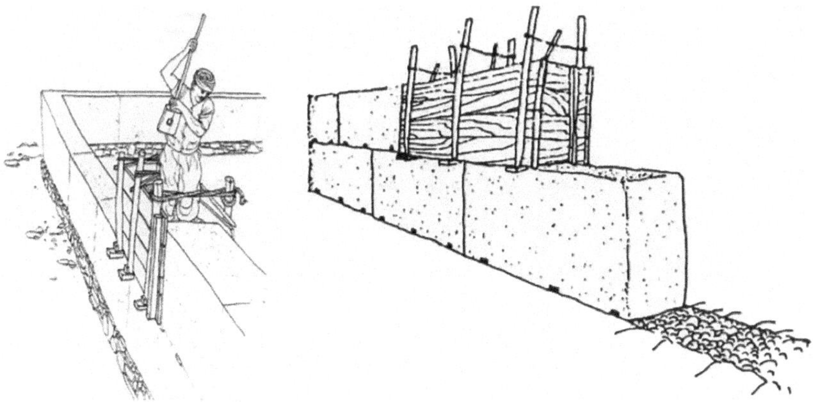

FIG. 4.7 Rammed-earth construction procedure with a mobile formwork (Borreguero Sánchez, 2015).

adaptation characteristics in the context of the sustainability challenges of the 21st century.

In Chile, this same technique is known as "tapial" or "adobón," because its disposition resembles large adobes (Karmelic, 2015). Despite being a widespread technique in countries with desert areas, it is used almost exclusively in the construction of partition walls to close orchards, courtyards, corrals, and other properties. Through the use of the material available in the territory, urban and rural landscapes were moulded with hundreds of kilometres of tapiales (Lacoste, Premat, Castro, Soto, & Aranda, 2012). Because of their low seismic resistance, this technique is not used in the construction of houses.

It is used mainly in the northern part of the country, in the villages of the Atacama Desert and in the central valley area, where they conform part of the landscape through the delimitation of roads.

Commonly, the structures of rammed earth are formed of three main parts: (i) the stone basement; (ii) the raw earth wall; (iii) the top. The latter is used mainly in the humid areas of the central valley (Lacoste et al., 2012) and can be constructed by clay tiles or dry cane (known as "barda" or "sombrerito") that grows in nearby plantations, see Fig. 4.8. In the northern part of the Atacama Desert, besides not having tiles at the top, they usually lack stone foundations, which often leads to problems of erosion at the base of the walls (Jorquera, 2012).

Currently, the construction of rammed earth is scarce, because rapidly installed industrialized materials have replaced it. Nevertheless, there are some initiatives that attempt to promote their preservation, through onsite training construction and raising awareness among the communities, some of them promoted by the Aldea Foundation and the School of Earthen Construction ECoT (Fig. 4.9).

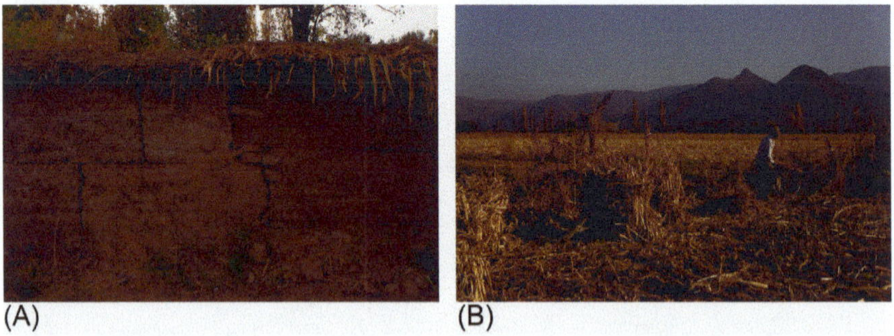

FIG. 4.8 Bucalemu, Aconcagua Valley, Chile. May 2019: (A) rammed earth with dry cane on the top; (B) harvesting of "curahuilla cane" for tapial finishing. Aldea Foundation https://www.somosaldea.org/.

FIG. 4.9 Onsite training Tapial construction, November 2019. Aldea Foundation https://www.somosaldea.org/.

In Bolivia, a clear example of the historical use of tapial is in the religious architecture of Santa Cruz. The most remarkable example of the architecture of Santa Cruz is the Jesuit Missions of Chiquitos, declared World Heritage by the United Nations Educational, Scientific and Cultural Organization in 1990. The construction of the churches of Chiquitos, in a mix of timber and tapial, was also an example of the participation of the local indigenous communities, whose influence is noticeable in the construction techniques used and in the design (see Fig. 4.10).

FIG. 4.10 Jesuit Missions of Chiquitos, Bolivia (https://jesuitas.lat/es/noticias/).

The use of the tapial technique in Peru has developed since antiquity, mainly in the vernacular construction of houses in rural areas located in the Andean region of the country. The material is inexpensive; tapial homes turn out to be three times cheaper than adobe homes due to the rapid construction (ININVI, 1989). Also, this material is renewable, leaves no residue, is a good temperature isolator, and has great historical value in the country, as it is part of the population's culture. Tapial construction has been revealed to be, however, highly vulnerable to earthquakes (ININVI, 1989). One of the leading causes for this vulnerability is the lack of appropriate out-of-plane performance of the walls, which is caused by inefficient connections between orthogonal walls and the poor mechanical properties of the material. Since the 1970s, research has been carried out to improve the response of these structures to earthquakes. Eucalyptus wood placed inside the walls (ININVI, 1989) or reinforced plasters with polyvinyl chloride meshes are just a few examples of the techniques investigated (Silva, 2013).

4.2.3 Wattle and daub

Wattle and daub represent a mixed system that takes advantage of some mechanical properties of fibrous materials (such as timber or bamboo) when covered with one or several layers of earth. This disposition enhances the structural system while combining the mechanical compression capacities of the earth with tensile-stress capacities of fibrous materials.

Bahareque, as it is known in Mexico, allows diverse configurations in the function of the size of the fibrous components and how these elements are connected. In Mexico, there are several regional varieties of this technique, also revealing the vast possibilities in terms of sections and scale (Guerrero Baca, 2017). Some examples located in Cueva del Maguey (Durango), in Fig. 4.11, are proofs of the durability of bahareque in demanding conditions (Punzo Díaz, 2016).

FIG. 4.11 Cueva del Maguey, Durango (Urbina, 2019).

Another remarkable example of the bahareque tradition is the Mayan houses (Fig. 4.12). These buildings have not substantially changed in the past centuries and are still a convenient production in the context of the challenging weather of the Yucatan peninsula. Up to date, the Mayan house is a material testimony of a way of life and social dynamics (Sánchez Suárez, 2006). Most of Mayan houses share the same pattern, revealing a compromise with a long constructive tradition in which every procedure, material, and system has a name and a sequence (Comán-Kalisch & Pinón-Jiménez, 2019).

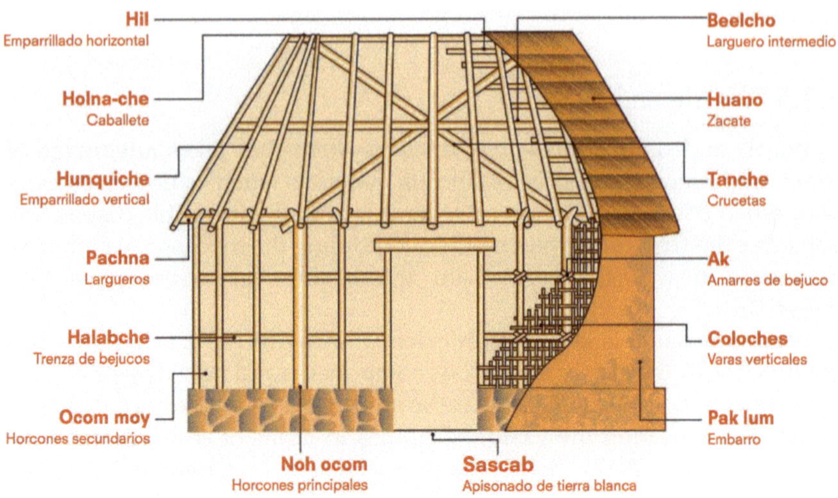

FIG. 4.12 Parts of traditional Mayan houses, as explained in Íñigo Dehud (2013).

Bahareque technique represents a light, easy, and economic strategy, even when compared with adobe and rammed earth. Some recent projects have explored contemporary designs with localized bahareque solutions, as the building for the Mexican Institute for Community Development (Instituto Mexicano para el Desarrollo Comunitario) in Barranca de Huentitán, Guadalajara (Fig. 4.13). This building combines a concrete structure based on frames with nonstructural walls of bahareque, providing adequate thermic isolation. In fact, those isolation properties permitted the development of specific typologies like the Cuexcomate (Fig. 4.14), a traditional barn design that has been used and replicated along the last 2000 years, at least (Uruñuela y Ladrón De Guevara & Plunket, 2012).

The mixed technique of wattle and daub is known in Chile, Peru, and Bolivia as Quincha, which in the quechua or ruma-sino language means perimeter fence (Marussi, 1989). This word, still used nowadays in all the Andean region, goes back to the pre-Columbian times. It is important to remark that pre-Columbian Quincha walls, differently from the contemporary ones, were not timber frame-based. Instead, local materials like organic chaff, small tree branches, bamboo sticks, and trunks of trees with the appropriate diameter were used to build them. Those fibers were mixed with animal leather or flexible branches, as in the case of Bolivian highlands. Once the structure was built, the fabric was whipped with mud to obtain a uniform 10- to 16-cm thick layer (Tomasi & Bellmann, 2018). This same technique is still used today. However, fibers have been replaced by timber framework and nails, in the case of joints. Following Kościuk and Kogut (2019), it is possible to distinguish four types of pre-Columbian Quincha walls (see Fig. 4.15).

Due to natural reasons, most of the organic materials are rarely preserved on archaeological sites, being difficult to determine precisely

FIG. 4.13 The building of the Mexican Institute for Community Development (Franco, 2015).

FIG. 4.14 Example of Cuexcomate (Ahrens, 2013).

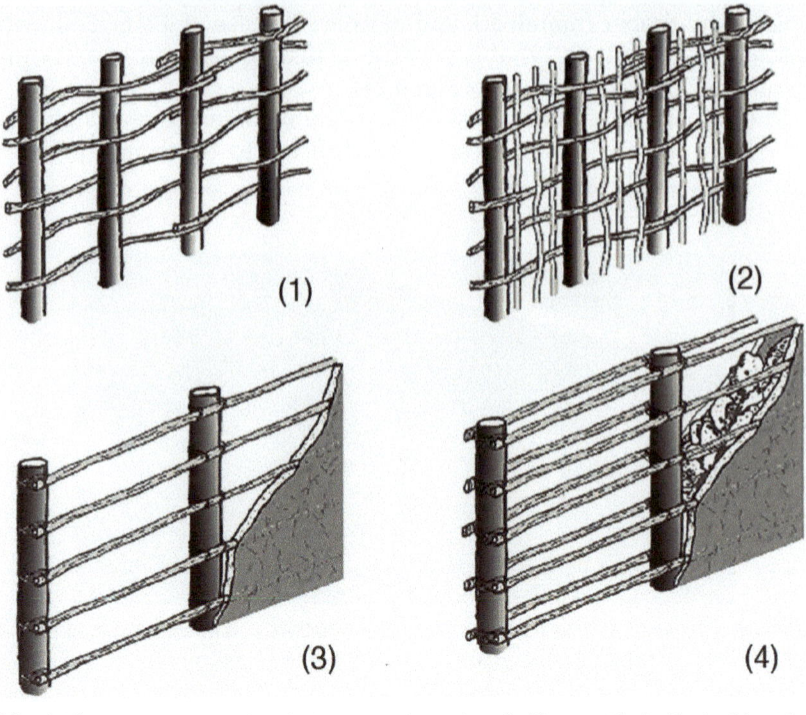

FIG. 4.15 (1) Single-interlaced, (2) crossed interlaced, (3) one-sided, (4) double-sided, often combined with an infill of small stones or adobe (Kościuk & Kogut, 2019).

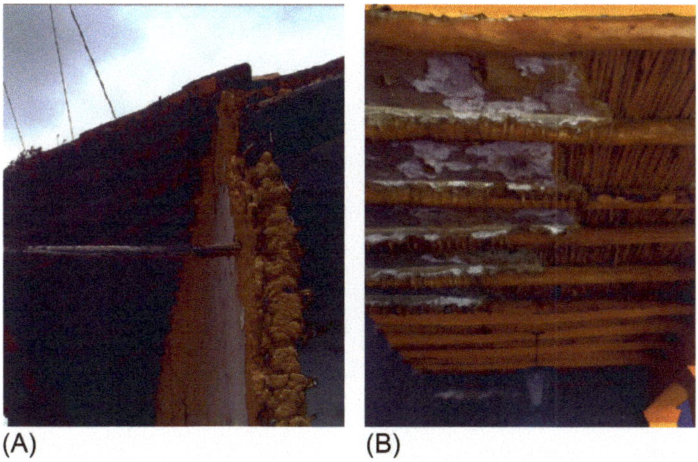

FIG. 4.16 (A) Quincha wall of a house in Bolivia (Tomasi & Bellmann, 2018), (B) Interior Quincha slab of colonial house in La Paz, Bolivia (Photo of F. Navia).

which kind of species were used before the Spanish conquer. Furthermore, pre-Columbian Quincha walls remain poorly studied.

During the Colonial period in Bolivia (from the 16th to 17th century), the use of Quincha wall arrived at the main cities, having been used mainly in the upper floors of the Spanish and creole aristocracy houses, due to their lightweight, Fig. 4.16A. Quincha technique was also used for roofs and slabs, where the fiber network was denser, forming almost a mat. In the case of roofs, the mud layer was mixed with high amounts of straw to make the mixture lighter and impermeable, see Fig. 4.16B. Quincha applied in roofs is still a widespread technique in the rural areas of Bolivia.

In Chile, Quincha has been used since pre-Columbian times by native people in the regions between Atacama and O'Higgins, which are characterized by the temperate climate and by the abundant presence of wood and limes (Jorquera, 2014). There is evidence that Quincha was the primary construction technique of the Diaguitas people (Rivera, 2017), see Fig. 4.17.

A notorious example of the exploitation of local resources was the use of "brea" as a secondary structure in Quincha constructions (Fig. 4.18). Located in the "Norte Chico," "brea" is a shrub that grows from the north to the center of Chile, and it can be mainly found on riverbanks or in soils with saline concentration (Cortez, 2014).

Quincha is still in usage in several rural towns of the transverse valleys of the Norte Chico area, where it is used for housing, warehouses, and perimeter fences of agricultural lands. The technique consists of three elements: (i) the main structure, which can be composed of logs or pieces of

FIG. 4.17 Quincha of the "Norte Chico" with medium size branches as structure (Rivera, 2017).

FIG. 4.18 Quincha with historical and contemporary "brea" (Cortez, 2014).

carved wood; (ii) a secondary structure conformed by branches, reeds, slats, or wires; (iii) the earth infill, that is received by the secondary structure and varies depending on the characteristics of the soil where the work is located (Jorquera, 2014).

The use of Quincha in Perú goes back to pre-Columbian times before the Spanish conquest of the Inca Empire. Few houses from the

FIG. 4.19 Houses with walls closed with wood and canes in Piura, Perú (Carbajal et al., 2005).

pre-Hispanic period remain, because the enclosed materials are susceptible to rotting when exposed to the elements. There is some evidence in archaeological finds, and from descriptions of unidentified chroniclers from the Spanish period, that indicates that early Quincha technology was rudimentary and was used only to construct houses for the lowest social classes. In contrast, the monumental constructions, such as the ceremonial centers, altars, and the houses of the curacas (Indian nobility or ruling class) were constructed with tapial or adobe (Carbajal, Ruiz, & Schexnayder, 2005).

The pre-Hispanic forms of Quincha consisted of frames formed from either branches or trunks of small trees. Similar to the Bolivian technique, frames were tied together with vegetal fibers. On these frames, the canes were braided to form the walls, and then a light layer of mud was applied. (Marussi, 1989).

With the arrival of the Spanish conquerors in 1532, other construction systems were tested and developed. In 1666, Quincha began to be used in monumental buildings, namely in domes and vaults. As the main advantages, quincha was a lightweight, economical, and quick to execute construction technique. This construction system is still used nowadays in the rural zones of the north coast of Peru, in the desert areas, as well as in some cities, such as Lambayeque, Piura, and Tumbes, see Fig. 4.19.

4.3 Vulnerabilities of traditional systems and reinforcement strategies

Throughout the years, the use of earth system construction techniques has been gradually replaced by industrial technologies. Among the various reasons that can be pointed out as the cause of this change, a general sense of insecurity resulting from the poor seismic performance of traditional construction techniques is undoubtedly one of the most important ones. Using the example of Chile, after the earthquake of Chillán, in 1939, the city collapsed, leaving 5648 dead according to official numbers (see Fig. 4.20). This event encouraged the development of an earthquake-resistance code that does not contemplate constructions in raw earth, contributing to its disappearance. Nevertheless, despite the occurrence of intense earthquakes, many earth buildings are still standing in the north and center of Chile, representing approximately 40% of the buildings declared national heritage (Karmelic, 2015). This demonstrates the effectiveness of various earthquake response strategies that have been adopted over the years.

The resistance of many earth constructions after the, 2010 Chile earthquake, considered as one of the 10 largest earthquakes in the world history, strengthened the interest of citizens, professionals, and local authorities in the conservation and restoration of earth buildings. Since then, numerous public initiatives with universities and local communities have been developed to preserve these buildings.

Unreinforced earth-based constructions are naturally vulnerable to weather and seismic events. This natural condition of materials urged to find strategies to reach better performance in terms of durability, resistance, resilience, and comfort. Most of these vulnerabilities are common to the diverse earth-based construction systems. In terms of seismic response, the most relevant properties are a high mass (a high density), low tensile strength, fragile behavior, softening behavior, and cumulative damage (Oliveira, Varum, & Vargas, 2012). To address such vulnerabilities, earth buildings in Mexico are usually retrofitted through the addition of fiber-based materials. Empirical experience showed that the presence of fibers in the earth components help to provide a higher mechanical resistance. This can be explained because of the tensile-strength properties of fibers that provide a certain ductility and dissipation of energy.

FIG. 4.20 Streets of old Chillan after the 1939 earthquake (https://www.csn.uchile.cl/efemerides-sismicas-terremoto-de-chillan-1939/).

Another essential strategy to guarantee good seismic performance is an adequate relation between the thickness of the walls and the height of the building. The most widely adopted slenderness in the north of Chile is 1:7. It is therefore frequent to find thick walls, being 50 cm the minimum thickness found in traditional architecture, to reach a height of 3.5 m. In monumental buildings, like churches or haciendas, it is possible to find greater heights with walls of 90, 120, or 150 cm thickness, measures obtained from the modulation of the adobe brick (Rivera, 2017).

Despite using these retrofitting strategies for centuries, several attempts with contemporary materials have been carried out with different grades of success. The adequate interpretation of crack patterns, deformations, and superficial pathologies may help to understand how and where mechanical capacities of earth-based walls have been exceeded. However, general reinforcements can be added to improve the global ductility of constructions.

In general terms, several reinforcement actions have been proposed as a consequence of the studies carried out after seismic damages in earth-based constructions (González Herrera, Aguilar Carboney, & Gómez Soberón, 2008). Among those, the most common alternatives usually involve: (i) internal cane mesh reinforcement; (ii) external wire mesh reinforcement; (iii) external polymer mesh reinforcement.

The internal cane mesh reinforcement consists of vertical cane rods anchored to a concrete foundation and placed inside the adobe walls. The adobe block layout defines the distance between the vertical cane rods at 1.5 times the thickness of the wall. As illustrated in Fig. 4.21, horizontal layers of crushed canes are placed for every few adobe rows and tied to the vertical cane reinforcement. Finally, this internal cane mesh reinforcement is tied to a wooden crown beam (Blondet & Aguilar, 2007).

This reinforcement system has demonstrated an excellent response in full-scale shaking table tests (Blondet et al., 1988). Fig. 4.22 shows the collapse of a full-scale adobe house model subjected to simulated seismic

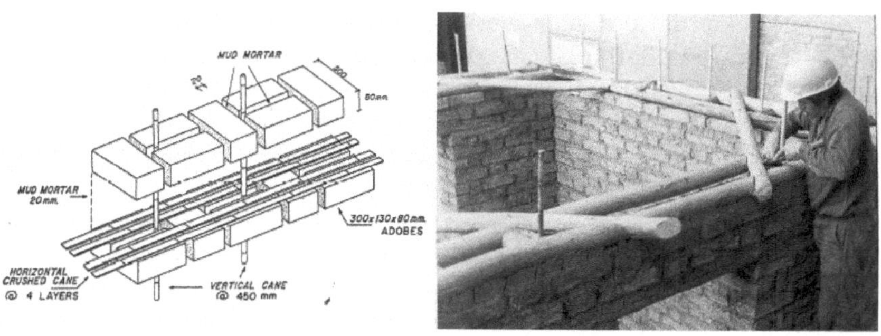

FIG. 4.21 Internal cane mesh reinforcement (Blondet & Aguilar, 2007).

FIG. 4.22 Internal cane mesh reinforcement (Blondet & Aguilar, 2007).

motion (left). During a similar test, the model reinforced with an internal cane mesh suffered significant damage but did not collapse. One of the limitations of this reinforcement system is the fact that cane is not available in all seismic regions or it is practically impossible to obtain the required amount for a massive construction or reconstruction program (Blondet & Aguilar, 2007).

In turn, the external wire mesh reinforcement consists of nailing wire mesh bands placed in horizontal and vertical strips against the adobe walls and then covering them with cement mortar. This technique provides significant additional strength to adobe models under earthquake simulation tests. For example, during a moderate earthquake in Peru in 2001, while surrounding houses were severely damaged or destroyed, houses reinforced with this system did not suffer any damage and were used as shelters (Fig. 4.23).

Wire mesh and cement are prohibitively expensive for the inhabitants of earthen houses in developing countries. External reinforcement with welded mesh could cost up to US $200 for a typical one floor, two-room adobe house, which exceeds the economic capacity of most Peruvian adobe users (Zegarra, Quiun, San Bartolome, & Giesecke, 2001).

Finally, external polymer mesh reinforcement study was performed using several similar full-scale adobe housing models with different amounts and types of polymer mesh on the PUCP unidirectional shaking table. The reinforcement consists of bands of polymer mesh tied to both sides of the walls with plastic string threaded through the walls (Blondet & Aguilar, 2007), see Fig. 4.24.

The first models were reinforced with different amounts of geomesh, having shown good dynamic response during the earthquake simulation tests: although the adobe walls suffered some damage, the collapse was avoided even during strong shaking. As expected, however, the amount and spread of damage on the adobe walls increased as the quantity of

FIG. 4.23 Reinforced house after Arequipa earthquake in 2001 (Blondet & Aguilar, 2007).

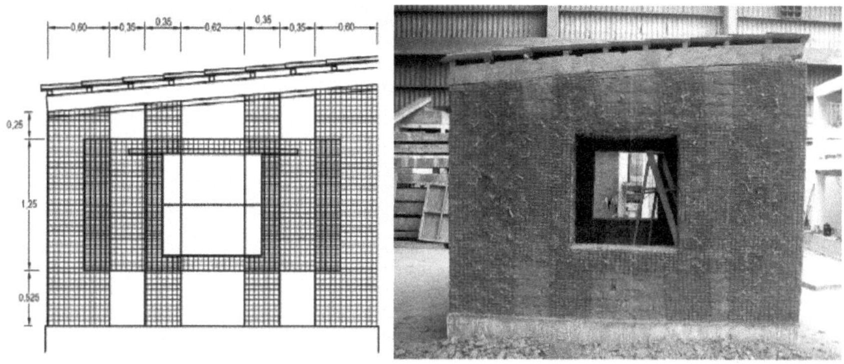

FIG. 4.24 External geogrid mesh reinforcement (75% of walls covered) (Blondet & Aguilar, 2007).

polymer mesh reinforcement decreased. Because geogrid polymer mesh is quite expensive in Peru, it was decided to study the use of a cheaper plastic mesh, usually employed as a soft safety fence in construction sites (Blondet & Aguilar, 2007).

The Getty Conservation Institute (GCI) has coordinated different projects aimed at developing technical procedures to prevent the structural instability of historic adobe buildings during earthquakes in Peru. During the years, 2010 and 2011, and as part of the Seismic Resistant Stabilization Project (SRP), several case studies, including the Santiago Apostle Temple of Kuñotambo (Kuñotambo Temple, in Fig. 4.25), were investigated by GCI staff, associates, and advisors (Cancino & Lardinois, 2013).

The Kuñotambo Temple was built in 1681 with thick walls and adobe buttresses seated on an overgrowth of irregular stone masonry, under a gabled wooden deck armour, called "par y nudillo" (pair and knuckle). This wooden reinforcement is composed of a discontinuous collar beam

FIG. 4.25 South interior of the Church of Santiago de Kuño Tambo in Peru, with its 17th-century wall paintings. Photo: Mario Santana, for the GCI.

at the top of the adobe walls, on which the pairs are seated, and by seven tie-down beams or braces, which do not have connections to the appropriate adobe walls, see Fig. 4.26.

The preliminary diagnosis was based on data obtained by qualitative methods, including historical research and direct observation. However, these points were confirmed with static tests of the different structural components of the temple as well as dynamic and sound tests.

These tests provided valuable information on the structural behavior of the building and allowed the validation of finite element models (FEM), which were used to simulate the current state of the building, as well as its seismic behavior before and after reinforcement (Vernaza, Cancino, & Rainer, 2018).

The main objective was strengthening the structure to improve its lateral resistance and its ability to dissipate energy. As the final result,

FIG. 4.26 Structural components of the temple. Adobe buttresses and wooden roof. Photos: C. Cancino and A. Ferrón, 2010 GCI.

reinforcement actions were divided into two groups (Vernaza et al., 2018): (i) techniques to strengthen the walls by adding buttresses; (ii) techniques to connect or reconnect the walls by adding wooden elements (internal reinforcements: collar beams and corner keys; external reinforcements: tie-down beams or braces and cotter pins).

4.4 Final remarks

The importance of the earthen-based techniques lies in the important amount of cultural assets that involve this material as well as its contemporary role to satisfy housing needs. Furthermore, this revival of earthen-based architecture is framed in the concerns about sustainability and environmental impact of constructions.

The existence of similar features in diverse places of Latin America is not only explained because of similar access to natural resources, but it is also a relative confirmation of success or "natural selection" that encourages the survival of the best strategies in terms of materials, human force, and durability.

To analyze and understand the intangible values that existing cultural assets offer (mostly in terms of technology) may be useful for their preservation as well as a base for contemporary innovation. Most of the challenges that current constructions must face have been traditionally well known and solved. Thus, it becomes essential to establish bridges between modern science and technology and ancient traditional knowledge.

This approach permitted, for instance, to experiment structural strengthening techniques with innovative materials, obtaining interesting and promising results. This exploration demonstrates that the use of earth-based architecture is more than a legacy, but a living tradition in constant enrichment that may be helpful to provide answers for 21st-century challenges.

References

Ahrens, C. (2013). *Coexcomate of temezcal? Deciphering the circular architectural features at Angamuco, Michoacán, Mexico.* Fort Collins, Colorado: EEUU: Colorado State University.

Blondet, M. (2010). Mitigation of seismic risk on earthen buildings. In E. H. Gökçekus, U. Türker, & J. LaMoreaux (Eds.), *Survival and sustainability: Environmental concerns in the 21st century.* Heidelberg, Berlin: Springer.

Blondet, M., & Aguilar, R. (2007). *Seismic protection of earthen buildings.* Lima, Peru.

Blondet, M., Ginocchio, F., Marsh, C., Ottazzi, G., Villa Garcia, G., & Yep, J. (1988). Shaking table test of improved adobe masonry houses. In *9th World conference on earthquake engineering, Tokyo-Kyoto, Japan.*

Blondet, M., Vargas, J., & Tarque, N. (2008). Low-cost reinforcement of earthen houses in seismic areas. In *14th World conference on earthquake engineering, Beijing, China.*

Blondet, M., Villa Garcia, G., & Brzev, S. (2003). *Earthquake-resistant construction of adobe buildings: A tutorial.* Oakland, California; Tokyo, Japan: Earthquake Engineering Research Institute and IAEE.

Borreguero Sánchez, L. (2015). *Arquitectura low tech. Refugio de Emergencia con Palets. Trabajo de fin de grado.* Valladolid: Universidad de Valladolid.

Canadell Ruiz, S. (2014). *Estudio estructural de domos realizados con la técnica de falsa cúpula y superadobe.* Barcelona: Universitat Politecnica de Catalunya.

Cancino, C., & Lardinois, S. (2013). *Proyecto de Estabilización Sismorresistente: Estudio de edificaciones tipológicas.* Los Angeles, California: Getty Conservation Institute.

Carbajal, F., Ruiz, G., & Schexnayder, C. (2005). Quincha construction in Perú. *Practice Periodical on Structural Design and Construction*, 56–62.

Catalán Quiroz, P., Moreno-Marínez, J. Y., Galván, A., & Arroyo Matus, R. (2019). Obtención de las propiedades mecánicas de la mampostería de adobe mediante ensayes de laboratorio. *Acta Universitaria, 29*, 1–13.

Comán-Kalisch, M. A., & Pinón-Jiménez, A. (2019). Cambios y permanencias de la tecnología constructiva de la casa maya en Mérida, Yucatán. *LEGADO de Arquitectura y Diseño, 14* (25), 72–83.

Cortez, D. (2014). *Sistemas constructivos tradicionales como sistemas de conocimiento local. Quincha con brea en totoral, Atacama, Chile.* Universidad de Chile, Facultad de Arquitectura y Urbanismo.

Ferrada, K., & Segovia, D. (2007). *Agua, Tierra y Paja. Construcciones de adobe en el Limarí.* Santiago: CNCA.

Franco, J. T. (2015). *Guadalajara, México: un edificio comunitario de muros de bahareque y celosía de carrizo.* Recuperado el 16 de 10 de 2019de Plataforma Arquitectura.(2015). https://www.plataformaarquitectura.cl/cl/762081/guadalajara-mexico-un-edificio-comunitario-de-muros-de-bahareque-y-celosia-de-carrizo.

Gandreau, D., & Delboy, L. (2010). *Patrimoine Mondial, Inventaire et Situation des Biens Construits en Terre.* Paris, France: CRAterre-ENSAG.

González Herrera, R., Aguilar Carboney, J., & Gómez Soberón, C. (2008). Vulnerabilidad de viviendas de adobe en Chiapas y alternativas de reparación. In: *XVI Congreso Nacional de Ingeniería Estructural*, Veracruz: Sociedad Mexicana de Ingeniería Estructural, pp. 1–21.

Guerrero Baca, L. F. (2007). Arquitectura en tierra. Hacia la recuperación de una cultura constructiva. *Apuntes, 20*(2), 369–383.

Guerrero Baca, L. F. (2017). Pasado y porvenir de la construcción con bajareque. *Gremium, 4* (08), 69–80.

Guerrero Baca, L. F. (2019). Comportamiento sísmico de viviendas tradicionales de adobe, situadas en las faldas del volcán Popocatépetl, México. *Gremium, 6*(1).

Hastings García, I., & Huerta García, G. (2015). *Reconstrucción y mejoramiento de la vivienda de adobe en la montaña de Guerrero, México. 15° SIACOT—ECUADOR Seminario Iberoamericano de Arquitectura y Construcción con Tierra*: (pp. 1–11). Cuencia: PRO_TERRA.

Houben, H., & Guillaud, H. (1994). *Earth construction: A comprehensive guide.* London, UK: ITDG Publishing.

Íñigo Dehud, L. S. (2013). *Tecnologías de la construcción. Diseño Arquitectónico 1.* Mexico, Mexico: Ediciones ECA.

ININVI (1989). *Construcción con tapial.* Lima, Perú: En I. N. Vivienda.

Janusek, J. W. (2009). Residence and ritual in Tiwanaku. In *Studies in Latin American ethnohistory & archaeology* (p. 159).

Jorquera, N. (2012). *Culture costruttive in tierra e rischio sismico. Conoscenza delle architetture cilene e valutazione della loro vulnerabilità al sisma.* Università degli Studi di Firenze.

Jorquera, N. (2014). Culturas sísmicas: estrategias vernaculares de sismorresistencia del patrimonio arquitectónico chileno. *Arquitecturas del Sur, 32*(46), 18–29.

Karmelic, L. (2015). *Terremoto Chile 2010. Reconstrucción/Restauración Patrimonial: proceso de validación de la tierra como material vigente.* Balance de las Intervenciones en la VI Región Universidad de Sevilla.

Kościuk, J., & Kogut, J. (2019). Numerical modelling of hypothetical quincha walls on Samaipata rock. In *Structural analysis of historical constructions* (pp. 835–843). Cham: Springer.

Lacoste, P., Premat, E., Castro, A., Soto, N., & Aranda, M. (2012). Tapias y tapiales en Cuyo y Chile (Siglos XVI-XIX). *Apuntes: Revista de Estudios sobre Patrimonio Cultural, 25*(2), 182–199.

Marussi, F. (1989). *Antecedentes históricos de la quincha.* ININVI.

Oliveira, C. F., Varum, H., & Vargas, J. (2012). Earthen construction: Structural vulnerabilities and retrofit solutions for seismic actions. In *15 World conference on earthquake engineering, Lisboa.*

Punzo Díaz, J. L. (2016). Revisando la arquitectura en el valle de chalchiuiteña Guadiana, Durango. El caso del sitio arqueológico de la Ferrería. Cuicuilco. *Revista de Ciencias Antropológicas, 23*(67).

Rivera, A. (2017). *The Chilean adobe as a seismic vernacular technology, the study of the "Norte Chico" area:* (pp. 675–680). https://doi.org/10.1201/9781315267739-118.

Sánchez Suárez, A. (2006). La casa maya contemporánea. Usos, costumbres y configuración espacial. *Península, 1*(2), 81–105.

Silva, E. (2013). *Ensayos en Tapial Mejorado.* Lima, Peru: SENCICO.

Takahashi, K. (2014). *Variedades de El Niño:* (pp. 4–7). Boletín Técnico, Instituto Geofísico del Perú.

Tomasi, J., & Bellmann, L. (2018). La quincha y los sistemas de entramados. *Estructuras, 1*(2).

Tsai, H. (2012). *Adobe bricks and labor organization on the north coast of Peru.* (Vol. 10). Andean Past.

Urbina, O. N. (2019). *Tesoros rodean la ciudad:* (p. 11). El Siglo de Durango.

Uruñuela y Ladrón De Guevara, G., & Plunket, P. (2012). De tronco-cónicos a cuexcomates: hacia la exhibición del almacenamiento a finales del Formativo. In E. S. Bortot, D. Michelet, & V. Darras (Eds.), *Almacenamiento Prehispánico del Norte de México al Altiplano Central* (pp. 41–60). San Luís Potosí: Centro de Estudios Mexicanos y Centroamericanos.

Vernaza, C., Cancino, C., & Rainer, L. (2018). *Proyecto de Estabilización Sismorresistente: Informe sobre el análisis de condiciones, diagnóstico y pruebas de protección para las pinturas murales–Templo Santiago Apóstol de Kuñotambo.* Los Angeles, California: Getty Conservation Institute.

Viñuales, G. M. (2007). Tecnología y construcción con tierra. *Apuntes, 20*(2), 220–231.

Zegarra, L., Quiun, D., San Bartolome, A., & Giesecke, A. (2001). Behavior of reinforced adobe houses in Moquegua, Tacna and Arica during the June 23, 2001 earthquake. In: *XIII National congress on civil engineering, Puno, Peru.*

CHAPTER

5

Nondestructive testing, assessment, and strengthening for reducing the seismic vulnerability of masonry structures

Tiago Miguel Ferreira, Nuno Mendes, Rui Silva, and Paulo B. Lourenço

Department of Civil Engineering, ISISE, Institute of Science and Innovation for Bio-Sustainability (IB-S), University of Minho, Guimarães, Portugal

5.1 Introduction

Masonry is one of the oldest and most broadly used building techniques, which makes it an essential part of the world's cultural heritage. The safeguard and protection of masonry buildings against aging processes and natural hazards, assuring their transmission to future generations, are thus our responsibility as a society.

As discussed by Vicente, Lagomarsino, Ferreira, Cattari, and Mendes da Silva (2018), aging processes, which are typically slow and inevitable, might affect the structural stability of masonry building due to: (i) material degradation, (ii) anthropic modifications, particularly in urban environments; (iii) climate and environmental changes. The capacity to develop accurate and reliable inspection and testing programs is a fundamental requirement in all three cases, both through advanced instrumentation techniques and qualified visual inspections, to detect when interventions are needed.

In what regards to the prevention from natural hazards, it requires a preventive assessment aimed at the specific vulnerabilities of the structure to that particular hazard (earthquake, fire, flood, etc.). In the particular

https://doi.org/10.1016/B978-0-12-821087-1.00008-9

case of the seismic hazard, which still represents the leading cause of damage to masonry structures, it is fundamental to have the ability to develop and implement preventive measures based on reliable vulnerability assessments to minimize earthquake damage and consequent restoration and reconstructions needs.

This chapter provides a general framework on the critical issue of the seismic vulnerability assessment of masonry structures by reviewing the state-of-the-art nondestructive inspection and testing procedures, as well as some of the most widely used vulnerability assessment approaches at different scales, and successful traditional and innovative solutions for strengthening existing masonry structures.

5.2 Inspection and testing

As mentioned before, the evaluation of the seismic performance of masonry buildings requires a specific methodology in which the inspection and testing correspond to the first works and are fundamental to guarantee a reliable assessment of structural stability.

Inspection and testing procedures are aimed to know the characteristics of the structural scheme of the building, namely geometric properties, material properties, and constructive details (e.g., type and efficiency of connections between floors/roof and masonry walls, and connections between orthogonal walls). The inspection and testing are also relevant for defining the knowledge levels of the Eurocode 8—Part 3 (EN 1998-3, 2005), which establishes the applicable type of structural analysis and the confidence factor to be adopted for determining the design values of the material properties. It is noted that the available data (drawings and technical specifications) should also be evaluated. In the particular case of existing masonry buildings, the initial knowledge level is in general low and the techniques of inspection and testing are fundamental, namely the nondestructive and minor-destructive techniques. Thus, the main techniques used in the diagnosis of existing masonry buildings are detailed below.

The visible geometric properties can be obtained through drawings, when available and updated, and surveys based on direct and indirect techniques (Fig. 5.1A). In general, survey with direct techniques can be performed through easy-to-use equipment (e.g., tape measure and laser distance meter), require a high in situ workload, and involve a result processing of low complexity. On the other hand, the indirect techniques require equipment of high complexity and accuracy (e.g., total station and laser scanner), lower in situ workload, and advanced knowledge and high computational effort for the result processing. Photogrammetry is also an indirect technique that allows the digital reconstruction of the

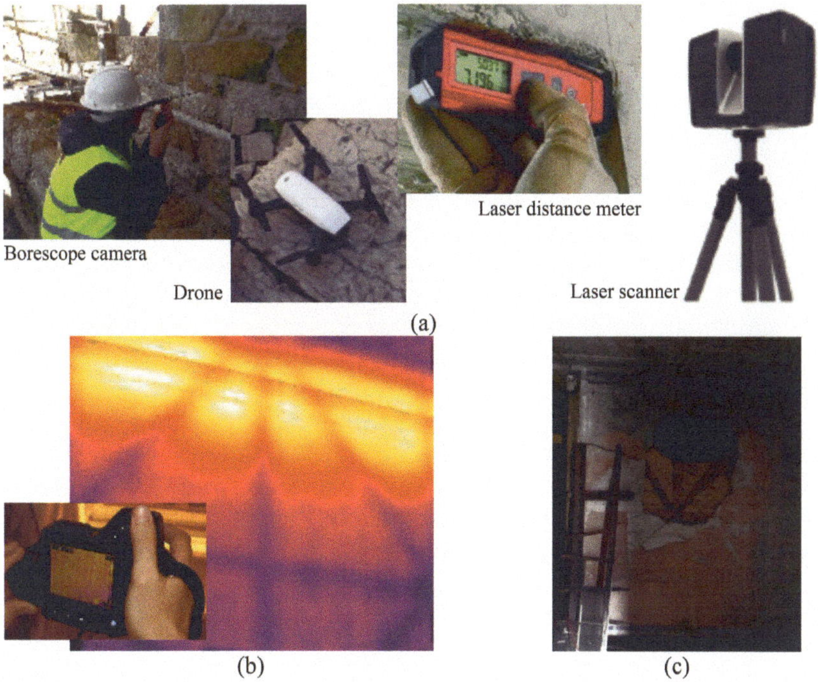

FIG. 5.1 Examples of techniques for the inspection: (A) equipment for geometry survey and inspection; (B) image obtained from thermal camera (detection of the timber structure embedded and not visible in a masonry wall with plaster); (C) destructive process of removing the plaster, allowing to identify the type of wall (masonry wall with embedded timber structure locally damaged due to an opening).

three-dimensional space through two-dimensional images, for example, obtained using digital cameras. This technique also requires advanced knowledge and high computational effort for the result processing. The indirect technique can also be used to evaluate deformations. The nonvisible geometric properties of the structural elements, such as foundations, can be estimated using trial pits and ground penetrating radar (GPR) testing. It is also noted that, in general, in a full structural stability assessment of a structure, it is necessary to know the properties of the foundation soil. These properties can be estimated through geotechnical testing, such as the standard penetration test and the cone penetrometer test.

The first phase of the works should include the visual inspection, aiming at performing the damage survey of the building, with particular attention for the structural damages that can affect the seismic behavior of the building, and characterizing the structural elements. The visual inspection should involve photographic records, which can be assisted by drone inspections. The visual inspection can also include observations with borescope camera, thermography, and opening of probes, aiming at obtaining information about the interior of the elements (Fig. 5.1). In the inspection with borescope camera, wall cavities can be used, or small diameter boreholes can be done to insert the borescope tube of the camera. This technique is of particular interest for inspecting the interior of masonry walls, for example, for identifying the number and the characteristics of the leaves of the walls, as well as in the assessment of the interior damage (Fig. 5.1A).

In addition to the detection of the presence of moisture and thermal bridges, the inspection with a thermal camera also allows to detect and distinguish structural elements, such as masonry walls with embedded timber elements covered with plaster (Fig. 5.1B), see structural details of floors and roof, and identify discontinuities, cavities, and anomalies (e.g., reduction of the cross-section of structural elements due to the presence of ducts). The opening of probes (Fig. 5.1C) allows to obtain additional information (e.g., evaluate the connection between structural elements) and validate the results obtained from the other techniques, such as the nondestructive and minor-destructive testing.

The nondestructive testing does not cause damage in the building and allows to identify: (1) nonvisible elements; (2) different types of materials and their conservation status; (3) damage; (4) interventions carried out in the past; (5) ducts and steel elements. Moreover, this type of tests allows for estimating material and geometric properties. Among the tests based on the propagation of elastic waves inside the element, sonic, ultrasonic, and impact-echo tests are highlighted. Sonic (20 Hz to 20 kHz) and ultrasonic (20 to 150 kHz) tests are based on measuring the velocity of the wave propagation from a transmitting transducer to a receiving transducer (e.g., instrumented hammer and accelerometer, in the case of sonic tests), through direct, indirect, or semi-indirect transmission (Fig. 5.2A). The propagation velocity of the wave is related to the elastic properties, allowing to estimate the dynamic Young's modulus of the material, as well as to identify heterogeneities. The impact-echo tests require a transducer only (accelerometer) and allow to estimate the thickness and identify damages in the interior of the element, based on the reflection of waves and result processing in the frequency domain. GPR tests (Fig. 5.2B) are based on the emission and measuring of the reflection of electromagnetic radiation and are also used in the diagnosis of masonry structures. This type of test allows to identify different types of materials, estimate the geometry of nonaccessible elements (e.g., foundations and embedded arches in walls) and partially accessible elements (e.g., the structure of timber floors and filling of vaults), and estimate the thickness of the leaves of masonry walls. Dynamic identification test using ambient vibration (Fig. 5.2C) corresponds to a nondestructive test that involves the measuring of accelerations using accelerometers of high sensitivity and allows to estimate the

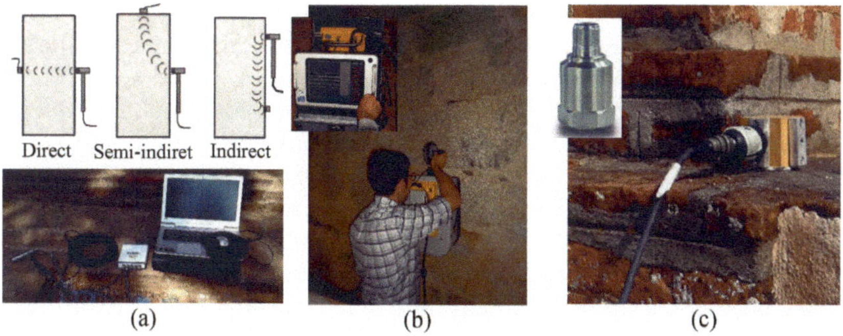

FIG. 5.2 Examples of nondestructive testing: (A) sonic tests; (B) ground penetrating radar test; (C) dynamic identification test (accelerometer).

dynamic properties of the building (frequencies, mode shape, and damping ratios). These results are of special interest for the calibration of numerical models developed for the structural analysis. Finally, the methods based on the surface hardness are referred, such as the tests with Schmidt hammer and pendulum rebound hammer, which allow to evaluate the homogeneity of masonry units (qualitative analysis) and to estimate the strength of the mortar of the joints, respectively.

In regard to minor-destructive testing and techniques, flat-jack tests, pull-out tests, and pull-off tests in masonry units, pull-out tests with a helix screw in the mortar, and the extraction of cores for strength tests in laboratory are highlighted. The flat-jack tests (Fig. 5.3A) require the execution of slots in the walls to insert the jacks and allow to evaluate the compressive stress level in masonry element (single flat-jack) and to estimate the material properties of the masonry (double flat-jack), such as Young's

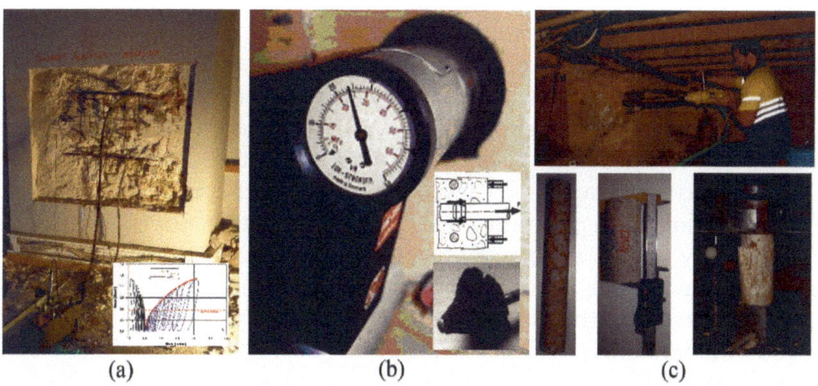

FIG. 5.3 Examples of minor-destructive testing and techniques: (A) double flat-jack test; (B) pull-out tests; (C) core extraction, sample, and compressive strength test.

modulus and the compressive strength. Pull-out and pull-off tests (Fig. 5.3B) can be used to evaluate heterogeneities in masonry units (qualitative analysis) and estimate their compressive strength, based on relations between the pull-out force and the compressive strength obtained from laboratory tests. On the other hand, the pull-out tests with a helix screw have a special interest for qualitative analysis and characterization of the mortar of the joints, based on the pull-out force applied in the screw inserted in the mortar. The cores extraction and the laboratory tests of samples made of masonry units and mortar (Fig. 5.3C) allow to estimate their compressive strength directly, and of the masonry indirectly through formulations based on the strength of each component. In the field of the minor-destructive testing, the tests with resistograph and Pilodyn are also mentioned, which allow assessing the integrity of timber elements of floors and roofs.

Finally, it is emphasized that the diagnosis of masonry structures should preferably involve prior knowledge on the type of building under analysis, including the type of materials and construction techniques usually used in its construction, and the integrated analysis of the results obtained from the inspection and testing, to guarantee reliable data for the seismic assessment of the structure. It is also recommended to estimate, whenever possible, the same parameter through different tests, aiming at increasing the confidence on the obtained results.

5.3 Assessment

The seismic vulnerability assessment of existing masonry structures is admittedly a difficult task due to two fundamental reasons: (i) the difficulty of acquiring as-built information on material parameters and structural details, as already discussed in the previous section; (ii) the still existing difficulties and limitations in interpreting and modeling the seismic response of the masonry buildings. As discussed by Ferreira, Mendes, and Silva (2019), the selection of the most suitable vulnerability assessment method to use must be based on a not always easy balance between the simplicity of the approach and the accuracy of the results. In any case, it is important to keep in mind that, regardless of the adopted approach, vulnerability results are always potentially by different types of uncertainty (related to the limitations of the model itself and to the inherent randomness of both the sample and the response) that should be acknowledged and addressed.

According to several authors, see for example Vicente et al. (2014) or Ferreira, Maio, Costa, and Vicente (2017), vulnerability assessment methodologies can be grouped into three main categories accounting for their

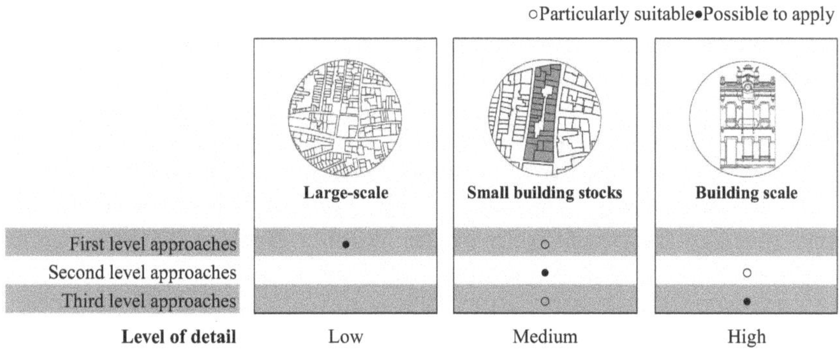

FIG. 5.4 Analysis techniques versus assessment scales from Ferreira et al. (2019).

level of detail, the scale of evaluation and use of data: i.e., first level, second level, and third level approaches, see Fig. 5.4.

First and second level approaches are typically based on the use of qualitative data, and therefore, they are particularly indicated for the development of large-scale analysis. Depending on the methodological formulation of each approach (some of them are much more quantitative, whereas others are more qualitative) and on the amount and the quality of the data available, first- and second-level approaches can be used to perform vulnerability assessments either at a large scale, an urban area or even a country, or at the scale of a single building. According to the literature (Ferreira et al., 2019), first- and second-level approaches can be further classified into four groups of methods: typological, indirect, conventional, and hybrid.

As described by Vicente et al. (2014), typological methods classify buildings into different classes depending on the materials, construction techniques, structural characteristics, and other features that directly or indirectly influence the seismic vulnerability of the building. In these approaches, vulnerability is defined as the probability of a structure to suffer a certain amount of damage for a given level of ground motion intensity. Evaluation of damage probability is based on observed and recorded damage after previous earthquakes and also on expert knowledge. For this reason, the results obtained from the application of typological methods must be treated in a statistical basis and are valid only for the area where the assessment was conducted, or for similar ones (regarding building typology and seismic hazard). Some examples of the use of typological-based approaches can be found in Whitman, Reed, and Hong (1973), Braga, Dolce, and Liberatore (1982), Di Pasquale, Orsini, Pugliese, and Romeo (1998), Dolce, Masi, Marino, and Vona (2003), and Lagomarsino and Giovinazzi (2006).

Regarding the indirect methods, they rely on the determination of a vulnerability indicator (a vulnerability index), followed by the establishment of relationships between damage and seismic intensity or peak ground acceleration. These methods have been applied extensively to assess the seismic vulnerability of large areas. Unlike the typological methods, whose vulnerability is inferred from the "average" features of a representative building typology, indirect methods can be used to determine the actual characteristics of the building stock under assessment. However, because they are defined empirically, the weights assigned to the parameters that compose the indices can be a potentially significant source of uncertainty and should therefore be taken into account. The "Vulnerability Index Method" proposed by Benedetti and Petrini (1984), the ATC-21 (1988) and the GNDT-SSN (1994) are three notable examples of indirect vulnerability assessment methods.

According to Ferreira et al. (2019), conventional methods are essentially heuristic, proposing a correspondence between a vulnerability index and a certain level of damage. Conventional methods can be divided into two categories: those that qualify the different physical characteristics of structures empirically—ATC-13 (1985) is one of the best known of examples of this first type; and those based on the criteria defined in seismic design standards for structures, evaluating the capacity-demand relationship of buildings—the HAZUS (1999) method is a well-known example of this second type of conventional methods.

Finally, hybrid methods seek to combine some of the features of the typological, the indirect, and the conventional methods. Some examples of the use of hybrid methods can be found in Faccioli, Pessina, Calvi, and Borzi (1999), where the characteristics of typological and indirect methods were put together to obtain an improved vulnerability index formulation, or in Ferreira, Maio, and Vicente (2017), where a hybrid approach is integrated into a GIS tool to evaluate the impact resulting from the adoption of different large-scale seismic retrofitting strategies. In the same line, Aguado, Ferreira, and Lourenço (2018) and Anglade et al. (2020) took advantage of a hybrid approach to discuss the potential benefits resulting from the application of different seismic retrofitting strategies, not only accounting for their effect on the reduction of individual and global damages resulting from different seismic scenarios but also regarding their global impact in terms of civil protection and urban accessibility (Ferreira et al., 2019).

As for the third-level approaches, refer to Fig. 5.4, they are specifically addressed to assess the vulnerability of individual buildings. As is it recognized, the assessment of the seismic vulnerability of existing masonry buildings is still a challenging task among other aspects, because the mechanical properties of their materials are highly variable (and their characterization is relatively costly), the accurate understanding of the building's change history is difficult to obtain, and also because most of the approaches prescribed in recent codes and guidelines are not applicable to these buildings. From a structural point of view, the evaluation of the seismic response of masonry buildings can be divided into two main types of behavior: out-of-plane and in-plane and gauged by the following features (Ferreira, Costa, & Costa, 2015):

- The material and construction quality of the load-bearing walls;
- The restricted slenderness of the walls, which rules their out-of-plane stability;
- The efficiency of the connections among walls and between the walls and the horizontal structures;
- The rigidity of the floor diaphragms;
- The presence of strengthening elements (such as ring beams, tie-rods, or buttresses).

In the last decades, several experimental and numerical studies on the in-plane and out-of-plane of existing masonry buildings were carried out, which allowed for significant improvement of the knowledge on its seismic performance. Furthermore, these studies helped develop new tools for the seismic assessment of existing masonry buildings. Because an exhaustive review of seismic vulnerability assessment approaches at the scale of the building would be too vast, the reader is referred to Ferreira et al. (2015, 2019), where the most relevant methodologies, techniques, and computational tools available to assess the out-of-plane and in-plane behavior of masonry buildings are reviewed and discussed.

5.4 Strengthening

A building system or an existing structure that is deemed unsafe under seismic loads, or that presents evident seismic limitations, can be improved to reduce its seismic vulnerability and allow safe use. This realization is patent through the history of humankind and led to the development of several traditional seismic-resistant building practices and intervention solutions. The development of these practices and solutions was initially based on the empirical knowledge gained through time by the local populations struggling against the repeating destruction of several earthquakes. The most successful practices and solutions are nowadays identified in many buildings around the world and are recognized as evidence of the local seismic cultures. According to Ortega, Vasconcelos, Rodrigues, Correia, and Lourenço (2017), the main factor that leads to the development of a seismic culture is earthquake hazard awareness. Furthermore, the impact of past earthquakes on the built-up environment is of great importance to shape the local seismic culture.

Thus, practices and solutions proved as dangerous after an earthquake are either abandoned or modified, whereas those withstanding the event are replicated in the reconstruction works.

More recently, the establishment of earthquake engineering brought scientific support on the application of the aforementioned traditional approaches. Recent developments also allowed the proposal of innovative seismic strengthening strategies based on advanced technologies and materials, which grant high structural efficiency.

In general, the traditional seismic strengthening solutions and practices follow five main principles:

- Improvement of the connections between structural elements;
- Stiffening of the horizontal structural elements;
- Consolidation of the vertical structural elements;
- Use of redundant structural elements as a backup for possible partial failure;
- Construction of additional structural elements contributing to the resistance of the building to horizontal loads.

The introduction of innovative strengthening solutions added two more principles, which are summarized as follows:

- Improvement of the mechanical properties of the masonry walls to increase the loading and displacement capacities;
- Reduction of the dynamic effects induced by earthquakes.

5.4.1 Traditional solutions

The floors and roofs of ancient masonry buildings, made typically of timber, are insufficiently stiff to provide to the structure the so-called "box-behavior" responsible for promoting the structural collaboration between the masonry walls. In the absence of such behavior, the global seismic response of a building becomes considerably limited (Lourenço, Mendes, Ramos, & Oliveira, 2011). Furthermore, the floor-wall, roof-wall, and wall-wall connections typically found in such buildings are poor, meaning that they also do not allow the structure to develop a satisfactory monolithic "box behavior" (Moreira, Ramos, Oliveira, & Lourenço, 2016). Thus, it is comprehensible that many of the traditional strengthening solutions are focused on improving these limitations of ancient masonry buildings.

Introducing ring beams between the walls and the roof constitutes one of the most effective solutions to ensure the structural integrity of masonry buildings. Traditional ring beams are built using a pair of longitudinal planks joined together by transversal elements in a ladder-like configuration (Vintzileou, 2008), as illustrated in Fig. 5.5. Rougher timberworks can

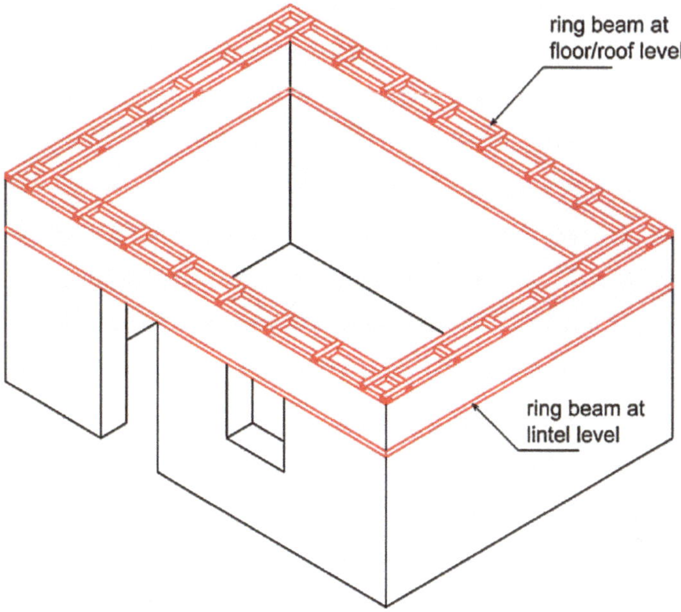

FIG. 5.5 Illustration of traditional strengthening with timber ring beams (Ortega et al., 2017).

also be used, such as a grid of horizontal timber trunks or tree branches lying longitudinally and transversally (Michiels, 2015). In the last few decades, ring beams were also built using reinforced concrete, but this procedure introduces compatibility concerns, such as stiffness and chemical incompatibilities.

Ring beams promote different types of improvement of the structural behavior, namely enhancement of the "box-behavior" (the increase of the loading and deformation capacity), prevention of the out-of-plane collapse of the walls, and tying of masonry leaves in multileaf walls. Thus, ring beams should be continuous around the entire building and present adequate connection at the intersections between walls and between the ring beam and the walls to provide the best performance. The main drawback of ring beams is its installation, because it may require removing and raising the roof.

Ancient buildings have also been traditionally strengthened with ties, which are typically composed of timber tie beams or steel rods connecting walls in opposing façades (Fig. 5.6). In this last case, steel anchor plates are used to fasten the tie rod conveniently to the masonry, meaning that this solution only works when the walls tend to move away from each other. Nevertheless, it allows enhancing the "box-behavior" of the building by avoiding the out-of-plane collapse of the façades and by constraining

FIG. 5.6 Example of strengthening with steel ties.

the floors and roof. Steel ties are introduced at the floor and roof levels, avoiding interference with the habitable spaces. The dimension of the anchor plates is another important aspect of this strengthening solution, as they should be sufficiently large to avoid shear failure of the masonry. In buildings with vaulted masonry floors and roofs, ties can also be used to limit relative lateral movements of the abutment walls caused by seismic loads. This type of strengthening promotes the preservation of the geometry of the vault, avoiding its collapse.

Timber wedges are also traditionally used to strengthen the connections of timber roofs and floors with the masonry walls. These elements ensure a tight mechanical connection between the joists and the walls (Fig. 5.7A). In case of absence of this solution in existing buildings, its implementation requires sufficient long resting lengths of the joists. Alternatively, metal brackets, steel straps, and ties can be more easily implemented to connect the timber floors and roofs to the walls (Moreira et al., 2016), as illustrated in Fig. 5.7B. Partial reconstruction can also be used to introduce or substitute timber wall plates, which ensure a better

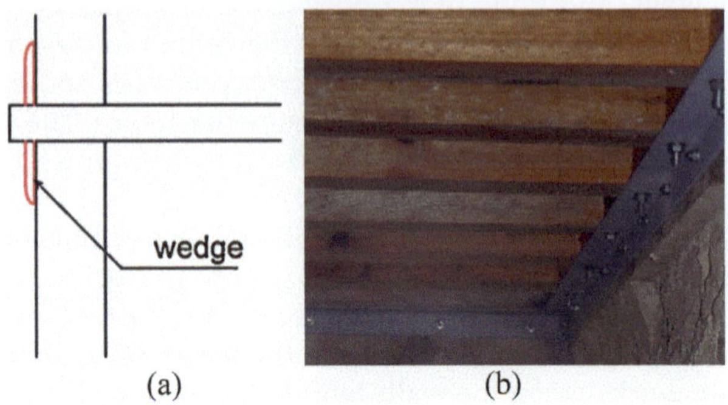

FIG. 5.7 Strengthening of the connection floor-walls: (A) schematic representation of timber wedges strengthening (Ortega et al., 2017); (B) with metallic elements.

transition of the roof or floor timber elements to the walls. As referred previously, proper roof-wall and floor-wall connections are fundamental to promote the "box-behavior" of the structure.

The reinforcement of the wall-wall connection is also a strengthening solution frequently used in existing masonry, namely at the corners of the building. This type of solution has been mainly implemented during the construction of the building by employing materials with better mechanical properties at these sections, which include high-quality stones, bricks, and timber elements. The strengthening of the wall-wall connection in existing buildings can also be achieved resorting to anchor systems, where a rebar is introduced in the masonry and is effectively bonded by means of grout injection and an anchor plate (Paganoni & D'Ayala, 2014), as illustrated in Fig. 5.8. The additional tensile strength provided by this strengthening solution improves the load transmission capacity between intersecting walls, enhancing the "box-behavior".

The diaphragmatic behavior of floors and roofs made of timber can be effectively improved by means of stiffening, which serves to improve the "box-behavior" of the existing structure. Stiffening can be achieved with diagonal bracings, triangulation, and by providing an additional layer of sheathing boards or timber planks placed perpendicularly to the existing ones (Senaldi, Magenes, Penna, Galasco, & Rota, 2014), as illustrated in Fig. 5.9. Stiffening can also be achieved by substituting the timber floors with slabs made of reinforced concrete or mixed concrete clay blocks, which require building-reinforced concrete beams on the walls to conveniently support the slabs. These beams should grant an adequate connection with the walls and provide uniform load transmission (Binda, 2006). On the other hand, this last approach is highly intrusive, which may deem it unacceptable in most of the situations.

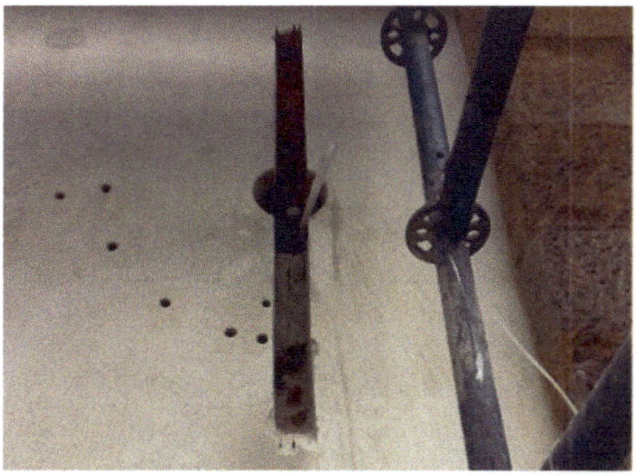

FIG. 5.8 Example of wall-wall connection strengthening with an anchor system.

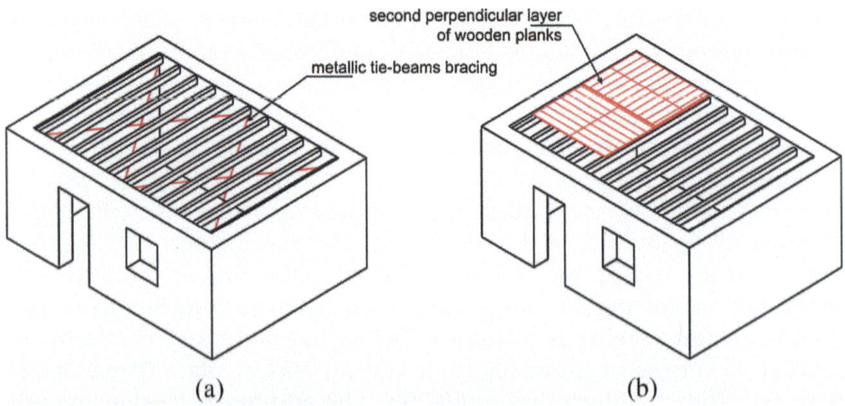

FIG. 5.9 Schematic representation of floor stiffening solutions (Ortega et al., 2017): (A) diagonal bracing; (B) additional layer of timber planks.

Consolidation is an intervention solution particularly addressed to masonry buildings with multileaf walls. Typically, seismic loads lead to an independent behavior of the different masonry leaves due to the lack of proper connection between each other, which makes them more vulnerable to separation and thus to out-of-plane failure. The connection between leaves is typically assured by through-stones, which consist of long stones placed through the full thickness of the wall. These elements serve as connectors between leaves and allow an enhanced load distribution (De Felice, 2011). Nevertheless, this connection can be further improved by means of transversal ties, where a rebar is introduced through the full thickness of the wall and is properly fixed with an anchor system and/or grout injection (Oliveira, Silva, Garbin, & Lourenço, 2012). Consolidation of masonry walls also involves repairing existing cracks (e.g., postearthquake cracks) that debilitate the in-plane and out-of-plane performance of the walls. Cracks can be repaired by means of partial reconstruction, soft stitching, and stapling (see Fig. 5.10). The first technique is labor intensive and intrusive, meaning that its use is only justified in the case of severe damage. Soft stitching consists of opening chases crossing the cracks, which are then filled with masonry units to create a mechanical bond. Stapling consists of applying staple shaped steel rebars transversally to the crack to connect the split sides (Hurd, 2006). Repointing (or deep repointing) is another technique used to consolidate masonry and consists in removing the bed-joint mortar to substitute it with a better one. This procedure allows for improving the accommodation between units and the bond in the masonry. In some situations, reinforcing rebars are also introduced with the mortar to mitigate other structural problems, such as creep damage (Valluzzi, Binda, & Modena, 2005).

Openings in masonry walls must have proper dimensions and characteristics to avoid collapse during seismic loadings. In this regard, lintels

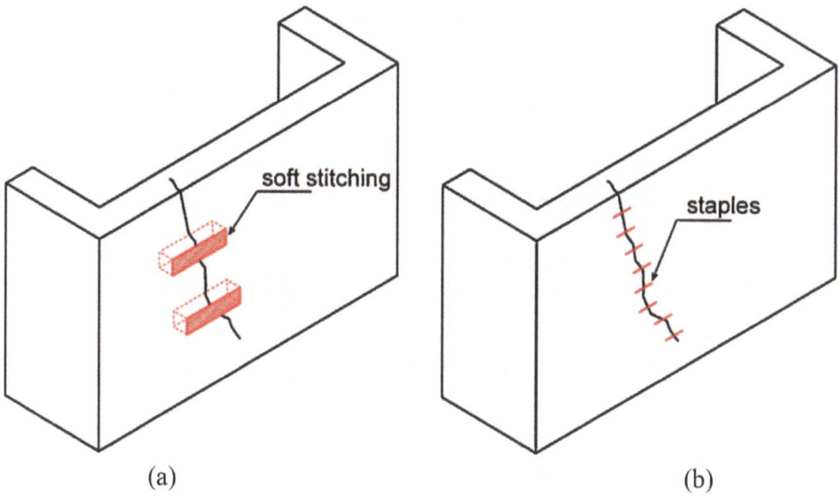

FIG. 5.10 Schematic representation of traditional crack repair solutions (Ortega et al., 2017): (A) soft stitching; (B) stapling.

are key elements that must grant proper redistribution capacity of the transmitted loads. Lintels can be constituted by single stones or timber elements with length longer than the span, and/or by discharging masonry arches. Thus, strengthening can be achieved by reinforcing the lintel resisting elements or by substituting them with stronger ones. Another option to reduce the seismic vulnerability associated with openings is to reduce their span or, as a more drastic solution, close them permanently (Ortega et al., 2017).

Build redundant structural elements in the existing building allow to avoid full collapse, even if the partial collapse of the most vulnerable structural elements occurs (Karakostas et al., 2005). This strategy allows preserving the life of inhabitants by avoiding falling parts of the building to the interior. To this purpose, an additional structure made of timber, reinforced concrete, or steel is built inside the existing building, which nowadays may constitute an intervention excessively invasive and likely unacceptable.

Building buttresses is a strengthening solution that serves to help the walls of a building counteracting the horizontal loads induced by earthquakes. Buttresses are massive masonry walls with typical triangular shape (see Fig. 5.11), which limit the out-of-plane movement of the adjacent walls due to the contribution of their heavyweight and additional shear stiffness and strength. Thus, buttresses and adjacent walls should be properly connected with cross ties to avoid independent behavior when both move apart, as undesirable impact damage may occur (Michiels, 2015).

FIG. 5.11 Example of a rammed earth building strengthened with buttresses.

5.4.2 Innovative solutions

The recent development of innovative seismic strengthening solutions originated new and efficient approaches for interventions in existing masonry buildings. Nevertheless, not in all situations the initial proposal of these solutions respected the main principles established in the conservation charters. Further scientific studies and experience allowed to improve this aspect, among which the fulfilment of compatibility requirements was the main focus. The subsequent paragraphs present a brief discussion on the main innovative strengthening solutions, which include grout injection, externally bonded composites, base isolation, and energy dissipation devices.

The strengthening of multileaf masonry walls with grout injection consists in filling the inner voids and cracks with a fluid mortar (see Fig. 5.12), which after hardening provides continuity and bond to the material (Toumbakari, 2002). Thus, the stiffness and strength of the injected walls

FIG. 5.12 Strengthening of a three-leaf wall with grout injection.

increase significantly (Oliveira et al., 2012), improving seismic behavior (Mazzon et al., 2009). On the other hand, grout injection presents important drawbacks, such as lack of predictability on the required volume of grout and irreversibility of the solution. The first is based on the impossibility of measuring adequately the ratio and interconnectivity of the voids, which compromises the cost control of the intervention. An intervention requiring the complete injection of all walls in a building can rise rapidly the costs, because, in general, voids represent an important percentage of the total wall volume (can go up to 30%). Alternatively, the injection can be prescribed only for the most vulnerable walls or sections. The irreversibility of this solution also raises concerns on the selection of the grout, meaning that the use of poor compatibility grouts should be avoided. In this regard, past interventions evidenced that epoxy- and cement-based grouts may introduce durability problems in ancient masonry walls. Nowadays, several commercial grouts specifically developed for this type of masonry are available and are mainly composed of lime and pozzolanas, which provide enhanced compatibility.

The strengthening of masonry with externally bonded composites has also been used to reduce the seismic vulnerability of masonry buildings, especially in Italy, where these solutions were mainly based on fiber-reinforced polymers (FRP). The popularity of FRP-based strengthening was driven by its ease of application and high efficiency in increasing the tensile strength and ductility of masonry walls while increasing the weight negligibly (high strength and stiffness to weight ratio). Although FRP-based strengthening possesses several drawbacks related with the use of organic matrixes (epoxy-based), such as poor fire resistance, lack of vapor permeability, low reversibility, high cost, and poor compatibility with masonry substrates (Papanicolaou, Triantafillou, Papathanasiou, & Kyriakos, 2008; Valluzzi, Modena, & De Felice, 2014). Thus, alternative composite solutions have been developed by integrating matrixes with enhanced compatibility, namely cement-, lime-, or clay-based mortars. The adequate embedding and bond of these mortars also required substituting the sheet textiles by mesh textiles or steel grids with adequate aperture sizes (see Fig. 5.13). These alternative composites are known as steel reinforced grout, fiber-reinforced cementitious matrix, or textile-reinforced mortar (TRM) (Ascione, De Felice, & De Santis, 2015; De Felice et al., 2014). Several commercial TRM systems are nowadays available in the market, from companies such as Ardea Progetti, BASF, Kerakoll, Mapei, Sika, and Simpson Strong-Tie. These systems are constituted by meshes of different fibers (steel, glass, carbon, basalt, and polyparaphenylene benzobisoxazole) and mortars of different compositions; nevertheless, they are in general high mechanical performance materials. Recommendations for designing FRP-based strengthening solutions are sufficiently documented in ACI 440 7R-10 (ACI, 2010) and CNR-DT 200

FIG. 5.13 Strengthening of a rammed earth wall with TRM.

R1/2012 (CNR, 2012) for concrete and masonry elements. These recommendations have also been used for designing TRM strengthening systems for masonry; nevertheless, they lack comprehensive instructions, in particular for what concerns application procedures, quality assurance of the materials, durability verification, laboratory and field-testing methodologies, and the effectiveness of the installation. More recently, the American Concrete Institute published the ACI 549.4R13 (ACI, 2013) to fill this gap.

The use of antiseismic devices to reduce the vulnerability of existing masonry buildings has been gaining attention in the last few years, especially in the seismic protection of buildings with significant cultural value. The most interesting types of devices are base isolators and passive dampers.

Base isolation consists in isolating the structure from the seismic vibrations by means of an interface with low horizontal stiffness and sufficient capacity to support the vertical loads, introduced at the foundations. The interface modifies the modes of the structure by increasing their fundamental periods and changing their configuration. The higher horizontal flexibility introduced by the isolators promotes a reduction of the accelerations in the structure (by avoiding resonance) and an increase in the total displacements. The structure behaves as a rigid body, which means that lower relative displacements of the structure are produced, and thus the associated damage is also lower. In fact, the total displacements concentrate at the interface of the system, which constitutes a limitation to the use of this solution if they may affect nearby buildings. Base isolation devices can be classified as elastomeric bearings and sliding bearings. Lead-rubber bearings are among the most established devices, whose installation in existing masonry buildings involves:

- Construction of reinforced concrete beams at the base of the walls;
- Construction of the base of the devices;
- Installation of the devices;
- Lifting of the structure and transfer of its dead-weight to the devices;
- Construction of a rigid diaphragm at the level of the beams to promote a joint operation of the full system.

The lifting-transfer phase is the most critical of the process, as it requires limiting the induced differential settlements to avoid causing damage to the masonry walls and other structural elements. This strengthening solution has an intrusive implementation and can be costly due to the complexity of the implementation process. Nonetheless, base isolation has been successfully applied in notable masonry buildings, such as the Salt Lake City and County Building in the USA (Bailey & Allen, 1988); see Fig. 5.14.

The installation of passive dampers aims at improving the capacity of the structure to dissipate the energy transmitted by an earthquake, which increases the seismic performance and reduces the induced damage. Dampers can be classified as: (i) viscous dampers; (ii) viscoelastic dampers; (iii) hysteretic metallic dampers; and (iv) friction dampers.

Viscous dampers are constituted by a hollow metallic cylinder filled with a silicon-based fluid and a piston that forces the fluid to pass through small holes at high speed. The movement of the fluid originates high forces that oppose to the relative movement of the damper and dissipates energy through heat, due to interparticle friction. The resisting forces are

(a) (b) (c)

FIG. 5.14 Example of a masonry building strengthened with base isolators (Bailey & Allen, 1988): (A) general view of the Salt Lake City and County Building; (B) location of the base isolators; (C) application of the base isolators.

proportional to the piston speed and depend on the viscosity of the fluid. This type of dampers allows increasing the structural damping significantly, leading to reduced deformations caused by the action of an earthquake.

Viscoelastic dampers are normally constituted by metallic sheets intercalated by deformable polymeric sheets. The shear deformation of the polymeric sheets due to relative movements of the damper extremities generates heat, which dissipates energy. The dissipated energy is a function of the elastic deformation and deformation speed. This type of dampers modifies the linear structural parameters for damping and stiffness.

Hysteretic metallic dampers dissipate energy by exploiting the plastic deformation capacity of steel, which can be achieved by using low yielding stress steels or cruciform cross-sections.

Friction dampers dissipate energy due to friction generated at the contact surface between two moving solid elements. These dampers are typically constituted by steel plates connected transversally by nuts and bolts fixed with a specific torque, which defines the friction force required to initiate energy dissipation.

Finally, passive dampers can be installed to strengthen specific structural elements, meaning that they constitute a solution less intrusive and costly than base isolation. The Siena Cathedral (Italy) is an example of a historical monument strengthened with dampers. In this case, two spring viscous dampers were installed to improve the out-of-plane behavior of the tympanum of the main façade (Castellano, Tosti, Bolletti, & Tosti, 2009), as illustrated in Fig. 5.15.

5.5 Final remarks

In this chapter, some of the most widely used approaches to test, assess, and mitigate the seismic vulnerability of existing masonry buildings were reviewed and discussed. Before making some final remarks, it is worth noting that, despite the authors efforts to compile a sample of works as relevant and comprehensive as possible, there are certainly several other important works that, due to space constraints or oversight, are not included herein.

From the review given in this chapter, it is possible to draw some important conclusions. The first has to do with the selection of the most appropriate seismic vulnerability assessment method to use. As it is discussed in the section dedicated to the assessment, the selection of the most appropriate method or technique to use must be made on the basis of a proper balance between simplicity of the approach and the accuracy of its results. In this sense, it is important to stress that vulnerability results can be strongly affected by several sources of uncertainty, which, for this reason, must be recognized and addressed.

The fundamental role of vulnerability assessment methods from the risk mitigation standpoint was also clearly demonstrated though the

(a)

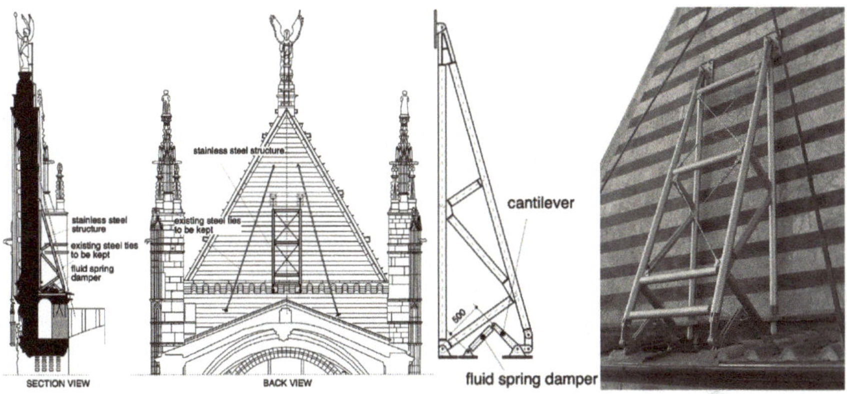

(b)

FIG. 5.15 Example of a masonry building strengthened with dampers (Castellano et al., 2009): (A) general view of Siena Cathedral; (B) details of the spring viscous damper installed at the tympanum of the main façade.

discussion presented herein. In fact, only on the basis of a comprehensive knowledge of the characteristics of the buildings and of their structural vulnerabilities, it is possible to select and implement proficient mitigation strategies and to outline strengthening interventions that can contribute to reduce their seismic vulnerability.

References

ACI (2010). *ACI 440.7R-10: Guide for the design and construction of externally bonded FRP systems for strengthening unreinforced masonry structures*. Farmington Hills, Michigan, US: American Concrete Institute.

ACI (2013). *ACI 549.4R13: Guide to design and construction of externally bonded fabric-reinforced cementitious matrix (FRCM) systems for repair and strengthening concrete and masonry structures.* USA: American Concrete Institute (ACI). ISBN: 9780870318528.

Aguado, J. L. P., Ferreira, T. M., & Lourenço, P. B. (2018). The use of a large-scale seismic vulnerability assessment approach for Masonry Façade Walls as an effective tool for evaluating, managing and mitigating seismic risk in historical centers. *International Journal of Architectural Heritage, 12*, 1259–1275. https://doi.org/10.1080/15583058.2018.1503366.

Anglade, E., Giatreli, A.-M., Blyth, A., Di Napoli, B., Parisse, F., Namourah, Z. , et al. (2020). Seismic damage scenarios for the Historic City Center of Leiria, Portugal: Analysis of the impact of different seismic retrofitting strategies on emergency planning. *International Journal of Disaster Risk Reduction. 44*, 101432. https://doi.org/10.1016/j.ijdrr.2019.101432.

Ascione, L., De Felice, G., & De Santis, S. A. (2015). A qualification method for externally bonded fiber reinforced cementitious matrix (FRCM) strengthening systems. *Composite Part B: Engineering, 78*, 497–506. https://doi.org/10.1016/j.compositesb.2015.03.079.

ATC-13 (1985). *Earthquake damage estimation data for California. Report ATC-13.* Redwood City, California: Applied Technology Council.

ATC-21 (1988). *Rapid visual screening of buildings for potential seismic hazards: A handbook.* Redwood City, California.

Bailey, J. S., & Allen, E. W. (1988). Seismic retrofitting: Salt Lake City and county building. *APT Bulletin: The Journal of Preservation, 20*(2), 32–44. https://doi.org/10.2307/1494249.

Benedetti, D., & Petrini, V. (1984). Sulla Vulnerabilità Di Edifici in Muratura: Proposta Di Un Metodo Di Valutazione. *L'industria delle Costruzioni, 149*(1), 66–74.

Binda, L. (2006). The difficult choice of materials used for the repair of brick and stone masonry walls. In *Proceedings of the 1st international conference on restoration of heritage masonry structures, Egypt, Cairo.*

Braga, F., Dolce, M., & Liberatore, D. A. (1982). Statistical study on damaged buildings and an ensuing review of the MSK-76 scale. In G. Athens (Ed.), In *Proceedings of the 7th European conference on earthquake engineering* (pp. 431–450).

Castellano, M. G., Tosti, G., Bolletti, G. P., & Tosti, M. (2009). Fluid spring dampers for seismic protection of the Cathedral of Siena. In *Proceedings of the conference protection of historical buildings—PROHITECH 2009, 669–674.*

CNR (2012). *CNR-DT200 R1/2012: Guide for the design and construction of externally bonded FRP systems for strengthening existing structures,* Italian Council of Research (CNR): Rome, 1–167.

De Felice, G. (2011). Out-of-plane seismic capacity of masonry depending on wall section morphology. *International Journal of Architectural Heritage, 5*(4–5), 466–482. https://doi.org/10.1080/15583058.2010.530339.

De Felice, G., De Santis, S., Garmendia, L., Ghiassi, B., Larrinaga, P., Lourenço, P. B. et al. (2014). Mortar-based systems for externally bonded strengthening of masonry. *Materials and Structures, 47*(12), 2021–2037. https://doi.org/10.1617/s11527-014-0360-1.

Di Pasquale, G., Orsini, G., Pugliese, A., & Romeo, R. (1998). Damage scenario from future earthquakes. In *11th European conference on earthquake engineering, 6–11 September 1998, Paris la Défense, France.*

Dolce, M., Masi, A., Marino, M., & Vona, M. (2003). Earthquake damage scenarios of the building stock of Potenza (southern Italy) including site effects. *Bulletin of Earthquake Engineering, 1*, 115–140. https://doi.org/10.1023/A:1024809511362.

EN 1998-3 (2005). *Eurocode 8: Design of structures for earthquake resistance—Part 3: Assessment and retrofitting of buildings.* European Committee for Standardization.

Faccioli, E., Pessina, V., Calvi, G. M., & Borzi, B. (1999). A study on damage scenarios for residential buildings in Catania city. *Journal of Seismology, 3*, 327–343. https://doi.org/10.1023/A:1009856129016.

Ferreira, T. M., Costa, A. A., & Costa, A. (2015). Analysis of the out-of-plane seismic behavior of unreinforced masonry: A literature review. *International Journal of Architectural Heritage, 9*, 949–972. https://doi.org/10.1080/15583058.2014.885996.

Ferreira, T. M., Maio, R., Costa, A. A., & Vicente, R. (2017). Seismic vulnerability assessment of stone masonry façade walls: Calibration using fragility-based results and observed damage. *Soil Dynamics and Earthquake Engineering, 103*, 21–37. https://doi.org/10.1016/j.soildyn.2017.09.006.

Ferreira, T. M., Maio, R., & Vicente, R. (2017). Analysis of the impact of large scale seismic retrofitting strategies through the application of a vulnerability-based approach on traditional masonry buildings. *Earthquake Engineering and Engineering Vibration, 16*, 329–348. https://doi.org/10.1007/s11803-017-0385-x.

Ferreira, T. M., Mendes, N., & Silva, R. (2019). Multiscale seismic vulnerability assessment and retrofit of existing masonry buildings. *Buildings, 9*, 91. https://doi.org/10.3390/buildings9040091.

GNDT-SSN (1994). *Scheda di esposizione e vulnerabilità e di rilevamento danni di primo e secondo livello (murata e cemento armato).* Roma: Gruppo Nazionale per la Difesa dai Terremoti.

HAZUS (1999). *Earthquake loss estimation methodology—Technical and user manual.* Washington, DC: Federal Emergency Management Agency.

Hurd, J. (2006). Observing and applying ancient repair techniques to pisé and adobe in seismic regions of Central Asia and trans-Himalaya. In *Proceedings of the Getty seismic adobe project 2006 Colloquium,* Los Angeles, CA: Getty Conservation Institute, 101–108.

Karakostas, C., Lekidis, V., Makarios, T., Salonikios, T., Sous, I., & Demosthenous, M. (2005). Seismic response of structures and infrastructure facilities during the Lefkada, Greece earthquake of 14/8/2003. *Engineering Structures, 27*(2), 213–227. https://doi.org/10.1016/j.engstruct.2004.09.009.

Lagomarsino, S., & Giovinazzi, S. (2006). Macroseismic and mechanical models for the vulnerability and damage assessment of current buildings. *Bulletin of Earthquake Engineering, 4*, 415–443. https://doi.org/10.1007/s10518-006-9024-z.

Lourenço, P. B., Mendes, N., Ramos, L. F., & Oliveira, D. V. (2011). Analysis of masonry structures without box behavior. *International Journal of Architectural Heritage, 5*, 369–382. https://doi.org/10.1080/15583058.2010.528824.

Mazzon, N., Valluzzi, M. R., Aoki, T., Garbin, E., De Canio, G., Ranieri, N. , et al. (2009). Shaking table tests on two multi-leaf stone masonry buildings. In *Proceedings of the 11th Canadian masonry symposium, Toronto.*

Michiels, T. L. (2015). Seismic retrofitting techniques for historic adobe buildings. *International Journal of Architectural Heritage, 9*(8), 1059–1068. https://doi.org/10.1080/15583058.2014.924604.

Moreira, S., Ramos, L. F., Oliveira, D. V., & Lourenço, P. B. (2016). Design parameters for seismically retrofitted masonry-to-timber connections: Injection anchors. *International Journal of Architectural Heritage, 10*, 217–234. https://doi.org/10.1080/15583058.2015.1113339.

Oliveira, D. V., Silva, R. A., Garbin, E., & Lourenço, P. B. (2012). Strengthening of three-leaf stone masonry walls: An experimental research. *Materials and Structures, 45*, 1259–1276. https://doi.org/10.1617/s11527-012-9832-3.

Ortega, J., Vasconcelos, G., Rodrigues, H., Correia, M., & Lourenço, P. B. (2017). Traditional earthquake resistant techniques for vernacular architecture and local seismic cultures: A literature review. *Journal of Cultural Heritage, 27*, 181–196. https://doi.org/10.1016/j.culher.2017.02.015.

Paganoni, S., & D'Ayala, D. (2014). Testing and design procedure for corner connections of masonry heritage buildings strengthened by metallic grouted anchors. *Engineering Structures, 70*, 278–293. https://doi.org/10.1016/j.engstruct.2014.03.014.

Papanicolaou, C. G., Triantafillou, T. C., Papathanasiou, M., & Kyriakos, K. (2008). Textile reinforced mortar (TRM) versus FRP as strengthening material of URM walls: Out-of-plane cyclic loading. *Materials and Structures, 41*(1), 143–157. https://doi.org/10.1617/s11527-007-9226-0.

Senaldi, I., Magenes, G., Penna, A., Galasco, A., & Rota, M. (2014). The effect of stiffened floor and roof diaphragms on the experimental seismic response of a full-scale unreinforced stone masonry building. *Journal of Earthquake Engineering, 18*(3), 407–443. https://doi.org/10.1080/13632469.2013.876946.

Toumbakari, E. (2002). *Lime-Pozzolan-cement grouts and their structural effects on composite masonry walls.* PhD thesis, Leuven, Belgium: Catholic University of Leuven.

Valluzzi, M. R., Binda, L., & Modena, C. (2005). Mechanical behaviour of historic masonry structures strengthened by bed joints structural repointing. *Construction and Building Materials, 19*(1), 63–73. https://doi.org/10.1016/j.conbuildmat.2004.04.036.

Valluzzi, M. R., Modena, C., & De Felice, G. (2014). Current practice and open issues in strengthening historical buildings with composites. *Materials and Structures, 47*(12), 1971–1985. https://doi.org/10.1617/s11527-014-0359-7.

Vicente, R., D'Ayala, D., Ferreira, T. M., Varum, H., Costa, A., Mendes da Silva, J. A. R. et al. (2014). Seismic vulnerability and risk assessment of historic masonry buildings. In *Structural rehabilitation of old buildings* (pp. 307–348). Berlin, Heidelberg: Springer. https://doi.org/10.1007/978-3-642-39686-1_11.

Vicente, R., Lagomarsino, S., Ferreira, T. M., Cattari, S., & Mendes da Silva, J. A. R. (2018). Cultural heritage monuments and historical buildings: Conservation works and structural retrofitting. In *Strengthening and retrofitting of existing structures* (pp. 25–57). Singapore: Springer Nature. https://doi.org/10.1007/978-981-10-5858-5_2.

Vintzileou, E. (2008). Effect of timber ties on the behavior of historic masonry. *Journal of Structural Engineering, 134*(6), 961–972. https://doi.org/10.1061/(ASCE)0733-9445(2008)134:6(961).

Whitman, R. V., Reed, J. W., & Hong, S. T. (1973). Earthquake damage probability matrices. In *Proceedings of the 5th world conference on earthquake engineering, Rome,* Vol. 2, (pp. 2531–2540).

Further reading

Salazar, L. G. F., & Ferreira, T. M. (2020). Seismic vulnerability assessment of historic constructions in the downtown of Mexico City. *Sustainability, 12,* 1276. https://doi.org/10.3390/su12031276.

Santos, C., Ferreira, T. M., Vicente, R., & Mendes da Silva, J. A. R. (2013). Building typologies identification to support risk mitigation at the urban scale—Case study of the old city Centre of Seixal, Portugal. *Journal of Cultural Heritage, 14,* 449–463. https://doi.org/10.1016/j.culher.2012.11.001.

CHAPTER

6

Geometrical, constructive, and mechanical characterization of the traditional masonry buildings in the historic city center of Leiria, Portugal

Tiago Miguel Ferreira[a], *Javier Ortega*[a], *and Hugo Rodrigues*[b]

[a]Department of Civil Engineering, ISISE, Institute of Science and Innovation for Bio-Sustainability (IB-S), University of Minho, Guimarães, Portugal, [b]RISCO, Department of Civil Engineering, University of Aveiro, Aveiro, Portugal

6.1 Introduction

The evolution of the historic city center can be considered a result of a long-term process dealing with ideological, ethnological, political, economical, architectural, and artistic values (Vicente, Ferreira, & Mendes da Silva, 2015). Its unique identity should be recognized and characterized by analyzing the urban morphology and by understanding the history and cultural background. Any potential intervention should be based on this knowledge and be a result of a systematic method of assessment and recording that respects the authenticity to ensure compatibility during the urban regeneration process.

The urban centers require information, wide knowledge that can support the rehabilitation of building stock, mainly because these buildings usually must comply with specific requirements for appropriate inclusion in the historic city center. The global survey of the main architectural and

Masonry Construction in Active Seismic Regions
https://doi.org/10.1016/B978-0-12-821087-1.00004-1

structural characteristics and the consequent elaboration of a diagnosis can represent an important tool to have a global image of the existent building stock. These works are a fundamental tool for practitioners to support the level and type of interventions and to define the most appropriate way of intervening in a building. The gathered information should also consider the materials properties, which, if possible, should be obtained with non-destructive technologies. As these techniques may not be fully available to all the cases, it is important to conduct tests on representative case studies that can be extended to the other buildings. This chapter intends to give an additional contribution to the evaluation of old city centers covering the building typology description regarding the main construction methodologies used through visual inspection and the mechanical characterization of a series of limestone masonry walls resorting to several minor and non-destructive evaluation techniques, namely sonic tests, ground penetrating radar (GPR), and results from flat-jack tests. The use of these techniques is typically limited to monumental structures, but the present work shows their application on traditional structures. A simple statistical approach is shown on how limited data extracted from a reduced number of buildings can be extrapolated to relatively homogeneous general built-up environments, aiming to support large-scale onsite inspections with quantitative data. The experimental data obtained from the tests were also complemented and inter-crossed with information gathered through a series of in-field inspections supported by checklists, especially prepared for the purpose. In the end, it is intended that this information can aid in the assessment of the structural safety of the buildings and the definition of optimized rehabilitation or retrofitting interventions.

6.2 The historic center of Leiria

Leiria is located in the center of Portugal and crossed by the Lis River and is one of the main cities between Coimbra and Lisbon. The historic city center is about 30 m above sea level and its landmark is the castle that has been protecting the city for years from the top of the hill. The proximity to the Atlantic Ocean influences its Mediterranean climate: ocean winds lower temperatures and increase humidity.

6.2.1 The history of Leiria

The city center area has been inhabited for a long time, although its early history is uncertain due to a lack of archaeological surveys. Little information and data have been collected from the era between the Roman dominion and the Christian reconquering. If prehistory and the Roman and Moorish rules are neglected, it may be said that the history of Leiria started in 1135 when D. Afonso Henriques built the castle (Mattoso, 1985). His aim was not only to defend Coimbra but also to plan the conquest of Santarém, Sintra, and Lisbon, commanding his army from this fortress (Zuqueste, 2003). In that period, castles boosted urban development of conquered areas and facilitated their social, economic, and administrative growth (Monteiro, 1999). According to Mattoso (Mattoso, 1985), the castle of Leiria may be considered an outpost of Soure, the southern borderline of the Portuguese realm. The fortress is also described as a starting point for potential incursion beyond the enemy lines because it was built beyond the internment castles within D. Afonso Henriques' territory (Nunes, 2005). On the contrary, Gomes (1995) argues that Leiria, Ourém, Pombal, and Porto de Mós are mainly defensive castles rather than offensive. After the Christian reconquering, there were Moorish incursions (between 1137 and 1145) that caused severe damage to the fortress and led it into ruins until the Portuguese conquered Santarém and Lisbon in 1147 (Mattoso, 1985).

Leiria was depicted by the historian Pedro Barbosa as a village almost uninhabited with a castle uphill (Barbosa, 1992). The end of Moorish incursions and the establishment of a tax-free zone in 1142 led to the demographic growth of the area (Gomes, 2005), which then underwent a massive urban change between the 13th and 14th centuries. Many existing buildings were demolished to make room for new structures (Carvalho & Inácio, 2011).

In the Middle Ages, the urban development of Leiria was linked to the castle and Lis River, which have influenced the daily lives of citizens (Gomes, 1993). The hill, west to east oriented, protects the southern urban areas not only from enemies but also from winds. The Lis River has flooded many times due to its sinuous path influencing the morphology of Leiria and the geological and climatic characteristics of the area. Initially, it flowed parallel to the street that today is known as Rua Direita, but then it was diverted to mitigate the risk of floods. However, the Lis River has boosted trade because it was navigable even though its mouth was often filled (Carvalho and Aveleira, n.d.). According to Gomes (1993), the hill influenced the urban layout that has developed in two different ways. The northern São Tiago district has evolved irregularly close to the bridge to Coimbra. In contrast, south of the hill, São Martinho has developed with a fishbone layout in the flat area (Gomes, 1993).

After becoming an administrative center in 1442 (Barbosa, 1992), Leiria updated its status to a city in 1545 when Pope Paul III declared it as a bishopric detached from Coimbra and Estremenha region in terms of both legal and administrative power. During the 16th and 17th centuries, the city of Leiria was developing as other European cities with farmlands or small vegetable gardens inside the urban area (Pousão-Smith, 2008). The courtyard was one of the main characteristics of urban structure regardless of social status.

During the Baroque period, some upgrades were carried out in the city center, such as the alignment of São Martinho square and the redirecting of Lis River, which was modified by Casa do Infantado between 1699 and 1702 to have a lower probability of floods (Carreira, 1995). In addition, Paço de São Simao was converted to the palace of bishop. The city has experienced many changes, but its medieval urban layout remains (Gomes, 1993).

In 1755, Leiria was affected by the Great Lisbon earthquake that caused major damage to the city, in particular to its castle and cathedral. The consequences on the urban layout should be taken into account when analyzing the urban growth, such as regarding public-private space ratio (Carvalho and Aveleira, n.d.). The city was also influenced by fires that were common and propagated rapidly because most of the structures were made of wood (Gomes, 1996).

During the 19th century, the urban layout was modified due to French invasions, which let the city fall into ruin. After the restoration, additional roads and buildings were constructed, but there were no heavy alterations to the medieval structure.

Existing buildings were aligned, and many streets were paved and redirected to make the transit easier. Moreover, existing bridges were restored, public areas were designed, and rural roads were enlarged to improve the network between the city and the surrounding areas.

During the 20th century, Leiria's urban area has undergone small changes such as alignments of buildings, façade renovation, and new constructions. Most notably, the urban area has sprawled outside the medieval profile because of the population growth. One of the main changes was the opening of Rua D. Afonso Henriques and Rua da Beneficência, which affected the medieval layout and led Rua Direita to lose its importance. From that point forward, the fishbone system had two parallel axes that go through the city center from north to east to south to west. These changes may be attributed to the issue of linking the center and the castle (Carvalho and Aveleira, n.d.).

In conclusion, the medieval area just in the south, south-east of the castle has developed harmoniously throughout the centuries, and in contrast, the urban area built between the 20th and 21st centuries has expanded following economic and political interests rather than architectural and urban quality (Carvalho and Aveleira, n.d.).

6.2.2 Urban layout

The development of the urban layout of the city center of Leiria during the centuries was affected by the geographical and typological features of the region, as discussed in the previous section.

The current layout of the historic city center has its focus on the Rua Barão de Viamonte that can be considered the fundamental axis of this area with 18 streets branching perpendicularly, 10 eastward and eight westward. The actual layout may be defined as the result of a long-term process of assembling several units with their own specific functions. This process affected the morphology and the dynamics of the urban area. It started during the Middle Ages between the 13th and 14th centuries, but the current layout is the result of the sprawl that occurred from the end of the 19th to 20th centuries (Gomes, 1993, 1996). It resulted in the actual urban fabric of the historic city center between Rua Barão de Viamonte and the Lis River consisting of blocks parallel or perpendicular to these two vectors.

6.2.3 Population

Most of the buildings in the historic center of Leiria are multi familiar or single familiar. However, many of these buildings have retail space on the ground floor and multi familiar or single familiar habitations on the upper floors. A total of 39.5% of the buildings in the city center are unoccupied. According to a socio-demographic study completed in 2006 and updated in 2011 census by the Municipality of Leiria (Dinis, 2006), the residents of the historic center are distributed in the following age classes: 2% in the age 0–4; 5% in the age 5–13; 11% in the age 13–24; 55% in the age 25–65; and 27% in the age over 65 years.

It is worth noting that the reliability of this demographic data is questionable as the total number of inhabitants of the historic city center of Leiria is reported as close to the number of buildings in this area. This means that only one or two people are living in each building. Even considering the number of unoccupied buildings in the city center, this number seems still too low.

6.3 Building characterization

The analysis presented in this section involved the investigation of 232 buildings, 45% of these are thought to have been built before 1945. A small percentage (1.7%) are public buildings and were not included in this analysis. The remaining 98.3% of private buildings were divided into three main typologies:

- Reinforced concrete (RC) buildings, which make up 31.1% of the analyzed building stock;
- Mixed structures, which make up 7.9% of the analyzed building stock, are mainly composed of external masonry walls and concrete slabs or

roofs in which the totality of the resisting system was replaced by an internal structural system (concrete or steel frame) and only the masonry façade wall remains for historic value;
- Masonry buildings, which make up 61% of the analyzed building stock, represent the focus of the present work.

The three typologies were identified by considering, in the case of RC buildings, the possible presence of columns or other hints that the structure is represented by an RC frame, or, in the case of the mixed buildings, the state of conservation. Any sign of recent restoration works and any sign of independent resisting system were used as identifiers.

Along with these three typologies of buildings, a few buildings were left out of the present investigation. These include some public and religious buildings, buildings in a bad state of conservation, and other buildings undergoing renovation works that were not possible to be evaluated because of the presence of scaffoldings or coverings.

6.3.1 Geometrical features

The historic city center of Leiria is composed of a fairly homogeneous building stock with common features that are easily identifiable. The focus of this section is to provide a general idea about the geometric features of typical buildings in the area.

Almost all the buildings are part of blocks organized in the layout described in "Urban layout." For this reason, at the time of their construction, all the buildings might have presented a fairly regular plan configuration, that is, square or rectangular, with common dimensions averaging from 3.5 to 23.5 m. Changes in ownership and necessities of the occupants likely generated the plan irregularity and the modification to the original floor plans.

The buildings in the city center of Leiria are mainly composed of three or four floors connected by a staircase system that, depending on the position of the building within the block, is generally concentrated on one of the sides of the building for small buildings, or, for wealthier and larger buildings, concentrated on the center of the plan to give access to both sides.

Most of the buildings bear the signs of later interventions that may be represented by superelevations of the upper floors of the buildings or the elimination or reduction of some masonry piers at the ground floors to create larger openings for shops installed later. These alterations and other characteristics such as the presence of galleries, the reduction of the plan area between the floors, or the presence of small dormers or towers are also worth mentioning. They represent a variation of mass and stiffness along the height of the building that can increase its vulnerability to horizontal actions beyond the intrinsic vulnerability of the façade walls.

In addition, the configuration of the façade walls needs to be discussed when trying to identify the typological characteristics of the building stock analyzed. First, in general, the façade walls assessed were found to measure from 0.4 to 1.0 m thick with an average of 0.7 m.

Second, it can be observed that the façade opening layouts represent a notable variability in the characterization of the building stock. It can be observed that the façades facing the main streets or the squares present a visible regularity in the layout. This regularity can also be seen in the dimensions of the openings, although an appreciable difference in the size of the openings is observed between the regular buildings and grander ones.

On the other hand, the secondary facades' openings do not present a comparable regularity either for layout or dimensions because the doors and the windows were generally created according to the internal plan rather than for external aesthetic purposes. Moreover, the opening layout and dimension regularity is affected by additions and interventions carried out on the buildings that may have led to the opening of additional doors or windows of differing dimensions and/or exceeding the regular grid of the existing ones.

6.3.2 Load-bearing walls

The external load-bearing walls of the buildings are mainly composed of stone masonry, which was identified to be composed of limestone units, according to its local availability, and bonded with weak hydraulic mortar joints. According to the observable wall sections, the walls are made with units of different sizes and sometimes mixed with units of other materials, for example, clay bricks and marlstones.

Fig. 6.1A illustrates the arrangement of the units of some of the masonry observed. It is possible to see that in most of the cases, the units have sub-horizontal mortar joints but that the alignment of the vertical joints does not indicate a good quality masonry (Borri, Corradi, Castori, & De Maria, 2015). In addition, there is large variability in the size and shape of the units.

As previously mentioned, the walls have a thickness between 0.4 and 1.0 m according to the visual inspections. Fig. 6.1B shows that the masonry walls can be defined as three-leaf walls with a weak inner core with no signs of stone connections between the two external leaves.

This process of masonry detection and identification led to the conclusion that the typical masonry walls of Leiria's buildings consist of stone masonry of irregular dimensions and rough surfaces. Although there

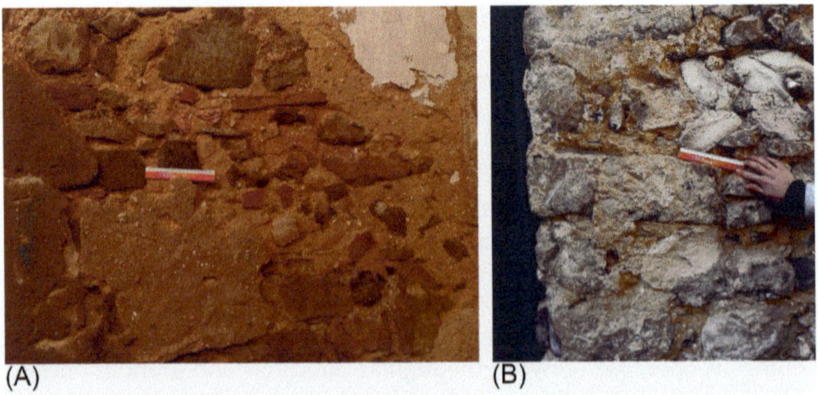

FIG. 6.1 Stone masonry with different material arrangement in elevation (A) and cross-section (B).

were no signs of cross or transverse connections or of metallic elements, in most cases, the units were adequately connected with mortar joints.

In several cases, the presence of clay brick masonry was observed. The brick units have different sizes from building to building and are in good condition, with mild erosion in some cases. On the other hand, as noted for the stone masonry, the mortar joints are in poor condition, which confirms the hypothesis of weak mortar joints.

The brick masonry is mainly located under window openings and could have been built to fill part of the wall, as this is a typical construction technique used on the buildings in Leiria.

The assessment of the connection quality between orthogonal façade walls is also a relevant aspect. No signs of tie rods used for this purpose were observed except for some rare cases. Thus, the connection performance is mainly entrusted to the quality of the stones placed at the corners of the buildings. Two different classifications of connections were observed:

- Good quality connections, achieved with the use of large, well-cut stones at the corners of the buildings, well interlocked with the surrounding masonry, usually not covered with any plaster or render;
- Bad quality connections accomplished with the use of smaller stones similar to the units of the surrounding masonry and not properly interlocked.

A good wall-to-wall connection was often evidenced by the presence of exposed large stones at the façade corners. Thus, when this feature was detected, the good quality of the wall-to-wall connection was generally assumed for the building.

It is important to note that one of the most important characteristics regarding the orthogonal wall connections is the presence of monolithic limestone post and lintels framing the facade openings. This is a clear example of how an architectural feature can affect the structural behavior of a building. It was noticed that sometimes the presence of openings close to the building corners causes a detachment of the orthogonal façade compromising the connection quality between the two walls, which is aggravated by the presence of these limestone elements.

6.3.3 Floors

Three different types of floors were observed in the historic city center of Leiria: solid concrete slab, metallic structure, and timber floors. The first is present in the buildings that have been restored recently, but their presence was generally an assumption based on the general state of conservation of the building and some information recorded by interviewing the owners.

The second type of floor was only observed in a few buildings. Steel beams support the wooden plank of the floor. The connection of this type of floor with the walls is assumed to be good because of the recent intervention.

The third type of floor, timber floors, is present in the large majority of the buildings. They are composed of pinewood and directly supported by the resisting wall, as shown in Fig. 6.2, where the location of the beams is clearly visible in an internal wall because the floor has been removed. However, in some buildings, the timber beams supporting the planks are supported by a primary order of timber beams, which assures the connection between the floor and the wall. No tie rods or metallic elements were observed to assure the connection between the floor and the wall.

Painted wooden planks or laths filled with mortar and coated with plaster to cover the floor structure were also observed in some cases. Because of these covers, the structural part of the floors was not always visible.

FIG. 6.2 Example of a timber floor found in the historic center of Leiria.

6.3.4 Roofing systems

All of the roofs investigated were made of timber elements. From the outside inspection, different typologies were observed such as pavilion, square, or flat roof. In some cases, half trusses were observed, but in most cases, king post trusses or simple trusses were used. In some bigger buildings, the trusses were more sophisticated, as illustrated in Fig. 6.3.

Such a diffuse use of timber trusses to support the higher layers of the roof led to the conclusion that, in most cases, the roofing systems generally do not exert improper thrusts on the building.

In all the inspected roofs, the truss structure was connected to the walls through a perimeter ribbon beam. This beam, illustrated in Fig. 6.4, is a mixed masonry-timber element, which indicates that the roof weight and thrust actions are properly distributed to the perimeter walls and a good connection between walls and roof is assured, although no tie elements were observed in the roofs.

(A) (B)

FIG. 6.3 King post truss (A) and simple truss (B).

FIG. 6.4 Perimeter ribbon beam supporting the roofing system.

Furthermore, the external roofing tiles are directly laid on the second-order rafters, without any insulation or impermeability system. Most of the time, they were fixed together with some mortar.

6.3.5 Nonstructural elements

Nonstructural elements connected to the façades are also worth mention herein, because they may represent debris in the streets even after low-intensity earthquakes and they embody an additional seismic vulnerability for the façade stability.

The widespread presence of large cornices was noticed protruding from the line of the façades facing the main streets and squares, whose connection to the structure itself was not always assured to be sufficient.

The building stock is additionally characterized by the presence of small stone balconies and balustrades, as well as lighter elements such as streetlamps and air conditioning systems.

6.4 Case studies

Representative case studies were selected from the city center to apply the nondestructive techniques and characterize the mechanical properties of the masonry walls.

The first case study was the "Casa dos Pintores," which is a representative traditional three-story building, dating probably from the 16th century. The building consists of outer walls of rubble masonry, as main load-bearing structural elements. The floors were in the original project made with timber elements, as well as the interior 'tabique' partition walls, see Fig. 6.5A and B. The building was rehabilitated in 2006, with the consolidation and strengthening of the existent masonry walls and the inclusion of a new steel structure to support the new floors. In the last years, the building presents some pathologies, namely diagonal cracking in the masonry walls due to soil settlements, see Fig. 6.5C and D. The NDT tests were performed in two representative walls located at the ground storey, see Fig. 6.6.

The second case study is also located in Leiria city center and is a habitational building, built in the 19th century, with a rectangular plan with three floors, which has not been used in the last years. The building is composed of load-bearing stone irregular masonry walls in the perimeter, whereas the inside partition walls are made of tabique. The floors and roofs are made with timber elements. The building was subject to flat-jack and sonic tests, see Fig. 6.7.

In a previous study, a single and double flat-jack test were performed to evaluate the young modulus and the compressive strength of the masonry (Pinheiro, Fernandes, Santos, & Rodrigues, 2017). In Fig. 6.8, the average

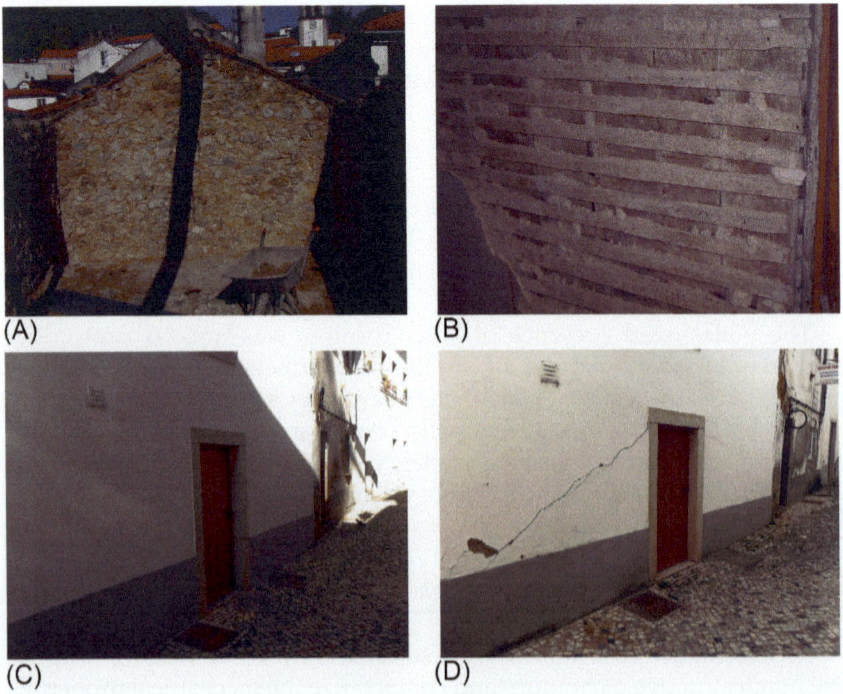

FIG. 6.5 Masonry walls from the "Casa dos Pintores": (A) masonry wall; (B) interior "tabique" partition walls; (C) façade wall in 2017; and (D) façade wall in 2020.

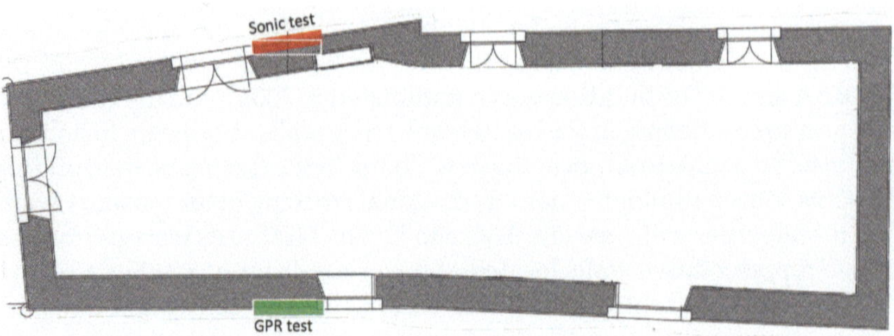

FIG. 6.6 Plan of "Casa dos Pintores" and location of NDT tests.

stress-strain results performed in the case study building are presented. The maximum stress of the masonry wall test ($\sigma_{max} = 0.721$ MPa) is compared with the in situ stress estimated in the single test ($\sigma_I = 0.137$ MPa). It can be observed that its current stress level is much lower when compared with the maximum capacity of the load-bearing wall.

FIG. 6.7 Housing building in the historic center of Leiria.

The evaluation of the elasticity modulus of the masonry walls tested was obtained through graphical analysis, estimating the stress-strain curves obtained in the double test. It was obtained with three different strategies according to the slope of the secant stiffness corresponding to different percentages of the maximum strength, which led to the following results: $E_0 = 3$ GPa; $E_{sec30\%}$ of 1.01 GPa; and $E_{sec30\%-60\%}$ of 0.35 GPa.

The third case study consists of a single wall, where the plaster was removed. It was used as a mere representative example of the wall typology available in the city center.

6.4.1 Nondestructive techniques for large-scale masonry characterization

Nondestructive techniques are a powerful tool for the diagnosis of the built heritage. Geophysical prospecting surveys based on wave propagation methods, such as sonic tests and GPR have been increasingly used to

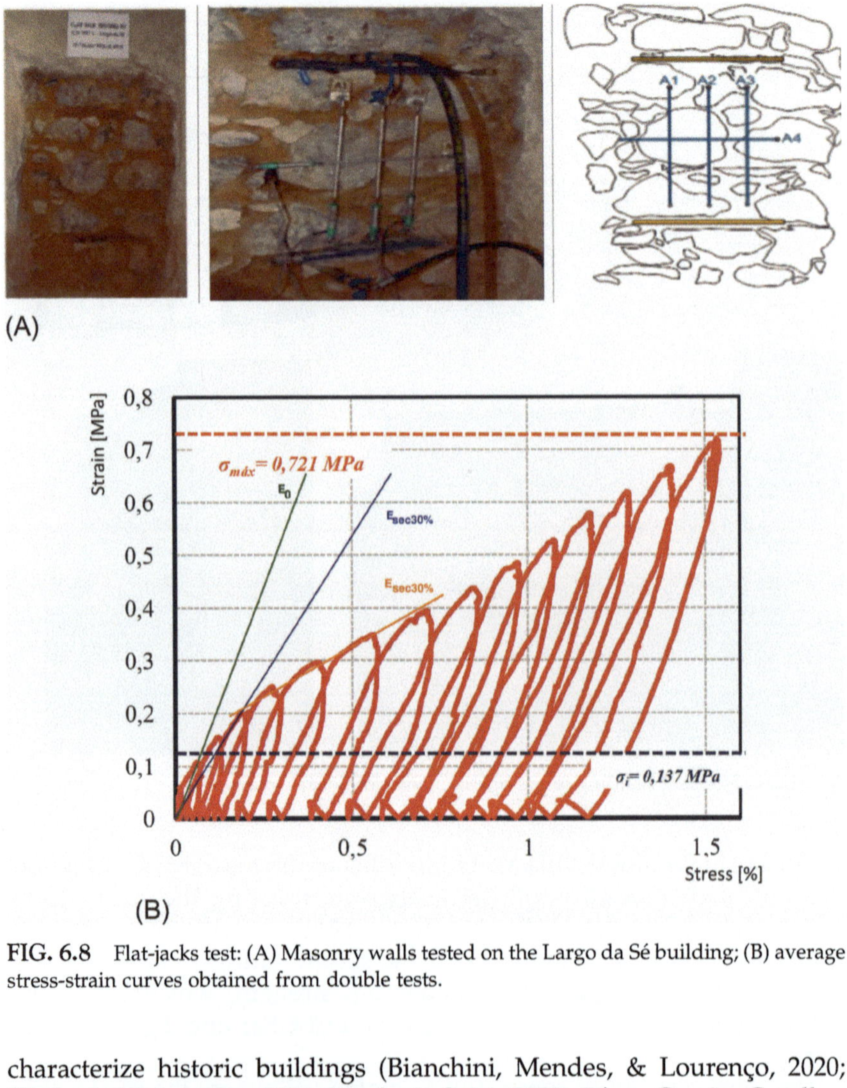

FIG. 6.8 Flat-jacks test: (A) Masonry walls tested on the Largo da Sé building; (B) average stress-strain curves obtained from double tests.

characterize historic buildings (Bianchini, Mendes, & Lourenço, 2020; Binda & Maierhofer, 2006; Binda & Saisi, 2009; Pérez-Gracia, Caselles, Clapés, Martínez, & Osorio, 2013; Ramos, Núñez García, Fernandes, & Lourenço, 2018). Sonic methods (ultrasonics, sonics, and impact-echo) can be used for the qualitative evaluation of the masonry, namely for the detection of voids, delamination, and weak areas (Sadri, 2003). However, there is a close relationship between the sonic wave velocities through the material and its physical and elastic mechanical properties (Everet, 2013). Thus, sonic methods can be used to estimate the masonry elastic properties. For the investigation of the built heritage, GPR is mostly used for the morphological characterization of masonry structures

(Riveiro, Solla, de Arteaga, Arias, & Morer, 2013) or to detect hidden elements, voids, and damage (Masini, Persico, & Rizzo, 2010).

Despite the unquestionable benefit of NDT for diagnostic investigations in historic masonry structures, their use is still limited and definitely not widespread among professionals in the field of conservation, which tend to rely more on literature, expert judgment, and others, given the typically high costs in terms of time and money of NDE methods (McCann & Forde, 2001). Moreover, these constraints have prevented their use in traditional and vernacular buildings, such as the ones that have been described in this chapter and compose the historic city center of Leiria. These buildings are typically not designed by specialists but are part of a process that involves many people over many generations and relies on empirical knowledge. Precisely due to its empirical and traditional nature, it is often seen as an obsolete and unsafe way of construction, which leads to its progressive abandonment. Therefore, there is a generalized lack of resources that can be assigned to its study and preservation. Most of the time, visual inspection is the only tool available to carry out the assessment.

The present section discusses how a reduced well-selected use of NDT that can be carried out in a one-day campaign can significantly enhance the knowledge of the structures under investigation and support visual inspection. The data extracted from NDT tests carried out in a few constructions can be later extrapolated to the general built-up environment through statistical approaches. The suggested approach that is presented here can be used to characterize the masonry of relatively homogeneous areas, such as Leiria historic center, in statistical terms, which can be particularly useful for large-scale assessments and can strengthen their reliability by adding quantitative data to approaches that typically rely solely on a visual inspection and qualitative assessments.

6.4.2 Onsite experimental campaign

The onsite experimental campaign carried out in Leiria included sonic tests and GPR inspections of masonry walls from three buildings. Sonic testing consists of measuring the velocity of the sonic waves propagating within the material under evaluation. Sonic waves are emitted by an instrumented hammer and received by an accelerometer. Direct tests aim to measure the velocity (V_P) of primary waves (P-waves) by placing the hammer and accelerometer in line at opposite sides of the wall. Indirect tests can measure both the velocity of P and R-waves (V_R) (Miranda, Rio, Miranda Guedes, & Costa, 2012). These velocities are dependent on the physical and elastic properties of the analyzed solid, namely Poisson's ratio and dynamic modulus, correlated through the following expressions:

$$\frac{V_P}{V_R}=\sqrt{\frac{2(1-\upsilon)(1-\upsilon)^2}{(1-2\upsilon)(0.87-1.12)^2}} \tag{6.1}$$

$$V_P=\sqrt{\frac{E}{\rho}\frac{(1-\upsilon)}{(1+\upsilon)(1-2\upsilon)}} \tag{6.2}$$

Where υ is the Poisson's ratio, E is the dynamic modulus, and ρ is the density of the material. Even though these expressions were developed for solid, elastic, isotropic, and homogeneous materials, they are adequate to provide an approximate first estimation of the mechanical properties.

In total, three sonic direct tests were performed on the case studies previously described, one of them at the *Casa dos Pintores* and two of them at the second house in the center. The wall tested at the *Casa dos Pintores* was plastered and the morphology was thus unknown, see Fig. 6.9. A 3×3 grid of 1×1 m with 0.5 horizontal and vertical spacing between the points was prepared. The thickness of the wall was measured onsite at the window, being 0.45 m. The values of the velocities shown in Fig. 6.10 reveal low variability and a relatively uniform masonry type. This can be also observed in Fig. 6.9, which shows the contour map with the wave velocities. The average velocity is 1140 m/s with a coefficient of variation (cov)

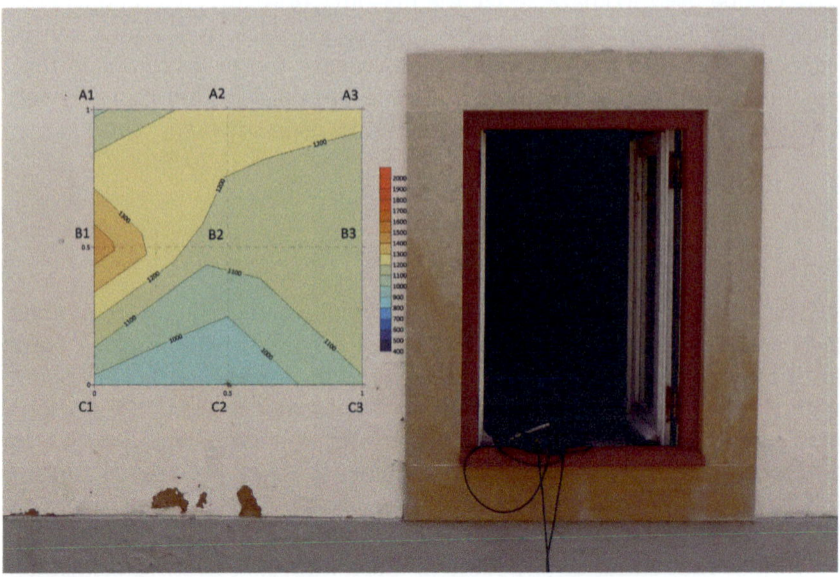

FIG. 6.9 Primary wave velocity (V_P) contour map at the wall tested in *Casa dos Pintores*.

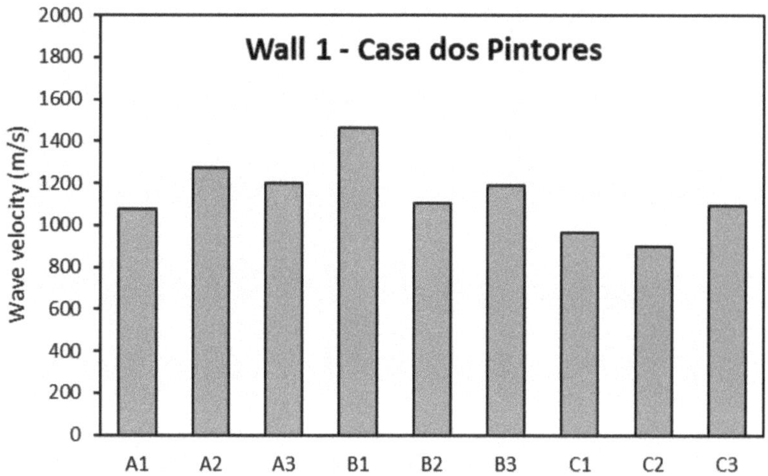

FIG. 6.10 Primary wave velocity (V_P) distribution obtained at each point of the grid in the wall tested in *Casa dos Pintores*.

lower than 15% among the different points. Ancient stone masonry can show a wide range of velocities from 350 m/s, corresponding to an irregular masonry with a high percentage of voids, to 2500 m/s, corresponding to a regular solid masonry wall (Binda, Saisi, & Tiraboschi, 2000; Lourenço, Trujillo, Mendes, & Ramos, 2012). The velocities obtained are typically over 1000 m/s, indicating the absence of significant voids throughout the cross-section.

The walls tested at the second house had an opening in the plaster that allowed observing the external morphology of the stone masonry. Irregular uncut stone masonry units of various sizes compose the masonry wall (Fig. 6.11). The largest dimension of the biggest stones observed is approximately 0.25 m, whereas there are smaller stones with dimensions of 5 cm or less. This type of masonry is likely to present voids within the cross-section because of the difficulties of filling the irregular spaces with the mortar. Two direct tests were performed at this location because the wall was accessible at both sides. The thickness of the wall was measured onsite as 0.6 m (without the particularly thick inner plaster, which was approximately 8 cm). The values of the velocities shown in Fig. 6.12 reveal a notable variability, reflecting the irregularity observed from the exterior. This can be also observed in Fig. 6.11, which shows the contour maps with the wave velocities. The average velocity of the first area is 820 m/s with a cov of approximately 30% among the different points. The average velocity of the second portion of the wall tested is 720 m/s with a cov of approximately 25% among the different points. In combination, the average velocity observed at this wall is 790 m/s with a cov slightly below 30%.

FIG. 6.11 Primary wave velocity (V_P) contour maps at the walls tested in the second house.

The velocities obtained are typically below 1000 m/s, and many of them are close to 500 m/s, which indicates the possible presence of significant voids throughout the cross-section. Moreover, the inner thick plaster must have a definite influence on the results, by decreasing the velocity of the waves in the second test.

In the second house, an additional sonic indirect test was performed at the location shown in Fig. 6.11 (left) on the interior side of the house. The hammer and the accelerometer are placed on the same face aligned vertically with the 0.5-m spacing (distance between points A and B in Fig. 6.11—left). An average P-wave velocity of 1050 m/s and an average R-wave velocity of 550 m/s were estimated, which resulted in an estimation of a Poisson's ratio of 0.28. The velocities obtained in the indirect tests were higher than those obtained in the direct tests. Because indirect tests measure the velocity through one leaf of the wall, these values indicate the probable existence of voids between the external leaves of the wall. Another indirect test was performed in the third case study, at the external lead of the masonry wall whose morphology could be observed from the exterior (Fig. 6.13).

Because only the exterior side of the wall was accessible, only indirect tests could be performed. The hammer and the accelerometer were placed on the exterior face aligned vertically with 0.5 distance in four points. An average P-wave velocity of 960 m/s and an average R-wave velocity of 540 m/s were estimated, but with a high cov of 40% and 60%, respectively. This high variation responds to the high irregularity of the wall tested, with irregular uncut stones of different sizes, lack of respected horizontal joints, and visible lack of mortar in some areas. The Poisson's ratio estimated is 0.21.

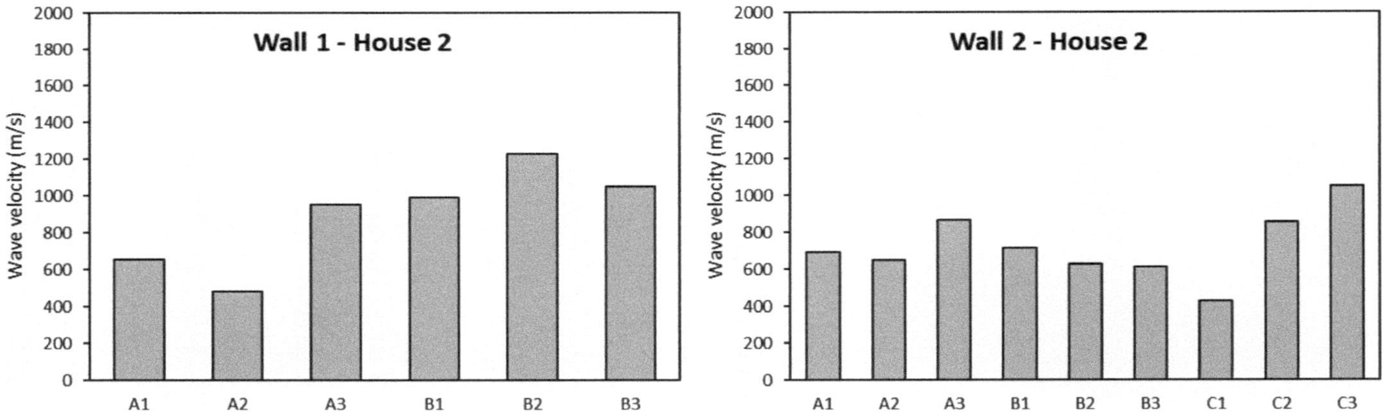

FIG. 6.12 Primary wave velocity (V_P) distribution obtained at each point of the grid in the walls tested in the second house.

FIG. 6.13 Stone masonry wall from the third house evaluated.

Using Eq. (6.2), the dynamic modulus (E) can be estimated for the different walls. Table 6.1 shows the values obtained assuming a range of typical values for the density of masonry walls between 1800 and 2200 kg/m^3. The variations shown in Table 6.1 are notable, with values of E ranging from 0.94 to 3 GPa. A high variability could be expected when addressing irregular stone masonry constructions with varying construction quality. Therefore, the relatively homogenous historic city center constituted by buildings with the thick limestone masonry walls previously described actually may differ significantly between them in terms of mechanical properties. This is a representative characteristic of historic masonry buildings within the same area, which were in fact built by different masons at different times. The number of resources invested in each building may also differ significantly, leading to a varying masonry quality within the center.

TABLE 6.1 Dynamic modulus (E) estimated at each wall with the sonic testing campaign.

House	Wall #	Type of test	V_P, m/s (cov)	V_R, m/s (cov)	ν	E (GPa)
Casa dos Pintores	1	Direct	1140 (15%)	–	–	2.34–3.00
#2	1	Direct	890 (30%)	–	–	1.44–1.84
#2	2	Direct	720 (25%)	–	–	0.94–1.20
#2	2	Indirect	1070 (6%)	550 (6%)	0.28	1.61–2.06
#3	1	Indirect	960 (40%)	540 (60%)	0.21	1.50–1.91

It should be also noted that these values of dynamic modulus cannot be directly compared with the values of elasticity modulus obtained from the flat-jack tests presented in Fig. 6.8. The empirically observed inequality between static and dynamic modulus of elasticity of the masonry has been noted in several studies (Makoond, Pelà, & Molins, 2019; Totoev & Nichols, 1997). However, most studies available relate to rock properties, within the geophysical context (Ahmad, Pamplona, & Simon, 2009; Mashinski, 2003). Definite correlations for masonry are not established. Commonly accepted ratios between the dynamic and the static modulus may vary between 1.2 and 1.3 (D'Ambrisi, Mariani, & Mezzi, 2012; Ripepe et al., 2014), but some standards establish ranges varying from 1.1 to 2 (BSI, 2004). Taking the former reflection into account, using the values of dynamic modulus (E) from Table 6.1 and assuming a ratio of 1.25, the static modulus of elasticity obtained for house #2 may vary between 0.78 and 1.72 GPa. The elasticity modulus $E_{sec30\%}$ obtained from the flat-jack tests lies within this range of properties.

The sonic testing campaign was carried along with a GPR investigation of the walls, intended to provide a qualitative characterization of the masonry morphology and cross-section. The same walls were thus also investigated with the GPR. The GPR can provide information about the level of homogeneity of masonry walls, the presence of voids, and structural configuration, such as the detection of multiple leaves in stone masonry walls. The method is based on the emission of short electromagnetic pulses that are transmitted into the building elements. The pulses are reflected when there is a change in the dielectric properties of the medium, which happens when they reach an interface between different materials, such as the end of elements (backside of walls), metallic elements, and voids. The antenna can be moved along the surface of the element under investigation, producing radargrams, which are charts that show the reflection amplitudes of the radar waves against receiving times at each position of the antenna. If the velocity of propagation of the radar waves within the media is known, the location and depth of the different interfaces can be estimated. A typical average velocity of radar waves within masonry is 14 cm/ns (Daniels, 2014; Riveiro et al., 2013).

Fig. 6.14 shows the radargram obtained at a wall from *Casa dos Pintores*. A clear two-leaf wall can be observed. The back of the wall is estimated at a depth of 0.55–0.65 m approximately. A central continuous reflection at a depth of 0.25–0.35 m shows the clear division of the wall between two leaves. Strong reflections can also be observed at a depth of 2–3 cm, indicating the end of the plaster. The strong reflections obtained at some locations indicate the lack of mortar at those areas and possible gaps at the interface between the leaves.

The GPR investigations at the second house show a similar wall typology with two clear leaves. However, the radargram is notably affected by

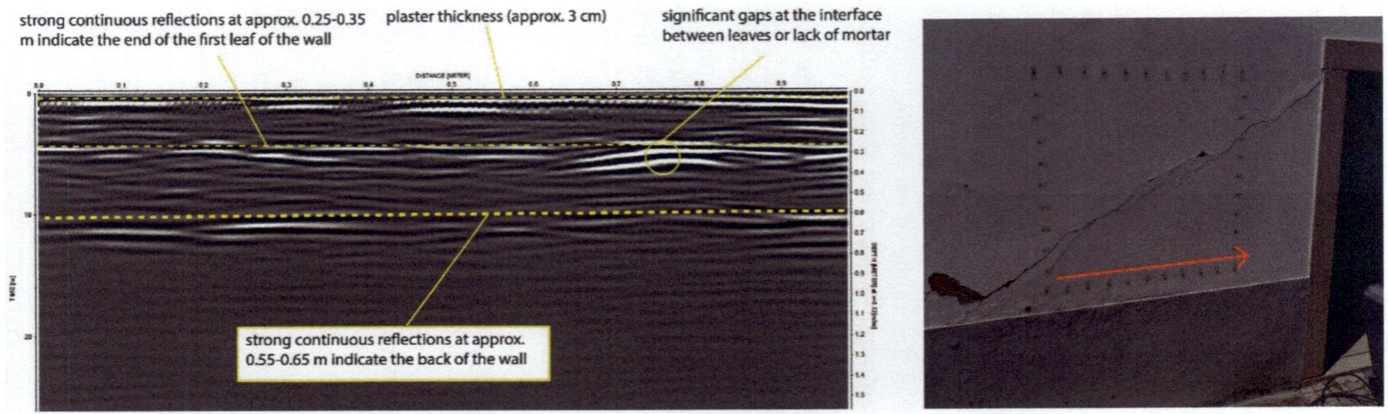

FIG. 6.14 Horizontal radargram obtained at an exterior wall from *Casa dos Pintores*.

the thick plaster layer of approximately 8 cm in the interior of the house (Fig. 6.15). The strong reflection at the end of the plaster layer highly attenuates the signals collected beyond. Nevertheless, clear discontinuous reflections can be observed in the middle of the wall (at a depth of 0.3–0.35 cm) and at the end of the wall (at approximately 0.7 m). In conclusion, a two-leaf stone masonry typology with thick plaster layers is detected in both walls.

6.4.3 Statistical approach for the extrapolation of the results

The data collected from the NDT campaign are extracted from a limited sample of only three buildings. However, all three buildings are considered to be representative of ancient masonry buildings from the historic city center of Leiria. In this chapter, a simple statistical approach proposed to extrapolate NDT results obtained from a limited sample of buildings to the global built-up environment within a reasonably homogeneous area is presented.

The approach consists of creating a probabilistic distribution of the dynamic modulus (E), based on the NDT results. According to the values shown in Table 6.1 from five sonic tests performed in three different buildings, a mean value of 1.8 GPa with a standard deviation (σ) of 0.56 GPa can be calculated. A histogram with values grouped in bins of 0.25 GPa is presented in Fig. 6.16A. According to the skewness of the distribution, a lognormal distribution was fitted for the data, see Fig. 6.16B. The lognormal distribution is considered adequate because it is also the distribution adopted in current codes (CNR-DT 212/2013, 2014) and avoids possible negative values. A normal distribution is also given in Fig. 6.16B for comparative purposes.

It is acknowledged that the exercise presented here is performed with limited sample size. However, enlarging the population with more buildings can help to create a reliable probabilistic distribution that is able to characterize the masonry of historic city centers better than the general provisions of the code. This approach can be particularly useful to support large-scale vulnerability assessments, by aiding the visual inspection with quantitative data on the masonry mechanical properties.

6.5 Final remarks

As discussed in this chapter, the in situ inspections carried out in the historic center of Leiria led to the identification of typical building typologies and to the characterization of features such as the quality of masonry and the connections.

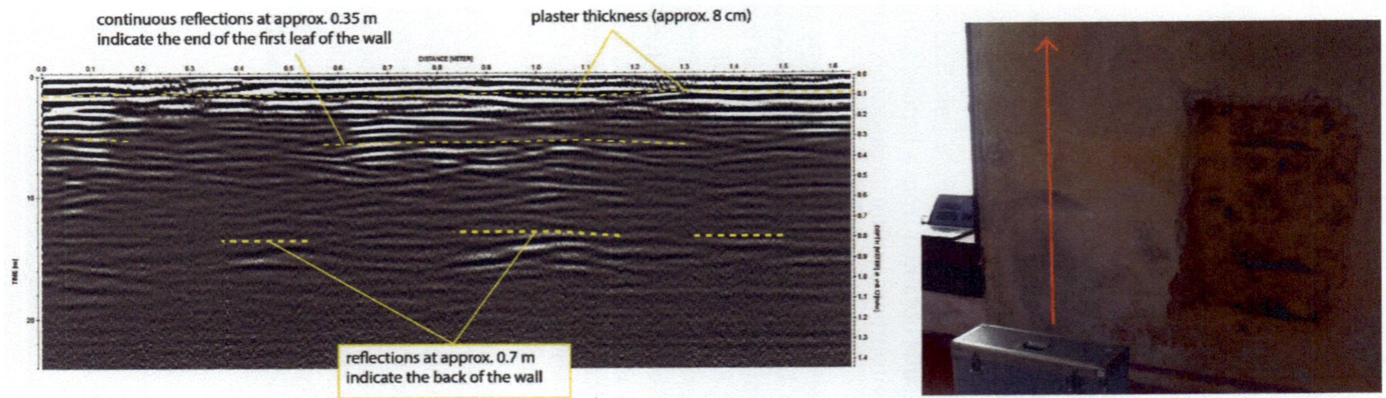

FIG. 6.15 Vertical radargram obtained at the interior side of a wall at the second house.

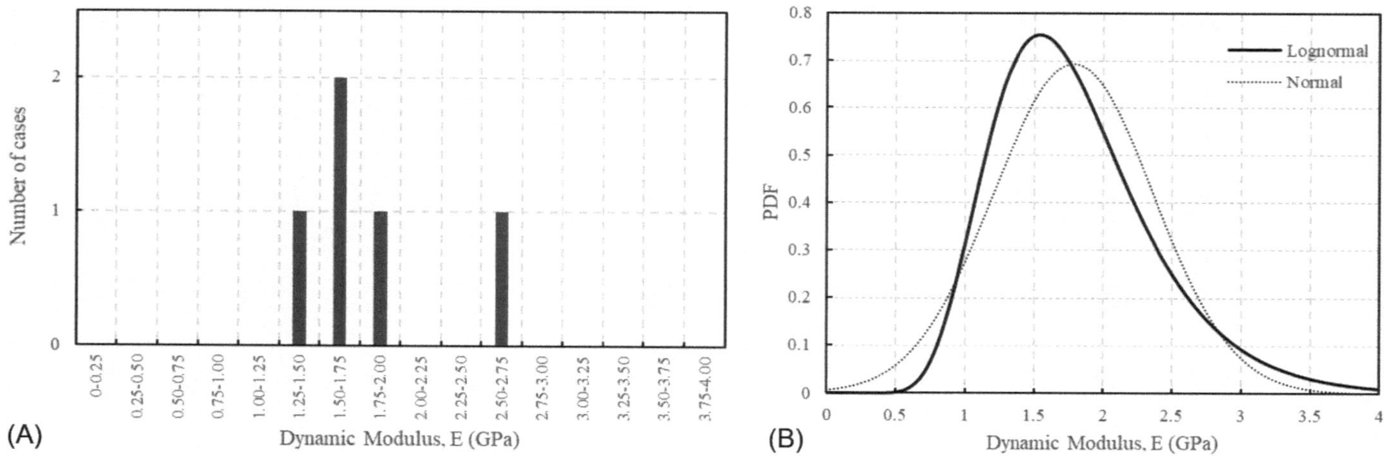

FIG. 6.16 (A) Histogram with the dynamic modulus (*E*) obtained from the sonic testing campaign; and (B) lognormal.

During the inspections, a number of features that may govern the seismic response of the buildings were observed. These are general construction features, such as the poor quality of masonry and, later interventions on buildings, typical in historic urban areas such as replacement of original timber floor with heavier concrete slabs and alterations on the building's regularity in height and openings. Another characteristic of the historic center of Leiria was the high rate of RC buildings that were alternating old masonry buildings and also a high rate of mixed structures. The interaction between the different typologies is an important issue that affects the vulnerability of the area.

This chapter has also presented the use of nondestructive techniques for the characterization of masonry elements in large-scale areas, such as the historic city center of Leiria. Sonic tests and GPR investigations were carried out in three different buildings, supporting the conclusions previously gathered by visual inspection on the type of masonry and its morphology. Moreover, quantitative data were extracted from the sonic tests, namely the Poisson's ratio and the dynamic modulus (E) of the masonry. A simple statistical study was then presented on how these data obtained at few constructions can be later extrapolated to the general built-up environment, which mainly consisted of the definition of a probabilistic distribution of the dynamic modulus (E) for the ancient masonry of the historic city center of Leiria. This approach is considered to be particularly helpful to support large-scale vulnerability assessment with quantitative data of masonry material properties obtained in situ.

References

Ahmad, A., Pamplona, M., & Simon, S. (2009). Ultrasonic testing for the investigation and characterization of stone—A non-destructive and transportable tool. *Studies in Conservation, 54*(Suppl 1), 43–53.

Barbosa, P. G. (1992). *Povoamento e estrutura Agrícola na Estremadura Central (Sec. XII a 1325)*: (p. 397). Lisbon: Instituto Nacional de Invstigação Científica.

Bianchini, N., Mendes, N., & Lourenço, P. B. (2020). Seismic evaluation of Bagan heritage site (Myanmar): The Loka-Hteik-Pan temple. *Structures, 24*, .

Binda, L., & Maierhofer, C. (2006). Strategies for the assessment of historic masonry structures. In M. Brusin (Ed.), *Proceedings of the 59th RILEM week* (pp. 103–124). Moscow, Russia: RILEM.

Binda, L., & Saisi, A. (2009). Application of NDTs to the diagnosis of historic structures. In: *Proceedings of the 7th international symposium on NDT in civil engineering (NDTCE 2009).*

Binda, L., Saisi, A., & Tiraboschi, C. (2000). Investigation procedures for the diagnosis of historic masonries. *Construction and Building Materials, 14*(4), 199–233.

Borri, A., Corradi, M., Castori, G., & De Maria, A. (2015). A method for the analysis and classification of historic masonry. *Bulletin of Earthquake Engineering, 13*(9), 2647–2665.

BSI (2004). *Natural stone test methods—Determination of sound speed propagation.* London, UK: British Standards Institution (BSI).

Carreira, A. (1995). Valores estético-urbanísticos de Leiria setecentista. In: *II Congresso Sobre História de Leiria e da sua Região*, pp. 85–94.

Carvalho, V., & Aveleira, A. (n.d.). "Enquadramento histórico e arqueológico." www.cm-leiria.pt/pages/405.

Carvalho, V., & Inácio, I. (2011). O projeto de investigação arqueológica do núcleo do castelo de leiria: enquadramento, objetivos e resultados. In: *I Congresso de História e Património da Alta Estremadura, October 28-30, 2011, Ourém, Portugal*, pp. 128–148.

CNR-DT 212/2013 (2014). *Guide for the probabilistic assessment of the seismic safety of existing buildings*. Italy, Rome.

D'Ambrisi, A., Mariani, V., & Mezzi, M. (2012). Seismic assessment of a historic masonry tower with nonlinear static and dynamic analyses tuned on ambient vibration tests. *Engineering Structures, 36*, 210–219.

Daniels, D. J. (2014). *Ground penetrating radar*. London, UK: The Institute of Electrical Engineers.

Dinis, C. (2006). *Estudo Sócio-Demográfico do Centro Histórico da Cidade de Leiria—Conclusão*. Leiria, Portugal: Departamento de Planeamento, Divisão da Habitação e Reabilitação Urbana, Câmara Municipal de Leiria.

Everet, M. E. (2013). *Near-surface applied geophysics*: (p. 415). New York, USA: Cambridge University Press.

Gomes, S. A. (1993). A Organização do Espaço Urbano numa Cidade Estremenha: Leiria Medieval, Separata A Cidade. In *Jornadas Inter e Pluridisciplinares. Atas II* (pp. 82–113). Lisbon, Portugal: Universidade Aberta.

Gomes, S. A. (1995). *Introdução à História do Castelo do Leiria*. Leiria, Portugal: Câmara Municipal de Leiria.

Gomes, S. A. (1996). Higiene e saúde na Leiria Medieval. In: *III Colóquio sobre História de Leiria e da sua Região*, Leiria, Portugal: Câmara Municipal de Leiria, pp. 9–43.

Gomes, S. A. (2005). *Noticias e Memorias Paroquiais Setecentistas*. Viseu: Palimage e Centro de História da Sociedade e da Cultura da Universidade de Coimbra.

Lourenço, P. B., Trujillo, A., Mendes, N., & Ramos, L. F. (2012). Seismic performance of the St George of the Latins church: Lessons learned from studying masonry ruins. *Engineering Structures, 40*, 501–518.

Makoond, N., Pelà, L., & Molins, C. (2019). Dynamic elastic properties of brick masonry constituents. *Construction and Building Materials, 199*, 756–770.

Mashinski, E. I. (2003). Differences between static and dynamic elastic moduli of rocks: Physical causes. *Russian Geology and Geophysics, 44*(9), 953–959.

Masini, N., Persico, R., & Rizzo, R. (2010). Some examples of GPR prospecting for monitoring of the monumental heritage. *Journal of Geophysics and Engineering, 7*, 190–199.

Mattoso, J. (1985). A cidade de Leiria na história medieval de Portugal. *Ler História, 4*, 2–19. Lisbon, Portugal.

McCann, D. M., & Forde, M. C. (2001). Review of NDT methods in the assessment of concrete and masonry structures. *NDT&E International, 34*, 71–84.

Miranda, L., Rio, J., Miranda Guedes, J., & Costa, A. (2012). Sonic impact method—A new technique for characterization of stone masonry walls. *Construction and Building Materials, 36*, 27–35.

Monteiro, J. G. (1999). *Os Castelos Portugueses dos Finais da Idade Media: presença, perfil, conservação, vigilância e comando*. Lisbon, Portugal: Edições Colibri.

Nunes, A. L. P. (2005). Dicionario de arquitectura militar. In *Caleidoscópio* (p. 364). ISBN 9789728801946.

Pérez-Gracia, V., Caselles, J. O., Clapés, J., Martínez, G., & Osorio, R. (2013). Non-destructive analysis in cultural heritage buildings: Evaluating the Mallorca cathedral supporting structures. *NDT and E International, 59*, 40–47.

Pinheiro, P., Fernandes, P., Santos, P., & Rodrigues, H. (2017). Mechanical characterization of masonry walls with flat-jack tests. Chapter 4In L. M. S. Gonçalves, F. Gaspar, & H. Rodrigues (Eds.), *Nondestructive techniques for the assessment and preservation of historic structures*. CRC Press. ISBN 9781138710474.

Pousão-Smith, S. (2008). *Rodrigues Lobo, os Vila Real e a estratégia de "Dissimulatio"* Lisbon, Portugal. ISBN 9789892008066.

Ramos, L. F., Núñez García, A. C., Fernandes, F. M., & Lourenço, P. B. (2018). Evaluation of structural intervention in the Quartel das Esquadras, Almeida (Portugal). *International Journal of Architectural Heritage, 12*(3), 465–485.

Ripepe, M., Coli, M., Lacanna, G., Marchetti, E., Cristofaro, M. T., De Stefano, M. , et al. (2014). Dynamic response of the Giotto's Bell-Tower, Firenze, Italy. *Engineering Geology for Society and Territory, 8*, 323–327.

Riveiro, B., Solla, M., de Arteaga, I., Arias, P., & Morer, P. (2013). A novel approach to evaluate masonry arch stability on the basis of limit analysis theory and non-destructive geometric characterization. *Automation in Construction, 31*, 140–148.

Sadri, A. (2003). Application of impact-echo technique in diagnoses and repair of stone masonry structures. *NDT&E International, 36*, 195–202.

Totoev, Y. Z., & Nichols, J. M. (1997). A comparative experimental study of the modulus of elasticity of bricks and masonry. In: *Proceedings of the 11th international brick/block masonry conference*.

Vicente, R., Ferreira, T. M., & Mendes da Silva, J. A. R. (2015). Supporting urban regeneration and building refurbishment. Strategies for building appraisal and inspection of old building stock in city centres. *Journal of Cultural Heritage, 16*(1), 1–14.

Zuqueste, A. (2003). *Monografia de Leiria: A cidade e o Concelho*. Leiria, Portugal: Folheto Edições & Design.

CHAPTER

7

The Borbone's antiseismic system: Historical, constructive, and structural analysis

Stefano Galassi[a], *Enrico Pagano*[b], *Nicola Ruggieri*[c], *Ginevra Salerno*[b], and *Giacomo Tempesta*[a]

[a]Department of Architecture, University of Florence, Florence, Italy,
[b]Department of Architecture, Roma Tre University, Rome, Italy,
[c]Superintendence of Archeology, Fine Arts and Landscape for the Province of Cosenza, Cosenza, Italy

7.1 Introduction

The high magnitude seismic shocks that hit the *Calabria Ulteriore,* in the South Italy, during the early months of 1783, caused the destruction of a large part of the Calabrian built environment. Furthermore, the quakes provoked a number greater than 35,000 victims, due also to the indirect effects aftermath the earthquake (Vivenzio, 1783). The measures put in place by the sovereign of the Regno di Napoli—Ferdinand IV whose government at that time was extended over the Calabria region—to relieve such a desperate situation were immediate and concerned socioeconomic aspects and, above all, the reconstruction of the destroyed countries. In this regard, immediately after the earthquake, an antiseismic code was drawn up—*Istruzioni per gli Ingegnieri commissionati nella Calabria Ulteriore*—with the aim of regulating the new building work and whose principles anticipate those on which modern seismic legislation relies (Ruggieri, 2017) (Fig. 7.1). It is the genesis of the Borbone's system, a complex synthesis of construction rules, whose main characteristic is the wall technique consisting of masonry with wooden frames reinforcement, which implemented advanced mechanical and technical principles in a

https://doi.org/10.1016/B978-0-12-821087-1.00003-X

heuristic way that would give shape to modern seismic engineering a century later (Reitherman, 2006). In fact, the choice to interconnect the load-bearing structures (Ruggieri et al., 2016), declared in the Borbone's code, shows a clear propensity to endow the construction with a box-like behavior and above all anticipates by a few centuries the concept of life-safety limit state, reaffirmed in the current Italian antiseismic legislation. In fact, even in case of filling masonry damage, the wooden frame continues to support the horizontal structures connected to it, thus avoiding the collapse of the entire building. Furthermore, the more usual organization of the wall with the wooden frame on the internal face of the perimeter walls suggests that, in addition to protecting the members from external agents, an attempt was made to "guide" the possible collapse of the masonry toward outside without affecting the interior of the house. This construction peculiarity allowed, on the occasion of the subsequent Calabrian earthquakes of 1894, 1905, 1907, and 1908, to limit the number of deaths considering the more or less extensive ruin of the walls of the buildings. In addition, the Borbone's code imposed the reduction of loads and, therefore, of the seismic mass, the limitation of the height of the buildings, a regularity of shape both in plan and in elevation, in addition to the introduction of further rules relating to the urban scale and concerning the distances between the buildings and the minimum dimensions of the roads (Stellacci et al., 2016), namely useful measures to guarantee safe escape routes and rescue operations following the earthquake.

Generally speaking, academic environments have a marked diffidence toward empirical knowledge (Ferrigni, Della Pietra, & Sorrentino, 2013), and this is even more valid in the context of modern seismic engineering, strongly based on mathematical models and on the measure of safety. Therefore the validation of a traditional construction technique if, on the one hand, requires an in-depth analysis of the construction details and their intimate function, on the other hand, requires scientific proof of its validity.

In other words, if you want to measure the structural safety of constructions of the Architectural Heritage in its modern sense and especially in light of current scientific knowledge, you need a mechanical characterization on theoretical and experimental bases, aimed at defining and calibrating predictive mechanical models (Galassi, Dipasquale, Ruggieri, & Tempesta, 2018; Galassi, Ruggieri, Dipasquale, & Tempesta, 2018; Galassi, Ruggieri, & Tempesta, 2015; Galassi, Ruggieri, Tempesta, & Zinno, 2014). To this end, both to understand the functioning of the existing Borbone's structures and to hypothesize their use in postearthquake reconstruction phases, or possibly to hypothesize their applicability to new buildings, much scientific work has been done. Starting from an experimental campaign that tested a sample of Borbone's wall reconstructed in the laboratory on a scale of 1:1 (Ruggieri, Tampone, & Zinno, 2015) under

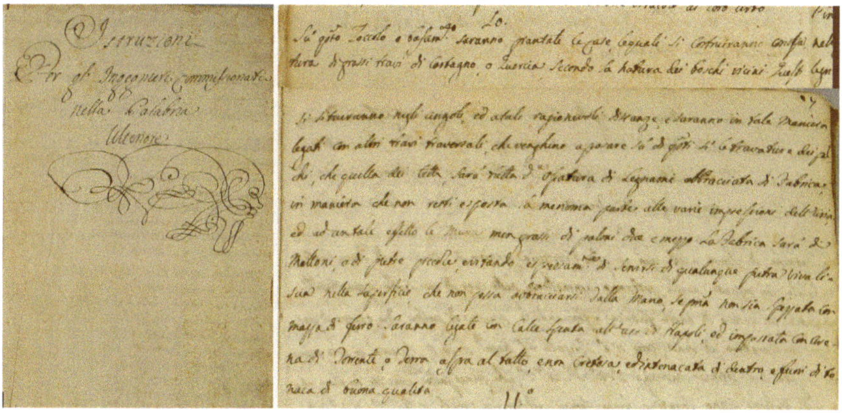

FIG. 7.1 *Istruzioni per gli ingegnieri commissionati nella Calabria Ulteriore*, the Borbone's code drafted at the end of the 18th century. The cover (on the right) and the article 10 (on the left).

quasistatic cyclic forces, two micromechanical models were developed (Galassi et al., 2014; Pagano, Ruggieri, Zampilli, & Salerno, 2019), both a basic model aimed at assessing the safety of an entire building.

7.2 The genesis

Empiricism, a fundamental concept in the general evolution of construction, had to play an important role in the genesis of the Borbone's system. In fact, this "... nasce dalle osservazioni fatte sul luogo ..." ("... arises from the observations made on the site ...") (Vivenzio, 1783), taking advantage of the *tests*, for dynamic actions, caused by the earthquakes that struck timber structures existing in the Calabria region before 1783 (Ruggieri, 2015). It should also be noted that the Calabria region had been characterized since ancient times for a thorough knowledge of woodworking, coinciding with an imposing wooded heritage. A material culture, implemented in the Middle Ages by the significant contribution derived from the Norman domination, which continued in the modern era and beyond, was grafted into the climate of the Neapolitan Enlightenment. Naples, capital of the Borbone's kingdom, in the second half of the 1700s represented a lively scientific environment. Neapolitan scientists were perfectly updated on the advancement of technology, and science in general, and actively participated in the debate about new discoveries that took place in Europe in diverse fields. The good practice spread among Neapolitan scholars to make numerous trips to France and Germany and, among nobles, to invite high personalities of science to the court and in the salons of the kingdom allowed the easy diffusion of culture. An extensive editorial production corresponded to this ferment, ranging from humanistic to scientific subjects with applications to the construction. The analysis of these latter volumes reveals a renewed interest in Vitruvius, whose precepts contained in *De Architectura* probably fed the theorization of the constructive system wanted by the Borbone's family.

Giovan Battista Mori, technician commissioned for the reconstruction of Reggio Calabria by King Ferdinand IV, in a 1789 report to the Vicar, General Pignatelli, attributed "... tutto il merito dell'invenzione ..." ("... all the merit of the invention ...") (Mauri-Mori, 1909) of the Borbone's constructive system to Francisco La Vega, director of the excavations of Pompeii and Herculaneum from 1780 to 1804 (Ruggieri, 2013). The excavation journals of the Spanish engineer (Pagano & La Vega, 1997) point out the discovery of wooden elements and in particular of framed walls of the *opus craticium* type, used in the buildings of the ancient Herculaneum and Pompeii. These discoveries lead us to postulate a possible influence of the Roman *parietes craticii* in the design of the framed walls made in the Calabria region. However, the contribution of Giovanni Vivenzio, an encyclopedic personality of the Borbone's court as well as the first doctor of King Ferdinand IV had to be decisive in theorizing of such a constructive technique. It is precisely Vivenzio who provides us with a testimony in this regard "... Tante distruzioni orrorose prodotte da' tremuoti ... da una punta del Mondo all'altra, spirano naturalmente il desiderio di ricercare i mezzi più efficaci per mettersi a coperto dè loro funesti effetti o piuttosto il mezzo di prevenirli. Da molto tempo ho rivolto le mie mire a quest'oggetto e credo di aver ritrovato un tal preservativo tanto desiderato. Molti dotti uomini a' quali ho comunicato le mie idee, le hanno accolte, e mi hanno indotto a pubblicarle ..." ("... So many horrendous destructions produced by tremors ... from one point of the world to another, naturally the desire to seek the most effective means to get covered by their dismal effects or rather the means of preventing them. For a long time, I have turned my sights to this object, and I think I have found such a much-desired protection. Many learned men to whom I communicated my ideas, accepted them, and led me to publish them ...") (Vivenzio, 1783).

7.3 The "istruzioni per gli ingegnieri commissionati nella calabria ulteriore"

The manuscript is made up of 12 chapters, in addition to the introduction that summarizes the content of the entire code. The chapter of greatest interest for the purposes of this contribution is the first, *Istruzioni sul metodo da tenersi nella riedificazione dè Paesi diruti della Calabria.* It consists of 26 articles relating to urban planning and sanitation, civil protection measures, in addition to those more properly referred to the construction. The recommendation contained in article 10 is the most identifying one of the Borbone's system. The wall has to be made "... *con ossatura di grossi travi di Castagno o Quercia ...Questi si situeranno negli angoli ed a tali ragionevoli distanze, e saranno in tale maniera legati con altri travi trasversali ... detta ossatura di legnami abbracciata di fabrica, in maniera che non resti esposta la menoma parte alle varie impressioni dell'aria ed ad un tale effetto non restino le mura men grossi di palmi due e mezzo...*" ("... with the framework of large chestnut or oak members ... These will be located in the corners and at such reasonable distances, and will thus be linked with other transverse beams ... this timber frame embraced by the masonry, so that no part is exposed to the various conditions of the environment and for such reason the walls are not less than two and a half spans ..."). A well-founded concern is inferred from this measure that the wood may be subject to degradation due to biotic agents if directly exposed to the air with high humidity values. The members will also be positioned in the corners, paying particular attention to the construction of the quoin, from now a consolidated construction rule. The provision in point 10 represents the fundamental requirement for achieving cohesion between the parties and ensuring "box-like" behavior, to be pursued to improve the seismic response of the building.

In addition, the code provides instructions on the maximum height, limited to 28 spans—about 7 m—with the possibility of derogating an additional floor, allowing to reach 38 spans—about 9 m—for the construction of wealthy people, while leaving the expertise of the engineer-in-charge of supervising the "sizing" and, in general, the technical characteristics of the masonry. With the aim of improving the behavior toward the seismic actions of the new churches, the rules differ from those imposed for ordinary buildings. In fact, for such a typology, the legislator, aware of the difficulty of supplying wooden elements of considerable height and high cross-section, in addition to the difficulty of execution, suggests, as an alternative to the wooden structure indicated in point 10, different construction measures. Specifically, article 18 recommends to follow criteria related to the *good rule of art* for the construction of religious buildings such as the execution of "... travi orizzontalmente lungo la sommità dei muri ..." ("... beams horizontally along the top of the walls ...")—a ring beam—and metal ties that can bond the walls counteracting their tendency to overturn. The *Istruzioni* prohibit without any misunderstanding the execution of vaults, recognizing the extreme vulnerability to horizontal actions of such a construction typology.

7.4 Constructive features

The Borbone's code contains indications not accompanied by drawings nor does the description include executive details that can guide the designer in the conception of a unique organization of the construction. Such a situation led to characterize the implementation of the regulations in the reconstruction for a heterogeneity of antiseismic solutions which, although compliant with the dictates of the code, differed in the configuration and in the geometric characteristics and, consequently, in the seismic response of the structure, as observed in various examples located in different Italian seismic areas (Pagano, Salerno, & Zampilli, 2018). The frame, in fact, could be positioned at different depths in the wall section; different types of bracing were made and the masonry was, in some cases, "reinforced" only by vertical wooden elements; in others, by a double frame with diatonic connecting elements. Another type is the "reduced" Borbone's system, applied in sporadic cases in the town of Reggio Calabria; it is distinguished by a first level in masonry and a first floor reinforced by wooden frames.

7.4.1 Foundations

The foundations, the main element on which the Vitruvian *Firmitas* and the 18th-century theory of construction relied, vary for the Borbone's antiseismic system, according to the two basic types, pile or shallow foundations.

Vivenzio (1783) recommends, in the description of *Case formate di legno* (houses made of wood), well-fixed wooden poles for a depth of about 3 m, placing himself in contrast with Milizia (1781) who in the same years, in a pioneering way, suggests to *isolate* the structure from the ground to improve the response of the building to seismic actions.

The drawing contained in the *Atlante* accompanying the book *Istoria dẽ fenomeni del tremoto avvenuto nelle Calabrie e nel Valdemone nell'anno 1783* (Sarcone, 1784) is interesting. Schiantarelli, the atlas drawings' author, with almost didactic intent, documents the various stages of the construction yard of a building in the town of Polistena (Rc) after the 1783 earthquake. The table shows a construction with the timber frame erected in the foreground and several stages for the execution of piling, the supply of wood, the perimeter of the foundation interposing arid material with respect to the planking level, and the execution of pointed poles stacked and ready to be stuck in the ground.

Article 9 of the Borbone's code orders the execution of the base of the building exclusively in masonry—a more or less deep foundation depending upon the consistency of the ground—from which the wooden framework stands out. It is a constructive measure aimed at isolating the wooden structure from the ground. A recommendation was anticipated by Vitruvius in relation to the *opus craticium* (liber II, capo VIII), which recommends to distance the frame from the ground, a possible vehicle of humidity, namely favorable conditions for biotic attacks that undermine the durability of the wooden elements. Such an arrangement can be found in the Palazzo Vescovile of Mileto (Vv), where part of the wooden framework is isolated from the ground by about 1 m of masonry without wooden elements. This base consists of stones of size, mineralogical composition, and organization different from those of the remaining facade. The foundations of Mileto's palace, at least as far as it is possible to see, would seem not to highlight any recess or extension to the base, contravening the precepts contained in the *De Architectura*. We are, therefore, in the presence not of a real foundation body but of a simple continuation of the above masonry.

7.4.2 The wall

The Borbone's antiseismic system is characterized by double framework, alternatively single inserted at different depths in the masonry (Fig. 7.2), based on examples described in the 18th-century literature and on constructed cases (Fig. 7.3). Differences are also recorded in relation to the method of stiffening in the plane, which could take place through crosses of Saint Andrew; alternatively, the deformability of the wooden frame was reduced by the masonry infill. Vivenzio recommends

FIG. 7.2 Borbone wall observed in the Bishop's Palace in Mileto, South Italy.

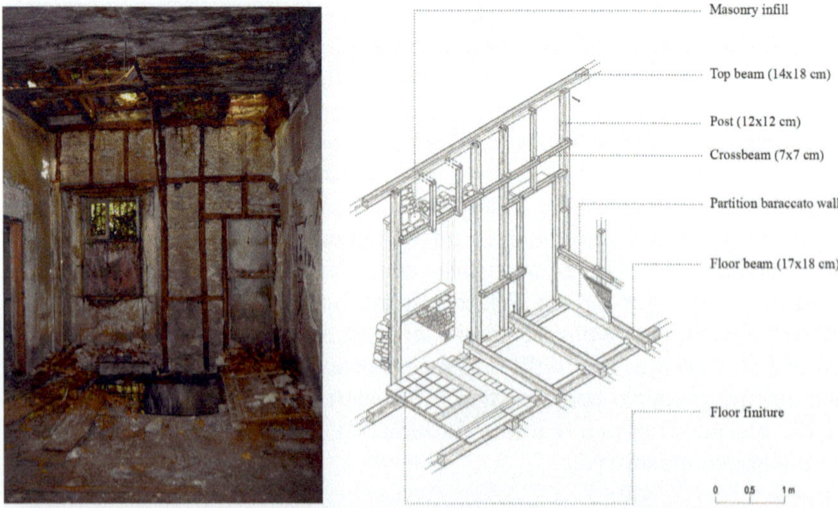

External timber framed masonry wall

Axonometric view of baraccato load-bearing structure

FIG. 7.3 *Case formate di legno* (house made of wood), drawings by Ferrarese.

a masonry fabric characterized by two parallel frames, suitably joined by means of wooden diatonic elements. A reinforced masonry is prescribed for public buildings, considering the significant economic and execution burdens (Fig. 7.4).

Therefore it represents a complex execution that motivates the scarce presence of buildings characterized by such an organization. Another type to be mentioned deals with only wooden pillars incorporated in the wall section; a configuration that, although it benefits from a lower seismic

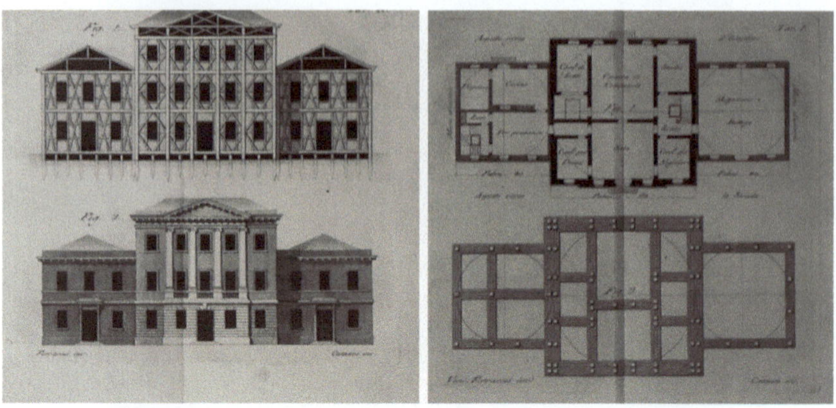

FIG. 7.4 Mileto (Vv), (end of the 18th century) The masonry reinforcement is constituted by only timber posts.

mass, thanks to the presence of the wooden elements, raises many doubts about the ability to improve the behavior of the masonry subject to horizontal cyclical actions. In fact, the wooden pillars could represent discontinuities in the wall and therefore the ability to counteract any overturning may be poor, given the action essentially concentrated in limited areas of the masonry; therefore they should rather be considered preferential ways of cracking in the masonry. In any case, the most common structural configuration is characterized by the masonry reinforced by wooden frames. A constructive technique whose notable example, slavish implementation of the Borbone's regulation, is represented by the *Palazzo Vescovile* built in Mileto (Vv) at the end of the 18th century. The Palace is a one-storey building beyond the basement and consists of a load-bearing structure characterized by masonry, with a thickness of about 60 cm, reinforced by timber frame with a more or less uniform interaxial distance. Other horizontal members are present inside the wall panel, however, irregularly organized and without any apparent order. There is a certain dimensional hierarchy with the top beam—a ring beam—measuring about 30 × 20 cm, while pillar and horizontal members have dimensions of 12 × 10 cm and approximately 7 × 7 cm, respectively. Some vertical wooden members continue beyond the attic of the ground floor to include the entire basement.

All the nodes between the various resistant elements are half-lap joints, connected with the help of a single metal nail. The internal partitions are smaller, about 25 cm and, given the small size, almost the entire section is occupied by the wooden frame. The uprights of these panels are simply juxtaposed to the orthogonal walls, while the constraint element is represented by the horizontal top member that connects the two orthogonal panels through a notch.

The masonry, stabilizing element in the plane of the frame devoid of Saint Andrew's crosses—even during the execution—is of the chaotic type with roughly squared stones, whose origin rock can be classified as calcilutite. However, there is a certain regularity given by the presence of pseudo-ashlars with variable size, stuffed with bricks in the joints, probably elements from buildings collapsed after the earthquake. In addition, the masonry fabric is characterized by the presence of not very regular courses. The mortar, arranged in beds of about 2–3 cm, is composed of lime and quartz-granitic aggregates, with dimensions between 2 and 5 mm. The connection between the pillars and the ring beam is obtained by means of a half-lap joint, stiffened by the presence of a pyramidal nail. Such a node presents technical devices aimed at withstanding a possible rotation of the wall both toward the outside and the internal direction of the building; the notches are in fact alternate in the two beam faces with a certain regularity.

7.4.3 Interfloor and roof slabs

The floors, strategic elements in the building's behavior under actions induced by earthquakes, can represent, if properly connected to the walls, constraints and so can contrast the overturning of the masonry panel.

The Palazzo Vescovile of Mileto has an interfloor slab characterized by beams, with defects—knots of large dimensions specifically—roughly squared measuring 25 × 20 cm and spaced about 80 cm, with the above boards secured to the beams by means of pyramid-type nails. The beams, arranged in a north-west, south-east direction and, therefore, perpendicular to the main facade, are connected to the perimeter wooden ring-beam by means of a half lap joint. A constructive peculiarity is represented by an inclined timber positioned at the lower edge of the floor near the corner, generating, albeit of a very low value, a reduction of deformability of the floor. There are several construction measures that pursue the reduction of seismic vulnerability also in relation to roofs. The trusses shown in the tables accompanying the Vivenzio's volume, signed by Vincenzo Ferraresi, are of the Palladian type, at least for the carpentry afferent to the two lateral bodies (see Fig. 7.3). The members' organization of such a structural unit consists of rafter, underrafter, kingpost, tie-beam, and top chord, from which does not arise outward thrust, hence the absence of horizontal component on the masonry support is ensured. The depiction in *Historia de tremuoti* of the carpentry roof of the main body is less explicit; however, the will to reduce the in-plane deformability by means of wooden bracing is nevertheless clear. Similar but with a less complex configuration is the roofing structure of the Palazzo Vescovile of Mileto composed of trusses of the "Mediterranean" type with purlins above and concluded at the top by minute slats—traditional organization in the Calabria region—which support the roof-tiles. Peculiar is the presence of wooden Saint Andrew's crosses in addition to other horizontal members placed perpendicular to the trusses with the evident intention of transversely stiffening the roof and limiting the tendency to pile up of trusses under horizontal action. It is, therefore, a measure aimed consciously at improving the behavior of the structural roofing system during the earthquake. In this regard, it should be noted that the roof structure consists of members in the minimum number essential to respond to gravitational loads and in the event of an earthquake, without structural redundancy, in some cases undersized, hence characterized by a modest seismic mass (Fig. 7.5).

The trusses, complete with struts and kingpost, highlight notches and nails at the ends with the aim of improving the connection between the members. There is metal hardware, for instance, a bracket that characterizes the king-post's foot and aimed at retaining the structural unit in the vertical plane. The tie beam to rafter node, with adequate dimension of the heel and stiffened by the presence of a metal plate, is translated by a few

FIG. 7.5 The roof carpentry of the Palazzo Vescovile of Mileto (end of the 18th century).

centimeters with respect to the wall support; however, the onset of shear stresses caused by the vertical component coming from the rafter is limited, thanks to the node almost centered on the wooden ring beam. These are, therefore, classifiable unions of gravity type, characterized, under modest seismic accelerations, by an appropriate behavior (Parisi & Piazza, 2000), which become more effective if the eventual vertical component of the earthquake is counteracted through the affixing of metal hardware, as in the case of the rafter-tie beam joint of the trusses of Mileto.

7.5 The evolution of the Borbone's system and the examples of the 20th century

The high-magnitude earthquakes that followed one another in the Calabria region—in 1832, 1835, 1836, 1854, 1870, 1905, and 1908—led to an increasing awareness of the need to build constructions that would oppose some resistance to the earthquake, at least in the years immediately after the telluric event. The experience gained with the Borbone's system, which had exhibited an appropriate behavior during seismic shocks compared to ordinary masonry constructions (Ruggieri et al., 2015), led to the promulgation of laws—in 1859 for the city of Norcia; in 1883 following the earthquake that hit the island of Ischia; in 1906 and 1909, following the Calabrian earthquakes of 1905 and 1908—which recommended the use of such a construction technique, reaffirming and taking up the indications contained in the *Istruzioni* (Ruggieri, 2016) (Fig. 7.6). That system, however, at least from the beginning of the 20th century (Barucci, 1990), underwent an evolution becoming more complex

FIG. 7.6 Typologies of masonry reinforced with timber frames of the 19th century in Calabria region, South Italy: Aprigliano (Cs) on the left; Cosenza, on the right.

in the organization, in parallel with the availability of cheap metal and the advancement of the science of building. The nodes, in fact, became more rigid with the use of various types of metallic fasteners, with a consequence also on the durability of the wood. In addition, the more precise woodworking, thanks to the advent of industrial production techniques, implied a further reduction of ductility compared to the Borbone's connections, in which the imprecision and craftsmanship guaranteed a greater generation of friction.

In any case, the extremely significant variation is to be recorded in the composition of the wall fabric more generally. In the Borbone's system, in the static field, also due to the considerable difference in elastic modulus between timber and masonry, it is the latter to bear almost entirely the

FIG. 7.7 Typologies of masonry reinforced with timber frames of the 20th century in Reggio Calabria, South Italy.

stresses, while in the event of an earthquake, the frame comes into play by exploiting the tensile resources offered by the wood. Such a structural hierarchy is reversed or cancelled in the examples created in the early 20th century. The wooden skeleton, betraying the original role, assumes the burden of absorbing the loads coming from the floors and the roof; instead, it is relegated to the masonry the function, together with other variously arranged wooden members, to brace the wall panel. However, it is worth to note that the typology made up exclusively of timber is customary; see the different examples in the town of Reggio Calabria (Fig. 7.7); a change that is also accepted in the evolution of the technical lexicon: if in the Borbone's documents known to us, the system does not have a precise denomination, at least since 1876 (Pesso, 1876), the construction technique is identified with the term *casa baraccata*. A neologism, evidently deriving from shack (baracca) whose semantics refers to a construction with a wooden structure, par excellence, at least in Italian culture, of a precarious and temporary nature.

7.6 Experimental investigation: Cyclic tests on Borbone walls

The mechanical analysis of the Borbone system has been calibrated from the results of the experimental test, carried out in the laboratory of CNR-Ivalsa of Trento in 2013, on an "imperfectly identical" copy of a Borbone wall extrapolated from the *Bishop's Palace* in Mileto (Vv) in South Italy. In particular, two timber-framed specimens, different from each other for the presence or not of the masonry infill, have been produced and subject to a cyclic test. In Fig. 7.8, the Borbone wall specimen is

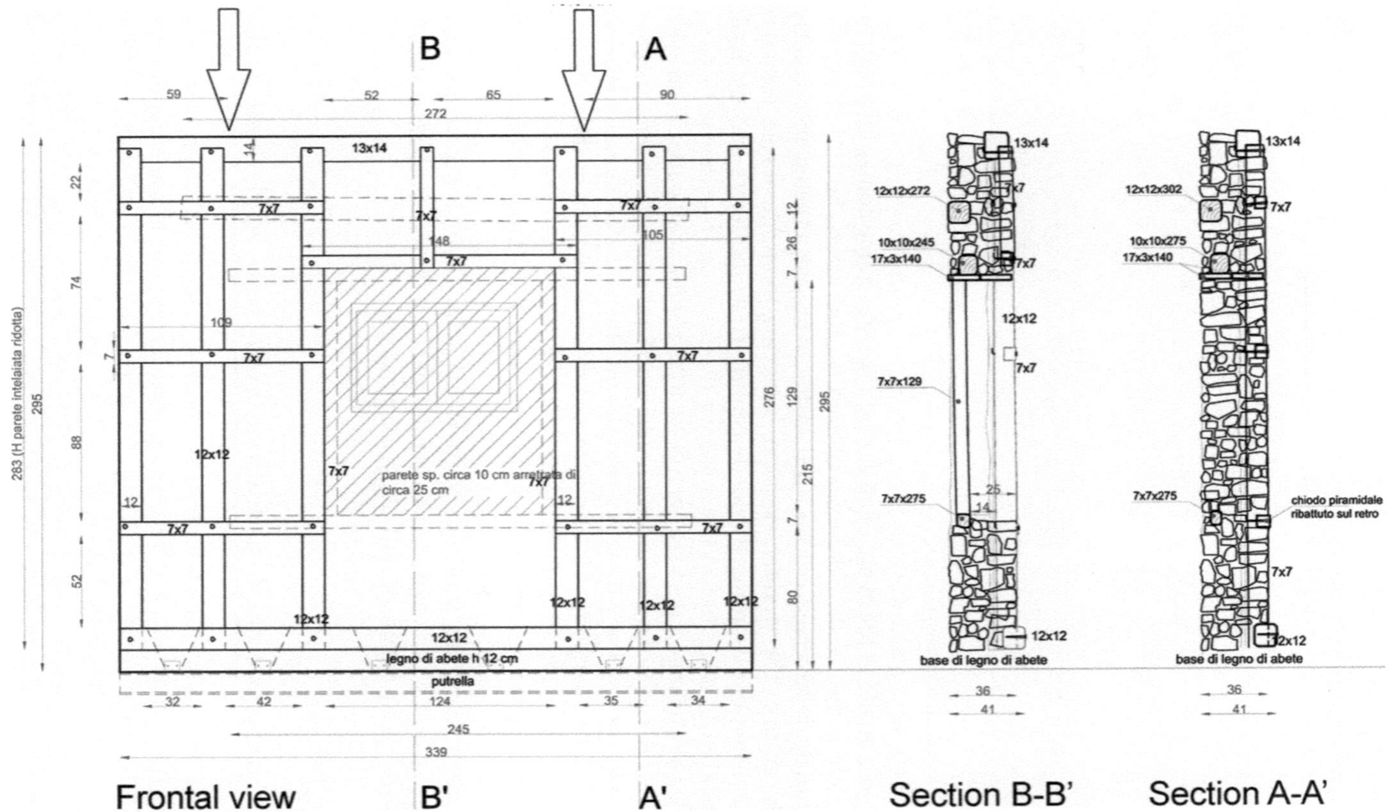

FIG. 7.8 The Borbone wall tested in the CNR-Ivalsa laboratory (Ruggieri et al., 2015).

represented, while more details about geometric and constructive characteristics are reported in Ruggieri et al. (2015).

The cyclic tests were carried out under horizontal displacement control, taking into account eight steps (displacements = 2.5, 5, 7, 10, 20 40, 60, and 80 mm), with a constant speed of 0.2 mm/s, up to the maximum displacement imposed of 80 mm. In particular, two loading phases were performed: one of *pushing*, with positive strength and displacement, and the other of *pulling*, with negative force and displacement. In addition, a uniformly distributed vertical load (18.6 kN/m) was applied to the specimens to replace the self-weight of the timber king post truss and the overall roof in general, discharging on the real wall.

To capture both the global behavior of the panel and the local effects, a network of several linear voltage displacement transducer (LVDT) was placed on the two samples, as shown in Fig. 7.9.

The analysis of the experimental results first highlights the strongly nonlinear behavior of the wall. In addition, the capacity curves of both specimens show a certain symmetry compared to the pushing and pulling directions.

Regarding the wooden frame specimen, the hysteretic curves show a high flexibility of the frame, reaching the ultimate displacement (80 mm), as a result of a force just over 6 kN. This response is due to the absence of a brace element for the frame, which in the Borbone system is ensured by the masonry infilled.

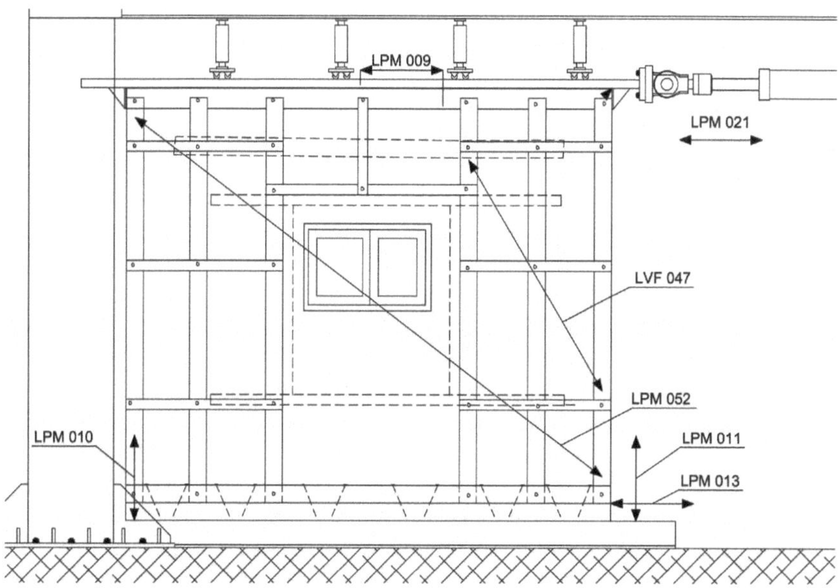

FIG. 7.9 Test layout (Ruggieri et al., 2015).

As a matter of fact, the envelope curve (Fig. 7.10A) shows an initial linear branch with stiffness equal to about 430 kN/m followed by a hardening branch that suggests the plasticization of some or all of the frame's joints, the only ones that can offer ductility to the system. As also indicated in the test report, the stiffness variation can be attributed to local mechanical phenomena affecting the half-lap joints, such as the poor grip between the contact surfaces of the elements. In this context, the hardening behavior of the wooden frame is related to an inelastic rearrangement of the geometric configuration of the joints, preceded and followed by a pseudo-elastic behavior. In fact, at the end of the test, both joints and wooden elements were not affected by ruptures.

With regard to the timber frame with masonry infilled, the results provide clear evidence of the mechanical behavior of the wall in terms of stiffness, strength, and ductility. The maximum imposed displacement is equal to 80 mm at which a horizontal force just over 100 kN was recorded both in pulling and pushing phases.

The envelope curve (Fig. 7.10B) shows that the wall had, excluding the first step with an imposed displacement of less than 1 mm, a linear elastic behavior in the first cycles, until the imposed displacement of 7 mm in which a horizontal stiffness equal to 6826 kN/m is recorded. Until the imposed displacement of 60 mm, the wall shows a plastic-hardening behavior, reaching the maximum strength in pushing (103.64 kN), while in the pulling phase, the maximum strength value (101.62 kN) is reached at the last step of the test, with an imposed displacement of 80 mm.

In general, the wall did not suffer any rupture, while slight damage was mostly limited in the interface joints between the wooden frame and the masonry.

In particular, the half-lap joints, while exceeding the ultimate displacement, continued to dissipate energy through the phenomenon of friction, contributing to the overall ductility of the system. On the contrary, the wooden elements did not exceed the elastic limit and did not suffer any plastic damage or deformation. The only joints affected by rupture are those between the corner posts and the lower current. This phenomenon can be attributed to an incipient rocking mechanism, causing a partial detachment of the posts from the lower current. In detail, the maximum uplift of the base of the corner posts recorded was equal to 3.2 cm in pulling and 2.6 cm in pushing. It's worth nothing that rocking mechanism of the wall is not properly configured as a rigid rotation because, from the first cycle, the specimens sample behaved as four no synchronous macroelements (see Fig. 7.11): two "piers" (1–2) at both extremities, a "spandrel" (3) in the upper part of the system, and a lower masonry block (4) devoid of internal timber skeleton. As evidence of the above, a kinematic analysis of the wall was conducted in the dual hypothesis of rigid rotation of the wall (Fig. 7.11A) and rotation of the macroelements. (Fig. 7.11B).

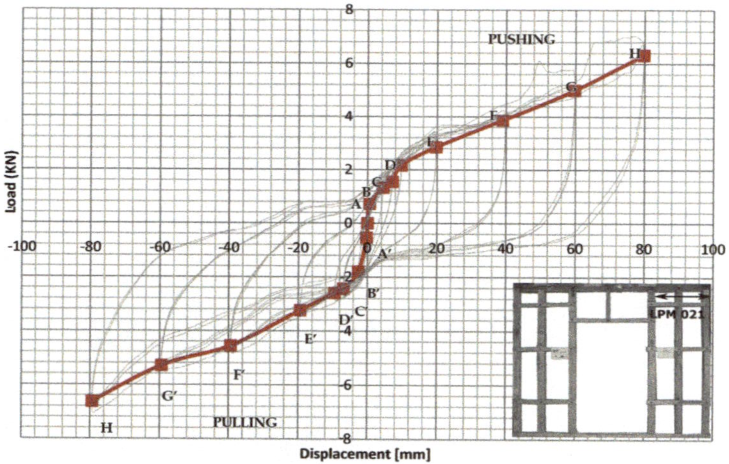

(A)

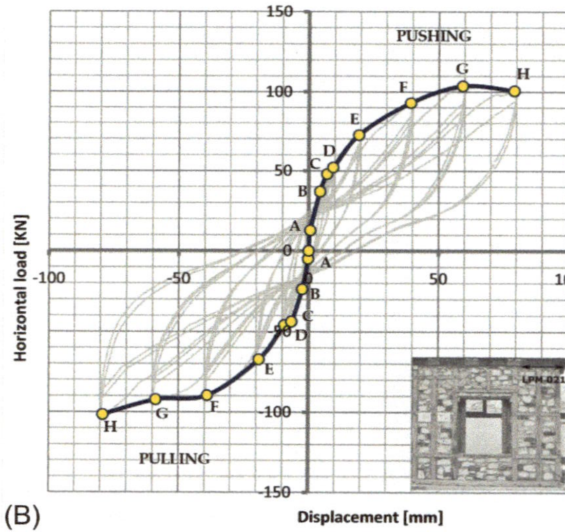

(B)

FIG. 7.10 Envelope curve obtained from the cyclic test on: (A) timber frame specimen; (B) timber frame with masonry-infilled specimen (Ruggieri et al., 2015).

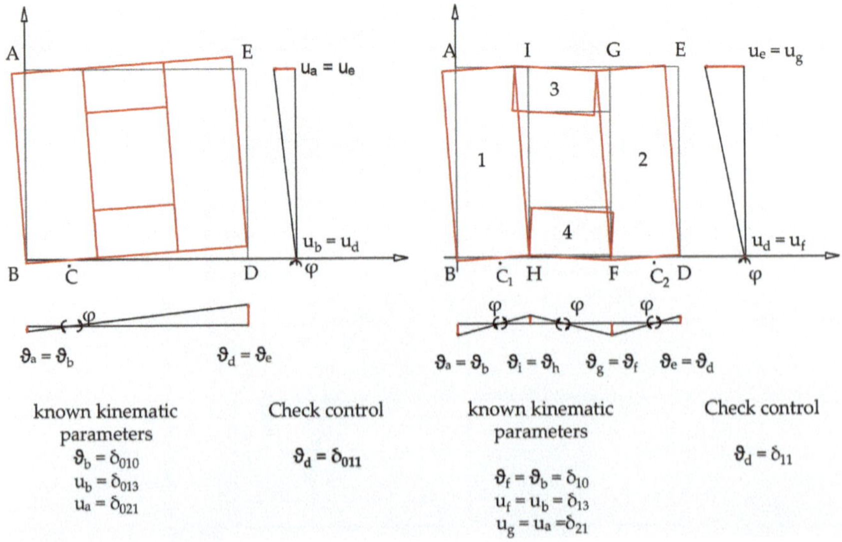

FIG. 7.11 Kinematic of the wall: (A) rigid rotation (rocking), on the left; (B) macroelements rotation, on the right.

For each step, a comparison between the numeric value of the vertical displacement of node D, considering the two hypotheses, and the vertical displacement measured by the LPM011 transducer during the experimental test was carried out. As result, it was possible to prove that the numerical displacement closest to the one measured by the transducer is the one referring to the hypothesis of rotation of macroelements. In addition, this kinematic check also made it possible to identify the yielding point of the envelope curve due to the plasticization of some elements that are likely to coincide with the interface joints between the wooden frame and the masonry, in the node piers-spandrel. This information has allowed to calibrate some of the constitutive laws of the springs that idealized, for example, the mechanical behavior of the joints.

Definitively, the analysis of the experimental results of the experimental suggests that the overall response of the wall is affected by a series of micromechanical phenomena, concerning the connections at the base of the wall and the interfaces between the timber frame and the masonry infilled. Such phenomena are extremely ductile and allow the wall to deform significantly before it is affected by ruptures. The damage, moreover, thanks to the structural configuration of the system, affects the resistant elements, according to a precise time order, as suggested by one of the most important criteria behind today's antiseismic design that is the concept of "hierarchy of resistance." The mechanical behavior of the Borbone system, in fact, before provides the activation of the masonry infilled, stiffer than the wooden frame, in which the ruptures are condensed. On the other hand, the timber frame, characterized by a higher elastic limit, remains in the elastic phase until the masonry reaches the ultimate damage and only later, in the presence of a prolonged stress over time, is called upon to contribute to the flexibility of the overall system.

In this context, it is evident that the collaboration between timber frame and masonry infill improves the mechanical performance of both. Timber frame encloses the masonry, forcing the latter to offer all its resistant capacity; at the same time, masonry, until it comes to rupture, stiffens the frame that consequently behaves elastically, avoiding fragile fractures of the wooden elements. After understanding these phenomena, two micromechanical models were developed to simulate and capture such mechanical behaviors.

7.7 Structural modeling

The mechanical modeling of masonry and, above all, of historic masonry is the subject of an international debate, because the seismic response of a historic building depends on so many global and local factors that many scholars are led to believe that the mathematical prediction models are one mere chimera and to prefer qualitative and descriptive approaches, which involve the use or restoration of some "rule of art." Precisely, this dialectical comparison, sometimes even very radical, becomes a stimulus and a challenge toward a continuous refinement of the mathematical models, with the support of the experimental tests and starting from the basic structural element of the historical building, namely, the wall. In the case of timber-framed masonry walls, the interaction between the masonry wall and the timber frame is an additional element of complexity. Therefore the experimental test carried out at CNR-IVALSA, in addition to having a value in itself, assumes a benchmark value for the validation of the structural models. As a matter of fact, today's paradigm of the mechanics of materials and structures is multiscale modeling (De bellis & Addessi, 2011; Pagano et al., 2019; Reccia, Leonetti, Trovalusci, & Cecchi, 2018), in which very detailed mechanical models of structural elements exchange information with coarse models, which, properly calibrated, aim to reconstruct the behavior of the building as a whole. Within a multiscale modeling of a timber-framed masonry wall, fine models are generally **discrete ones**, in which every single unit of the wall (a brick, a mortar joint, a timber joist and so on) finds its geometric and constitutive descriptions and directly compares itself with the experimental tests, with the ambition to reconstruct stiffness, strength, and ductility of the wall. On the contrary, the coarse models are often **continuum models**, predominating equivalent frame models (Pagano et al., 2018) or two-dimensional continuum (Masiani, Rizzi, & Trovalusci, 1995; Salerno & De Felice, 2009) interacting with the fine models to grasp constitutive information from them (Salerno & Felice, 2000; Uva & Salerno, 2006) and aiming at a simplified modeling of the whole building. Below, we present two fine numerical models, which compare with the experimental test, to reproduce the detailed behavior of the wall, in terms of configurations, damage evolution, stiffness, strength, and ductility. The definition of coarse models for the analysis of the entire building will be the subject of future work.

7.7.1 A first approach to the numerical modeling

The mechanical interpretation of the structural behavior of timber-framed masonry was conducted through the adaptation, to this specific case, of the original software *BrikWORK* (Galassi, Ruggieri, & Tempesta, 2019, 2020), developed by some of the authors for the analysis of a general masonry structure.

In the original procedure, the masonry arrangement is modeled as a collection of rigid elements (in this case stones) connected by mortar joints, where the elastic-brittle properties, which describe the behavior of the material, are concentrated. Consequently, the main type of damage mechanism considered in the mortar joints is a tensile failure: until such a failure does not happen, the joints are supposed to retain an elastic behavior.

Such an approach, therefore, involves that the masonry, as a whole, has a good capability to carry compression loads and, taking into account that the reference masonry is that of historical architecture heritage, the tensile strength of the material is limited to the poor cohesion between mortar and stone elements.

Under these assumptions, the mechanical characterization of masonry refers to a system of rigid blocks connected by unilateral contact and frictional links, which means that the mechanical model takes on the characteristics of a model with elasticity concentrated in the joints (Fig. 7.12).

Such a model, which considers the rigid blocks linked together by means of deformable joints with no-tension behavior, is especially effective to interpret the influences which the dimensions of the blocks and the joints' orientation have on the behavior of historical masonry buildings. Moreover, it is also capable to describe very clearly the progression of the damage of masonry under loading.

In turn, the joints are modeled through an interface device composed of two links orthogonal to the interface (n_1 and n_2), which transfer the axial force X_n, and two links arranged along the interface (t_1 and t_2), which transfer the shearing force X_t.

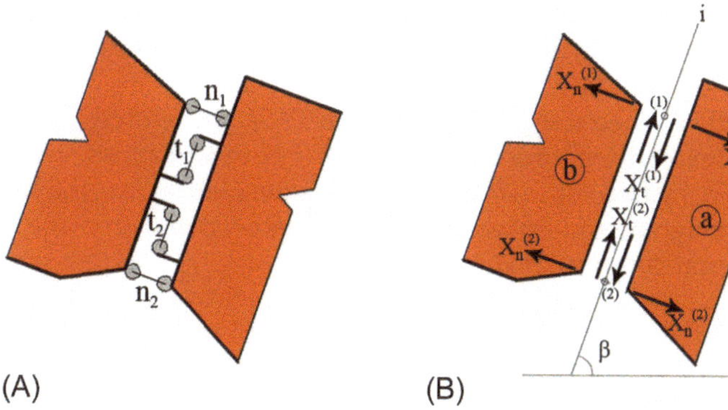

FIG. 7.12 (A) Discrete model of joint interface; (B) Axial and shear forces transmitted by the joint links.

$$\begin{cases} AX = F + \alpha\overline{F} \\ \tilde{A}x + KX = \Delta \end{cases} \tag{7.1}$$

In the equilibrium equation (7.1).$_1$, A is the geometrical configuration matrix, F is the vector of the dead loads, and $\overline{F}$ is the vector of the external loads that are increased by the factor α. In the specific case of the wall specimen tested at the CNR Ivalsa Laboratory, described in Section 7.6, $\overline{F}$ is the horizontal force provided by the hydraulic jack, which impressed an increasing horizontal displacement at the top of the wall, and α is the incremental factor.

In the elastic-kinematic equation (7.1).$_2$, $\tilde{A}$, the transpose of matrix A, is the kinematic equation, and K is the diagonal matrix whose entries are the deformability coefficients of the links of the joint device. Vectors X and x are the static and kinematic unknowns of the problem, the former being the vector of the forces in the interface links and the second one being the vector of the horizontal, vertical, and rotational displacements of the block centroids. Vector Δ is the "distortion" vector, and its use in the numerical procedure will be discussed later. As aforementioned, the failure criterion is assigned to the interface links. Stresses in the axial links X_n cannot overcome the threshold value of the tensile strength (which is zero in the case of a no-tension material). The solution of Eq. (7.1) is shown in Eq. (7.2), which provides the values of the forces of the links of the joint devices:

$$X = X_0 + \left(I - K^{-1}\tilde{A}\left(AK^{-1}\tilde{A}\right)^{-1}A\right)\cdot\Delta \tag{7.2}$$

where Eq. (7.3) is the linear-elastic solution:

$$\mathrm{X}_0 = \mathrm{K}^{-1}\tilde{\mathrm{A}}\left(\mathrm{A}\mathrm{K}^{-1}\tilde{\mathrm{A}}\right)^{-1}\bullet(\mathrm{F} + \alpha\overline{\mathrm{F}}) \tag{7.3}$$

The numerical procedure is iterative. At the beginning of the analysis, vector Δ is initialized to zero; therefore the initial solution corresponds to that of Eq. (7.3), that is, $X = X_0$.

Afterward, it is necessary to check, in Eq. (7.2), if the forces in the axial links do not exceed the threshold strengths. For this purpose, vector X is partitioned in the two subvectors X_n and X_t, whose entries are the forces in the links orthogonal to and along the joint, respectively, and the tensile failure criterion is formulated in Eq. (7.4):

$$X_n^{(j)} \leq \overline{\sigma}_n \cdot S^{(j)} \tag{7.4}$$

in which $\overline{\sigma}_n$ is the tensile strength of the joint, and S is the area of the cross-section of the jth link. If the solution provided by Eq. (7.2) does not agree with the constraint of Eq. (7.4), the analysis becomes nonlinear and the procedure must nullify, step by step, the inadmissible forces, and vector X in Eq. (7.2) is computed again each time.

In order to nullify the inadmissible force of a link, the numerical procedure introduces an "impressed distortion" into that link. The set of impressed distortions introduced into the links whose forces do not satisfy the conditions expressed by Eq. (7.4) is composed of the nonzero coefficients of vector Δ.

At each step of the iterative analysis, entries of vector Δ are unknown. To compute the entries of vector Δ to be input in Eq. (7.2), this equation has been rewritten, partitioning all matrices and vectors, considering the coefficients corresponding to the links with admissible forces (subscript "b") and the links with inadmissible forces (subscript "c") and putting:

$$C = \left(I - K^{-1}\tilde{A}\left(AK^{-1}\tilde{A}\right)^{-1}A\right):$$

$$\begin{bmatrix} X_b \\ 0 \end{bmatrix} = \begin{bmatrix} X_{0b} \\ X_{0c} \end{bmatrix} + \begin{bmatrix} C_b & B \\ \tilde{B} & C_c \end{bmatrix} \cdot \begin{bmatrix} 0 \\ \Delta_c \end{bmatrix} \tag{7.5}$$

The solution of Eq. (7.5) provides the subvector of the "effective distortions" (Eq. (7.6)) and the subvector of the final solution collecting the nonzero forces (Eq. 7.7):

$$\Delta_c = -C_c^{-1}X_{0c} \tag{7.6}$$

$$X_b = X_{0b} - BC_c^{-1}X_{0c} \tag{7.7}$$

Finally, to draw the final configuration of the structure, the effective displacement vector $\overline{x}$ and the vector $\overline{\Delta_c}$ that measures the depth and position of cracks in the mortar joints are computed, considering both the links that elastically deformed and those that failed. In this purpose, the second of Eq. (7.1) is partitioned, considering once again the links with admissible and inadmissible forces:

$$\begin{bmatrix} \widetilde{A}_b & K_b & 0 \\ \widetilde{A}_c & 0 & K_c \end{bmatrix} \begin{bmatrix} \overline{x} \\ X_b \\ 0 \end{bmatrix} = \begin{bmatrix} 0 \\ \overline{\Delta}_c \end{bmatrix} \tag{7.8}$$

The solution of Eq. (7.8) provides the vector x of the displacements of the block centroids (Eq. 7.9) and the vector $\overline{\Delta}_c$ that measures the depth and position of cracks in the mortar joints (Eq. 7.10):

$$\overline{x} = -K_b X_b A_b \left(\widetilde{A}_b A_b\right)^{-1} \tag{7.9}$$

$$\overline{\Delta}_c = \widetilde{A}_c \overline{x} \tag{7.10}$$

The appropriate adaptation of the original numerical model to the one used for the analysis of the masonry wall reinforced with timber frames has necessarily had to take account of a specific definition and calibration of the mechanical parameters to be assigned to the contact joints between stone and stone, wood and stone and for the contact joint between wood and wood, in its turn considered perfectly elastic .

For this purpose, several correlations were made in the comparison between the results of the numerical model and those deduced from the experimental tests described in Section 7.6. In the first place, the mechanical and geometrical characteristics have been defined to be assigned, at the joints of interface, to the discretized model with concentrated elasticity so as to reproduce the same field of deformation and displacements obtained by the experimental tests carried out on the single timber frame (Fig. 7.13). To define the elastic modulus to be considered

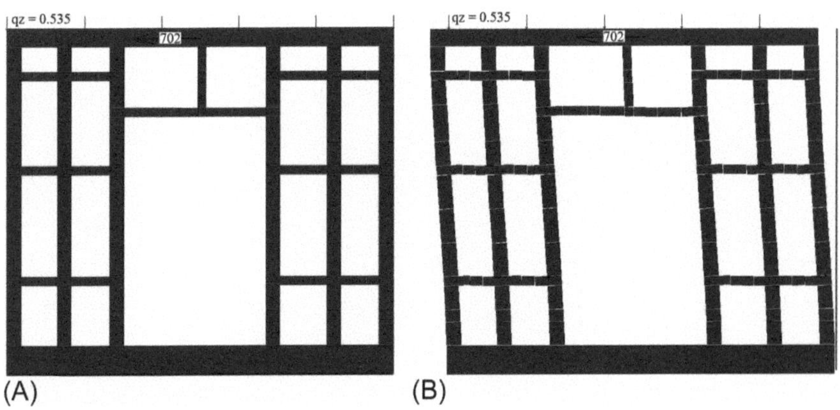

FIG. 7.13 Modeling of the timber frame specimen. (A) Original configuration; (B) Deformed configuration.

in this concentrated elasticity model, the comparison of the actual deformation values of the tested timber frame with those parameterized obtained by the numerical model was performed.

Starting from the observation of the actual behavior of the experimental model, in which even for the maximum value of the applied load the timber frame has never shown failure mechanisms in any section, the limits of resistance in the contact joints have been conventionally assumed equal to infinity for both tensile and compressive interactions.

It was subsequently defined the discretized model of the masonry infill, respecting as much as possible the original shape and the arrangement of the stone elements of the experimental model, as to reproduce, as closely as possible, also with regard to the thickness of the joints, the reality of the experimental model (Fig. 7.14).

As mentioned earlier, in accordance with the comparative values deduced by the comparison between the experimental results and those obtained using the numerical model, a limit of tensile strength equal to 0.4 MPa has been assumed for the joints between stone and wood, while for the joints between the stone and stone, an ultimate tensile strength equal to 0.8 MPa seemed to be the closest and most coherent with respect to the actual values deduced from the tested structure.

Regarding the boundary conditions of the mechanical model subjected to the numerical analysis, considering the choices made for the experimental tests, fixed supports have been assumed at the base and movable hinges at the top.

The calculation process adopted in the application of the numerical model has retraced the hysteretic cycle used in the experimental test, as reported in Section 7.6 (see Fig. 7.10).

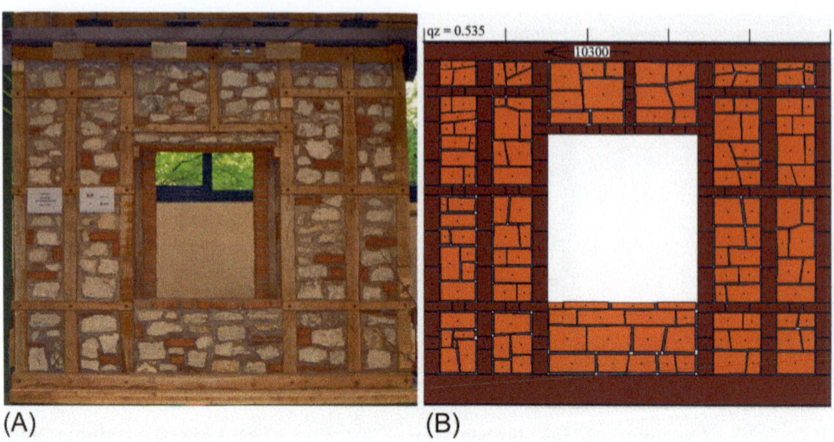

(A) (B)

FIG. 7.14 Mechanical modeling of the masonry wall made on the basis of the Borbone constructive system. (A) Experimental model; (B) Numerical model.

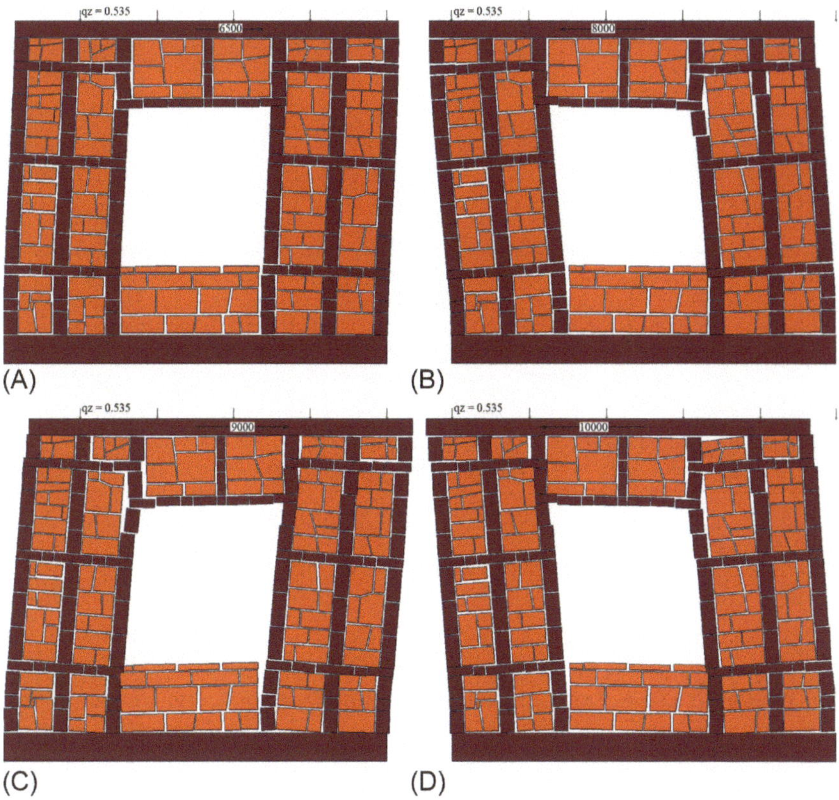

FIG. 7.15 Configurations in correspondence to a load of +65 kN (A), −80 kN (B), +90 kN (C), and −100 kN (D).

In this regard, the loading history followed a cycle of alternating loading steps (Fig. 7.15), first to the left and then to the right, with a sequence of values as much as possible corresponding to hysteresis loops from the tests: −50, +65, −80, +90, −100, and +100 kN.

At each assigned value of the load, before restarting with the next increase of the latter, an internal loop was performed, always with the same load, in which the joints were disconnected in which the two contacts orthogonal to the interface resulted zeroed, by the calculation algorithm, for having reached the threshold tensile strength.

The capacity curve represented in Fig. 7.16 shows the general behavior of the process.

The application of the numerical procedure, according to the modalities described earlier, has shown to catch with good approximation, in correspondence of each step of inversion of the direction of the load, the progressive decay of the stiffness of the structure, mainly due to the progressive detachment between the masonry filling and the timber frame at

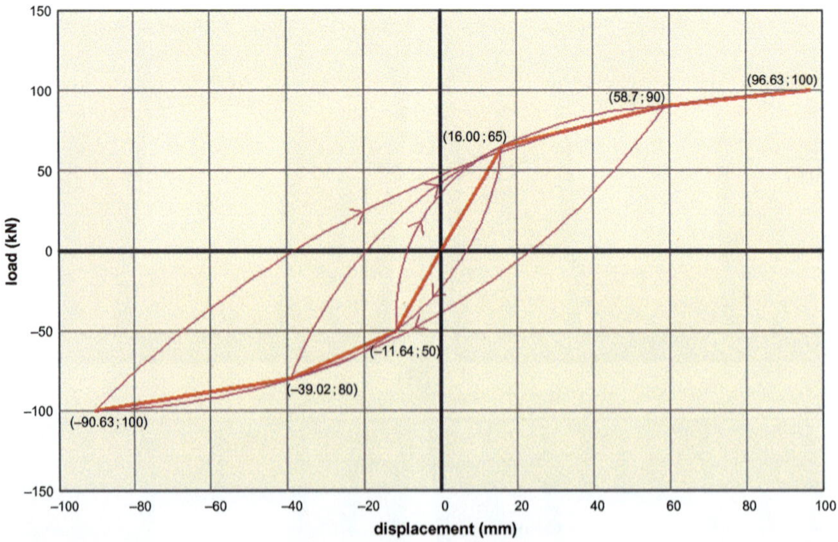

FIG. 7.16 Capacity curve obtained by the numerical procedure.

the contact joints between the two materials. The results obtained with the numerical modeling have provided an interpretation of the behavior of the wall very close, both quantitatively and qualitatively, to that of the experimental model. It is interesting to notice how, in terms of fracture and detachment, the crack pattern produced as the output of numerical analysis has shown a significant similarity with the real one.

7.7.2 Toward a simplification of the detailed model

This second model has the ambition to simplify many aspects of the model of Section 7.7.1 and of a model in literature (Salerno et al., 2016), which fully shared its modeling detail, while maintaining a dominant factor of the timber-framed masonry walls, that is, the interaction between the wooden frames and the wall panels.

For this purpose, the modeling has focused on the mechanical behavior of the timber frame, tested also separately in the experimental tests described in Section 7.6, of the wall panels, for which a simplified model provided by the literature has been proposed, and of the connections. For each of these structural elements of the wall, the elastic and inelastic properties have been defined (stiffness, strength, and ductility), referring to the Italian Code for structural design "NTC18" or through calibration from the results of the experimental tests of Section 7.6. In particular, the wooden elements were modeled as linear-elastic Timoshenko beams, while the half-lap joints were modeled by using continuity restraints of the nodal displacements, with the addition of rotational links, characterized by elastic-plastic hardening constitutive behavior. For the masonry panels, the macromodel developed by Caliò and Pantò (2014) for the in-plane analysis was chosen, with two diagonal shear springs, and further wood-masonry interface springs, characterized by Mohr-Coulomb behavior under shear and elastic-plastic behavior under compression. The aim was to evaluate both the overall response (stiffness, strength, and ductility) and the different mechanical phenomena occurred during the experimental test, such as the ground restraint rupture and the damage progression of the interface between timber frame and masonry infilled. Finally, the selected approach models the wall as an assembly of one-dimensional elements. Nonlinearity was considered by using of nonlinear Springs (Nlink), reproducing the timber frame joints, the connections between wooden frame and the ground system, the masonry and the frame-masonry interfaces.

7.7.2.1 Timber frame, half-lap joint, and ground joints.

Frame elements were modeled as Timoshenko beams, with elastic perfectly plastic behavior, while the half-lap joints were modeled by means of rotational links with elastic-plastic hardening behavior, as shown in Fig. 7.17A. The flexural stiffness was calculated considering that the relative rotation between post- and cross-beam generates in the node compressive stresses, as illustrated in Fig. 7.17B.

Then, the compressive force "N" has been calculated as reported in Eq. (7.11):

$$N = \frac{\varphi \cdot E \cdot \alpha^2 \cdot s}{2 \cdot l_{\text{mean}}} \tag{7.11}$$

where φ = relative rotation between the elements at the joint, E = elasticity modulus parallel to grain, a = length of the compressed section, s = thickness of the joint, and l_{mean} = mean length of the wood intersecting elements. Then, the flexural stiffness "K_φ" was put equal to $K \cdot \varphi = N \cdot b$. The mechanical parameters of the wood are reported in Table 7.1.

The values of the yield strength "$M_{y\text{ joint}}$," the plastic slope, and the ultimate rotation were empirically calibrated from experimental results. On the contrary, the ground restraints behavior was defined considering the in-plane gap rupture between wooden element and steel nail, as highlighted in previous tests conducted on TFM walls (Meireles, Bento, Cattari, & Lagomarsino, 2012; Ruggieri et al., 2015). This phenomenon has been idealized through an axial and transversal NLink. Stiffness and ultimate displacement values were empirically calibrated, while the ultimate axial strength was computed according to Eurocode 5,

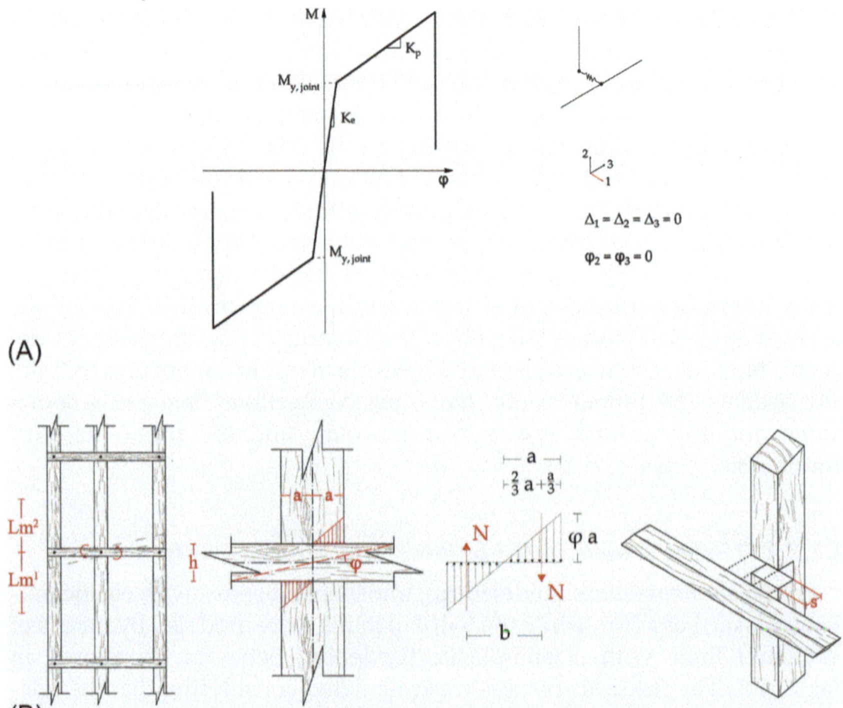

FIG. 7.17 Half-lap joint: (A) constitutive behavior law; (B) compressive stresses induced by relative rotation.

considering an uplift of the post. In compression, the ultimate strength of the joint was considered equal to the maximum compressive strength perpendicular to grain of the lower beam. In the transversal direction, the ultimate strength was empirically calibrated.

7.7.2.2 Masonry infill

Interaction between frame and masonry was simulated according to the macroelement developed by Caliò and Pantò (2014). It is constituted by an articulated quadrilateral with four rigid edges connected by four hinges and two diagonal nonlinear springs. Each side can interact with the timber frame by means of a discrete distribution of Nlink, denoted as interface with zero thickness. Each interface Nlink simulates the behavior of a "fiber" of masonry, with reference to a specific influence area, as illustrated in Fig. 7.18.

Axial and shear behavior of the interface springs were considered elastic perfectly plastic. In the axial direction, a material with no tensile strength was considered, while the compression behavior was defined

TABLE 7.1 Mechanical parameters of wood class D24 (chestnut).

	Strength properties (N/mm²)						Stiffness properties (kN/mm²)			
Wood class EN 338	**Bending**	**Tension parallel**	**Tension perpendicular**	**Compression parallel**	**Compression perpendicular**	**Shear**	**Mean modulus of elasticity parallel**	**Mean modulus of elasticity perpendicular**	**Mean shear modulus**	**Density (kg/m³)**
D24	$f_{m,k}$	$f_{t,0,k}$	$f_{t,90,k}$	$f_{c,0,k}$	$f_{c,90,k}$	$f_{v,k}$	$E_{0,\mathrm{mean}}$	$E_{90,\mathrm{mean}}$	G_{mean}	r_k
	24	14	0.6	21	7.8	4	10	0.67	0.62	485

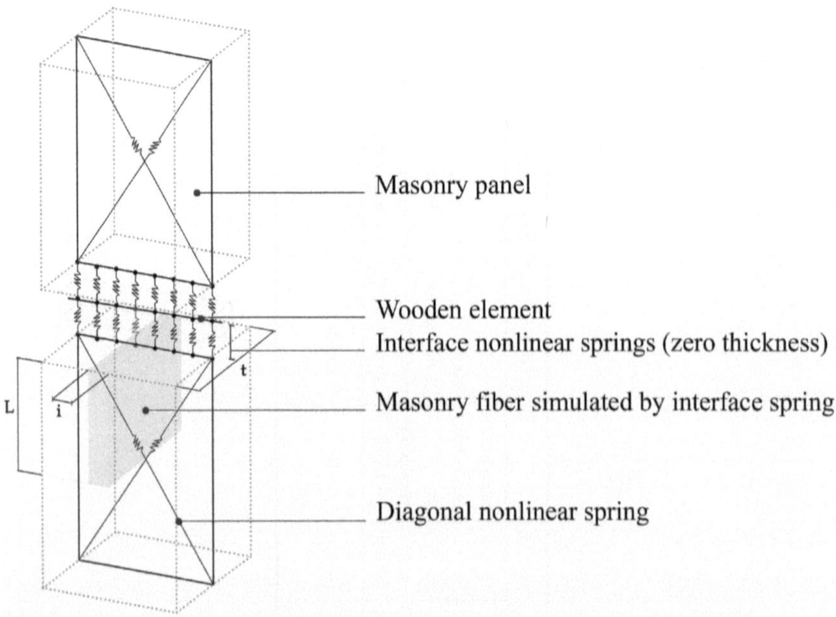

FIG. 7.18 Macroelement adopted to model the masonry infill and the interface.

by considering a constant distribution of the normal stresses on the interface.

In this schematization, the spring stiffness is put equal to:

$$K_{\text{comp}} = \frac{E \cdot (i \cdot t)}{L} \tag{7.12}$$

where: E = Young's modulus, i = interaxial distance between the springs, t = thickness of the macroelement, L = half-length of the macroelement.

The ultimate compression strength is equal to the product between the compression strength of masonry "f_m" and the influence area:

$$F_{c,y} = f_m \cdot (i \cdot t) \tag{7.13}$$

while no ultimate deformation was assigned. The shear behavior was defined according to Mohr-Coulomb law. The shear stiffness was put equal to:

$$K_{\text{sliding}} = \frac{G \cdot (i \cdot t)}{L} \tag{7.14}$$

where G = masonry shear modulus. The maximum shear strength was calculated as follows:

$$F_{\text{sliding},y} = \tau_0 \cdot (A_m + \mu \cdot N_{\text{comp}}) \tag{7.15}$$

where τ_0 = shear strength of masonry, extrapolated from Italian Code NTC18, referring to an "Irregular masonry type with soft stones," A_m = influence area of the spring; μ = coefficient of internal friction (tan $\alpha = 0.4$), and N_{comp} = average compression on the interface. Values of the ultimate deformation are the results of sensitive analysis. The mechanical parameters of masonry type extrapolated by NTC18 are presented in Table 7.2.

Diagonal shear collapse of the masonry panel was simulated by the two diagonal springs with elastic perfectly plastic behavior. According to the reference model, the elastic stiffness of each spring was obtained enforcing an elastic equivalence between the panel and the corresponding masonry wall, considered as a pure shear deformable homogeneous 2D continuum with shear modulus "G," cross-area section "A_t," and height "h." Spring stiffness is:

$$K_{\text{diag}} = \frac{G \cdot A_t}{2h \cdot \cos^2 \cdot \alpha} \tag{7.16}$$

where α = arctan (b/h), b being the base length of the panel. The yielding strength is:

$$F_{\text{diag},y} = \frac{\tau_u \cdot A_t}{2 \cdot \cos \alpha} \tag{7.17}$$

where $\tau_u \cdot A_t$ is the ultimate shear strength of the macroelement, whose value is the minimum between that calculated according to the Mohr-Coulomb criterion and that obtained in accordance with the Turnsek-Cacovic criterion. Finally, as proposed by Magenes and Calvi (1997), the ultimate displacement δ_u was associated to a specific value of the ultimate angular deformation γ_u, put equal to 0.5% with reference to the Italian NTC18 standards:

$$\delta_u = \gamma_{mu} \cdot h \cdot \cos \alpha \tag{7.18}$$

The so defined micromodel was implemented in SAP2000 software. Pushover analyses were carried out for the two specimens of the wall

TABLE 7.2 Mechanical parameters of masonry material.

Masonry type | Italian code NTC18	Average Young modulus E (MPa)	Average shear modulus G (MPa)	Average compressive strength f_m (MPa)	Average shear strength τ_0 (MPa)	Weight W (kN/m^3)
Irregular masonry type with soft stones	1.080	360	1.8	0.035	16

(unreinforced timber frame and timber frame with masonry infill), applying monotonically a horizontal displacement up to a maximum value of 80 mm. In addition, a previous load condition, representing both the self-weight of the wall and the vertical loads coming from the roof structure, was considered. Figs. 7.19 and 7.20 show the comparison between the numerical and experimental capacity curves.

The numerical capacity curve captures the transition from elastic (points A and B) to pseudo-elastic configuration (point C), validating the constitutive law defined for the half-lap joint, on which the ductile behavior of the entire timber frame depends. In particular, by observing the damage patterns, the ground restraints remain in the elastic phase, while all the other connections undergo inelastic deformations, according to the following progression: (A) The joints at the head of right pier are the first to overcome the elastic limits; (B) The head joints of the posts and those belonging to the center of the right pier overcome the elastic limit; and (C) In the end, symmetrically compared to right pier, the center joints of the left pier overcome the elastic limit. This configuration remains unchanged until the last imposed displacement (point D), at which no damage is recorded either to the frame member or to the joints. With reference to the timber frame with masonry infill model, Fig. 7.4 shows the damage progression recorded in SAP2000.

Interface springs and the ground restraints at the base of the piers are the elements most affected by plasticization and rupture. The left end ground restraint was affected by a rupture in the longitudinal direction, due to the uplift of the post, while the right one overcame the ultimate compressive strength perpendicular to the grain of the lower wooden beam. On the other hand, the springs that simulate the connections between wooden elements don't overcome the yielding point, less than what was recorded in the case of empty timber frame. Rotations close to the yield value were recorded only in the joints between the spandrel and the piers. This condition led to assume that the presence of the masonry infill operates a stiffening action for the joints, subject to low flexural stresses. Ultimately, the mechanical behavior recorded with the numerical model is compatible with the results of the experimental test. Other aspects don't replicate the experimental results, and they require further in-depth analysis with reference, for example, to definition of the ultimate displacement for several Nlinks.

7.8 Conclusion

The continuous earthquakes, to which Calabria was subjected, led, at the end of the 18th century, to develop construction rules for the realization of seismic-resistant constructions. The conception of the Borbone

Step	Displacement	BaseForce
	m	KN
0	0,000094	0
1	0,004094	1,15
2	0,008094	1,75
3	0,016094	2,35
4	0,024094	2,89
5	0,032094	3,44
6	0,040094	3,98
7	0,048094	4,53
8	0,056094	5,07
9	0,064094	5,62
10	0,072094	6,17
11	0,080094	6,71

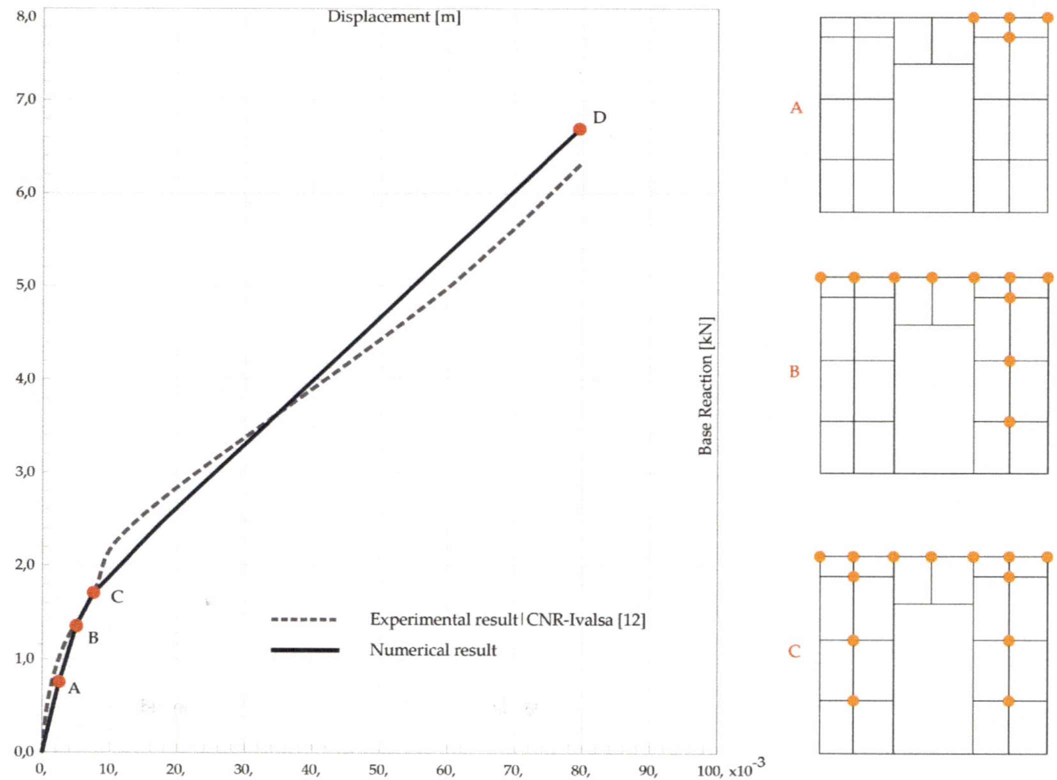

FIG. 7.19 Micromodeling: Experimental and numerical capacity curves (unreinforced timber frame).

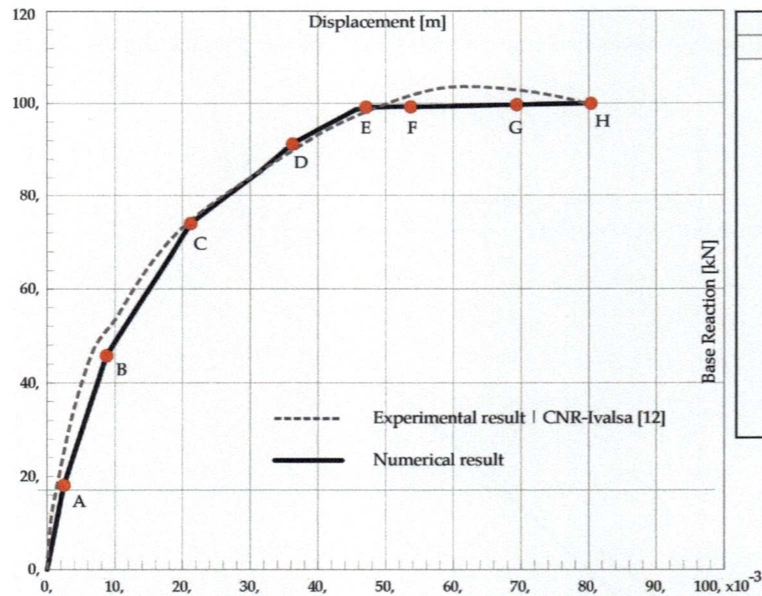

Step	Displacement	BaseForce
	m	KN
0	-0,00115	0
1	0,002499	18,120
2	0,008749	45,779
3	0,021249	73,970
4	0,031249	85,072
5	0,036249	91,338
6	0,045624	98,656
7	0,046874	99,010
8	0,048124	99,169
9	0,053124	99,318
10	0,054374	99,245
11	0,069374	99,657
12	0,079374	99,957
13	0,080729	99,998
14	0,080730	99,998
15	0,080731	99,999

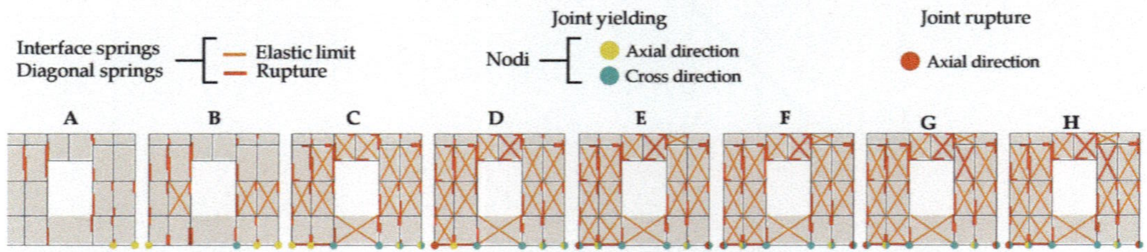

FIG. 7.20 Micromodeling: Experimental and numerical capacity curves (timber frame with masonry infill).

system, whose peculiar character is the masonry reinforced by timber frames, lays the foundations, for the principles pursued, for modern seismic engineering.

It, therefore, represents a significant chapter in the history of science that requires in-depth interdisciplinary knowledge, which includes, in addition to the historical and constructive aspect, analysis through mathematical models, adequately validated by experimental tests, aimed at reconstructing the dynamic behavior of buildings based on this constructive system.

On this purpose, detailed models have been used to carry out nonlinear analysis and rely on data deriving from an experimental campaign conducted on specimen in full scale, made on the basis of the Borbone constructive system.

A preliminary numerical approach was developed by means of the adaptation of the original software BrikWORK. The results obtained have provided an interpretation of the behavior of the wall very close, both quantitatively and qualitatively, to that of the experimental model. In fact, in terms of fracture and detachment, the crack pattern produced as the output of numerical analysis exhibited a significant similarity with the results of the experimental test.

A successive fine model has been developed, with the aim of simplifying the former by adopting a macromodel only for single portions of wall material. Also, in this case, the comparison between experimental and numerical results showed a good correspondence in terms of stiffness, strength, and ductility. As a matter of fact, this second model doesn't retrace the softening behavior of the shear wall, because no hysteretic behavior was defined for the nonlinear springs; however, it has allowed to retrace the progression of damage and to reconstruct the capacity curve of the timber frame and of the timber-framed masonry wall, objects of the laboratory experimentation.

Both models, now identified at the constitutive level, can allow the definition of even more simplified models that can be aimed at modeling an entire barracked building. But this will be the subject of future work.

References

Barucci, C. (1990). *La casa antisismica: prototipi e brevetti: materiali per una storia delle tecniche e del cantiere*. Gangemi.

Caliò, I., & Pantò, B. (2014). A macro-element modelling approach of infilled frame structures. *Computers and Structures, 143*, 91–107.

De bellis, M. L., & Addessi, D. (2011). A Cosserat-based multi-scale model for masonry structures. *International Journal for Multiscale Computational Engineering, 9*, 543–563.

Ferrigni, F., Della Pietra, A., & Sorrentino, M. (2013). Rafforzamento appropriato dell'edificato storico in zona sismica. Mid-term report in LAReHBA Project "Local Appropriate Retrofitting of Historial Built-up Areas" *Finanziato da EUR-OPA Major Hazards Program.*

Galassi, S., Dipasquale, L., Ruggieri, N., & Tempesta, G. (2018). Andalusian timber roof structure in chefchaouen, northern morocco: construction technique and structural behavior. *Journal of Architectural Engineering*, *24*(3), 04018012. 09/01; 2020/03.

Galassi, S., Ruggieri, N., Dipasquale, L., & Tempesta, G. (2018). Assessment of the Moroccan vernacular timber roof: a proposal for an eco-friendly strengthening system. *Journal of Architectural Conservation*, *24*(3), 224–248.

Galassi, S., Ruggieri, N., & Tempesta, G. (2015). Seismic performance evaluation of timber—framed masonry walls experimental tests and numerical modelling. In N. Ruggieri, G. Tampone, & R. Zinno (Eds.), *Historical earthquake-resistant timber frames in the mediterranean AreaCham* (pp. 95–103). Springer International Publishing.

Galassi, S., Ruggieri, N., & Tempesta, G. (2019). Ruins and archaeological artifacts: Vulnerabilities analysis for their conservation through the original computer program BrickWORK. In *Structural analysis of historical constructions* (pp. 1839–1848). Springer.

Galassi, S., Ruggieri, N., & Tempesta, G. (2020). A novel numerical tool for seismic vulnerability analysis of ruins in archaeological sites. *International Journal of Architectural Heritage*, *14*(1), 1–22.

Galassi, S., Ruggieri, N., Tempesta, G., & Zinno, R. (2014). Stability and stiffness contribution of the masonry in the borbone anti-seismic system. In: *Proceeding of 9th international masonry conference*.

Magenes, G., & Calvi, G. M. (1997). In-plane seismic response of brick masonry walls. *Earthquake Engineering and Structural Dynamics*, *26*(11), 1091–1112.

Masiani, R., Rizzi, N., & Trovalusci, P. (1995). Masonry as structured continuum. *Meccanica*, *30*(6), 673–683.

Mauri-Mori, G. (1909). *Riedificazione di Reggio Calabria dopo i terremoti del 1783: Appendice.* Nuova Antologia.

Meireles, H., Bento, R., Cattari, S., & Lagomarsino, S. (2012). A hysteretic model for "frontal" walls in Pombalino buildings. *Bulletin of Earthquake Engineering*, *10*(5), 1481–1502.

Milizia, F. (1781). *Principi di architettura civile.*

Pagano, M., & La Vega, F. (1997). *I diari di scavo di Pompei, Ercolano e Stabia di Francesco e Pietro La Vega: 1746–1810: raccolta e studio di documenti inediti. I diari di scavo di Pompei, Ercolano e Stabia di Francesco e Pietro La Vega.*

Pagano, E., Ruggieri, N., Zampilli, M., & Salerno, G. (2019). The borbone's system: A multiscale model for the building seismic analysis. In: *International conference on structural health assessment of timber structures (SHATIS19).*

Pagano, E., Salerno, G., & Zampilli, M. (2018). The baraccato anti-seismic constructive system: enhancement and preservation of the existing heritage in southern Italy. In: *HERITAGE 2018 - 6th international conference on heritage and sustainable development.*

Parisi, M. A., & Piazza, M. (2000). Mechanics of plain and retrofitted traditional timber connections. *Journal of Structural Engineering*, *126*(12), 1395–1403.

Pesso, L. (1876). *Sul consolidamento delle fabbriche nelle Calabrie contro i danni da terremoto.* Torino: L' ingegneria civile e le arti industriali.

Reccia, E., Leonetti, L., Trovalusci, P., & Cecchi, A. (2018). A multiscale/multidomain model for the failure analysis of masonry walls: a validation with a combined FEM/DEM approach. *International Journal for Multiscale Computational Engineering*, *16*(4).

Reitherman, R. (2006). Earthquakes that have initiated the development of earthquake engineering. *Bulletin of the New Zealand Society for Earthquake Engineering*, *39*(3). 09/30; 2020/03.

Ruggieri, N. (2013). Il sistema antisismico borbonico, muratura con intelaiatura lignea, genesi e sviluppo in Calabria alla fine del '700. *Bollettino degli Ingegneri*, *10*, 1–14.

Ruggieri, N. (2015). *L'ingegneria antisismica nel Regno di Napoli.* Roma: Aracne Editrice.

Ruggieri, N. (2016). Reliability of timber framed constructions in seismic prone areas in the practice codes issued from 18th to 19th. In *Historical earthquake-resistant timber frames in the mediterranean area* (2nd ed.).
Ruggieri, N. (2017). The borbone "istruzioni per gli ingegnieri": A historical code for earthquake-resistant constructions. *International Journal of Architectural Heritage, 11*(2), 292–304.
Ruggieri, N., Geremia, F., Pagano, E., Salerno, G., Stellacci, S., & Zampilli, M. (2016). The masonry timber framed load bearing structure of the Palazzo Vescovile in Mileto (Italy). In *Historical earthquake-resistant timber framing in the mediterranean area* (pp. 135–145). Springer.
Ruggieri, N., Tampone, G., & Zinno, R. (2015). In-plane versus out-of-plane "behavior" of an Italian timber framed system—The Borbone constructive system: Historical analysis and experimental evaluation. *International Journal of Architectural Heritage, 9*(6), 696–711.
Salerno, G., & De Felice, G. (2009). Continuum modeling of periodic brickwork. *International Journal of Solids and Structures, 46*(5), 1251–1267.
Salerno, G., & Felice, G. (2000). Continuum modeling of discrete systems: A variational approach. In: *European congress on computational methods in applied sciences and engineering, ECCOMAS 2000 01/01.*
Salerno, G., Geremia, F., Pagano, E., Zampilli, M., Ruggieri, N., & Stellacci, S. (2016). The masonry timber framed load bearing structure of "baraccato" system: a numerical model. In *Historical earthquake-resistant timber framing in the mediterranean area* (pp. 205–213). Springer.
Sarcone, M. (1784). *Istoria de'fenomeni del tremoto, avvenuto nelle Calabrie.*
Stellacci, S., Geremia, F., Pagano, E., Ruggieri, N., Salerno, G., & Zampilli, M. (2016). urban planning and building reconstruction of Southern Italy after the 1783 earthquake: The case of Mileto. In *Historical earthquake-resistant timber framing in the mediterranean area* (pp. 125–134). Springer.
Uva, G., & Salerno, G. (2006). Towards a multiscale analysis of periodic masonry brickwork: A FEM algorithm with damage and friction. *International Journal of Solids and Structures, 43,* 3739–3769.
Vivenzio, G. (1783). *Istoria e teoria de'tremuoti in generale ed in particolare di quelli della Calabria, e di Messina del 1783 di Giovanni Vivenzio.* Napoli: Stamperia Regale.

CHAPTER

8

Seismic vulnerability of historical constructions in Northern Southeast Asia (Indochina): Lessons learnt from recent earthquakes

Teraphan Ornthammarath[a], Saw Htwe Zaw[b], Nakhorn Poovarodom[c], and Yin Myo Min Htwe[d]

[a]Department of Civil and Environmental Engineering, Faculty of Engineering, Mahidol University, Nakhon Pathom, Thailand, [b]Myanmar Earthquake Committee, Yangon, Myanmar, [c]Department of Civil Engineering, Faculty of Engineering, Thammasat University, Bangkok, Thailand, [d]Department of Meteorology and Hydrology, Ministry of Transport and Communications, Nay Pyi Taw, Myanmar

8.1 Introduction

In the past decade, several moderate earthquakes ($Mw > 6.0$) that occurred in northern Southeast Asia (Indochina) caused significant loss attributed to building and infrastructure damage. Not only seismic damage to modern structures within epicentral area is commonly observed but also historical structure damage is quite common. However, the studies of these ancient structures due to earthquakes are quite limited because these heritage structures have no engineering design consideration. Without proper understanding of construction materials and techniques, the problems of cultural preservation are complicated and involve multidisciplinary consideration. Studies of the seismic vulnerability of historical buildings are needed to mitigate the associated hazard and to retrofit vulnerable parts are of great importance.

https://doi.org/10.1016/B978-0-12-821087-1.00007-7

In this context, an overview of historical structural damage that was observed due to 2016 Chauk earthquake in central Myanmar that took place 1 or 2 weeks after the earthquake is presented. After providing a general summary of the regional tectonic settings and seismicity, the observed ground motion is discussed, and the performance of heritage structures is evaluated.

8.2 Seismotectonic settings

On a continental scale, all earthquakes in Northern Southeast Asia (NSA) occur due to continuing collision process between two plate boundaries, the Eurasian and the Indian plates. As the Indian plate is colliding with Eurasian plate at a rate of 45 mm/year creating a counter-clockwise rotation, (Bilham, 2004), the Sunda Plate, comprising major parts of NSA, is moving eastward relatively to the Eurasian plate with a clockwise turning about 0.34-degree per million years (Simons et al., 2007). The Sunda Plate's western boundary is defined by active tectonic structures, including Arakan Thrust and Andaman Trench, as shown in Fig. 8.1. These north-south trenches have the capacity of generating gigantic events as the Mw 9.2, December 26, 2004, in Northern Sumatra. In addition, the dextral strike-slip Sagaing fault is believed to accommodate parts of the shear component of plate motion related to India-Eurasia convergence at 18 mm/year, according to Socquet et al. (2006). The greatest instrumental earthquake magnitude in NSA is the 1946 Mw 7.7 earthquake generating along the Sagaing fault. Almost all fault plane solutions along Sagaing fault exhibit dextral strike-slip faulting (e.g., November 11, 2012, Mw 6.8, Thabeikkyin earthquake).

The other main tectonic features in NSA due to the collision of the India and Eurasian plates result in predominantly NW-SE and NE-SW strike-slip faults. Some important and well-defined active faults are capable of producing earthquake greater than 7.0 based on fault length in the northern NSA, including the Dien Bien Phu fault, Three Pagoda fault zones, Red River fault, etc. For Northwestern Myanmar, the seismicity intensifies farther to the north of Myanmar, associated within subducting slab down to 150 km depth. Most intermediate and deep earthquakes have strike-slip or normal focal mechanism eastward, oriented down the dip of the seismic zone (e.g., Fitch, 1970). These tremors then occur within the Wadati-Benioff zone, representing its slab pull or negative buoyancy. The largest earthquake in this zone is the 1988 Mw 7.2 earthquake at 100 km depth. However, intraslab earthquakes are not normally observed below 18°N latitude, though intermediate and deep tremors have been shown to reappear at 15°N latitude propagate along the Andaman trench, and along the deformation from Nias Island north through the Andaman Islands.

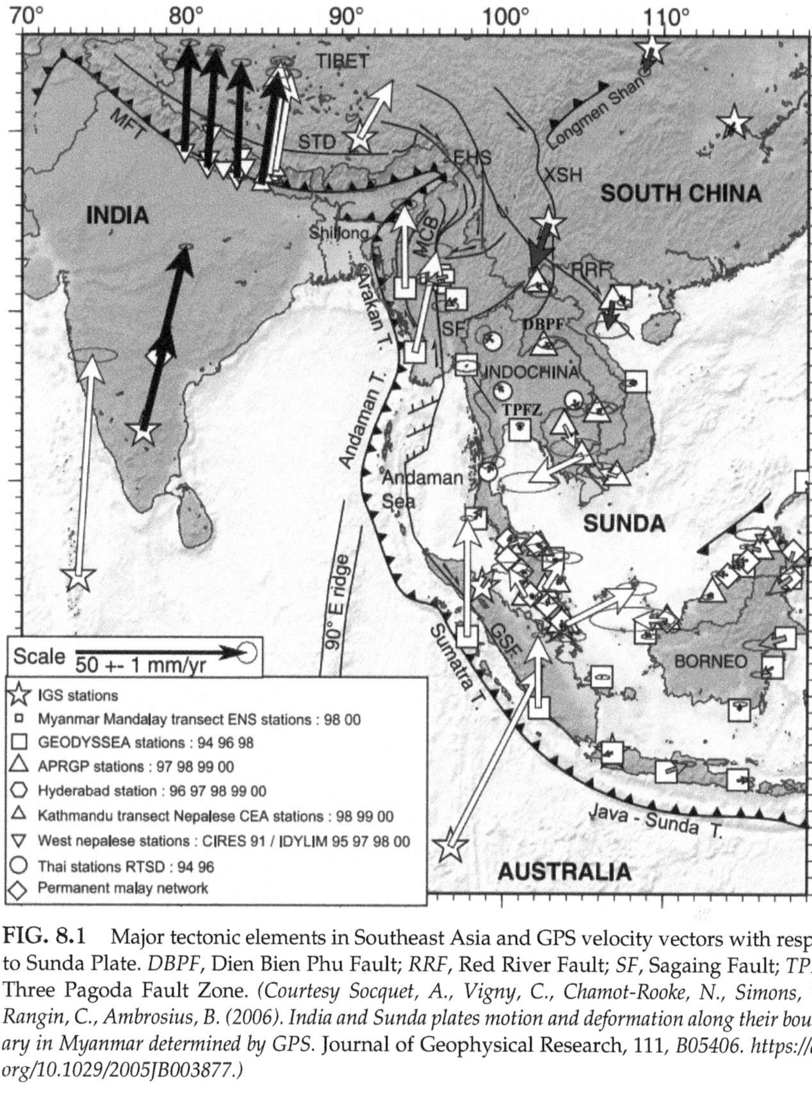

FIG. 8.1 Major tectonic elements in Southeast Asia and GPS velocity vectors with respect to Sunda Plate. *DBPF*, Dien Bien Phu Fault; *RRF*, Red River Fault; *SF*, Sagaing Fault; *TPFZ*, Three Pagoda Fault Zone. *(Courtesy Socquet, A., Vigny, C., Chamot-Rooke, N., Simons, W., Rangin, C., Ambrosius, B. (2006). India and Sunda plates motion and deformation along their boundary in Myanmar determined by GPS.* Journal of Geophysical Research, 111, *B05406. https://doi.org/10.1029/2005JB003877.)*

8.3 The August 24, 2016, Mw 6.8 Chauk earthquake

The kingdom of Pagan was one of the first empires in Indochina from 9th to 13th centuries that later on unify the modern-day Myanmar. Its capital city is located in Bagan, and during the peak period, Bagan's kings and their upper-class citizen constructed over 10,000 Buddhist temples and pagodas, with around 2000 religious monuments remain in presence. Located in central Myanmar at 160 and 200 km from Mandalay and Nay

Pyi Taw, respectively, the last historical structural damage in Bagan due to 1975 earthquake was reported here due to many ancient monuments within old Bagan city with an area of 100 km^2 along the Irrawaddy River. Based on historical stone inscriptions, Bagan suffered several historical earthquake damages (Swe, 2015). The 1975 Mw 6.8 event is the biggest tremor that occurred near Bagan in the past 100 years and is the most damaged one until 2016 event. Based on available seismicity, seismicity in and around Bagan is mainly controlled by intermediate and deep earthquakes.

The Mw 6.8 event on July 8, 1975, was located about 34 km northwest of Bagan with hypocentral depth of 100 km. More than 70% of historical religious monuments were affected, with at least 15 collapsed pagodas (Shwe Gaig Tha, 1976). Afterward, on August 24, 2016, at 10:34:54 UTC, an Mw 6.8 earthquake occurred at 85 km focal depth, 26 km west to Chauk, a rural town of 180,000 inhabitants in the Magway Division, Myanmar. This earthquake was widely felt throughout Myanmar, as well as people in high-rise buildings in Calcutta and Bangkok (at epicentral distance of 500 and 1000 km, respectively). The event took place in an area previously characterized as moderate seismic hazard. The seismic hazard of Bago and Chauk has been estimated with estimated peak ground accelerations around 0.2 g for 475-year return periods at bedrock conditions (Shedlock, Giardini, Grünthal, & Zhang, 2000). The 2016 event was located 50 km southern of the 1975 event. The rupture length of the 2016 event would then occur on different rupture zone compared to the one in 1975. The fault plane solution of the 1975 event shows that it resulted from in-slab event with strike-slip focal mechanism (Storchak et al., 2013). In contrast, the fault plane solution of the 2016 tremor shows that it resulted from in-slab event with thrust focal mechanism. The strongest aftershocks occurred on August 24, 2016, at 10:57:16 UTC ($mb = 4.5$), located 10 km east to the mainshock with 92 km focal depth. Because local buildings in epicentral area are largely not damaged, and situation in epicentral area appeared to return to normal after 2 weeks after the main earthquake, a significant amount of damage occurred only to a very large number of ancient monuments.

Ground motion records were obtained from the Department of Metrology and Hydrology (DMH), Incorporated Research Institutions for Seismology (IRIS), and the Thai Metrological Department (TMD) (Fig. 8.2). Different seismic instruments were used to record strong ground motion. All records were processed by applying the standard zero-order correction. Each recorded ground motion is visually inspected for obvious errors or multiple earthquakes. From August 24, 2016 event, the nearest accelerograph station is in Nyaung-U Seismic Station, located at a 48 km distance from 2016 epicenter. At this station, the observed PGA in NS, EW, and UD reach 0.08, 0.08, and 0.12 g, respectively. Fig. 8.3 displays the acceleration

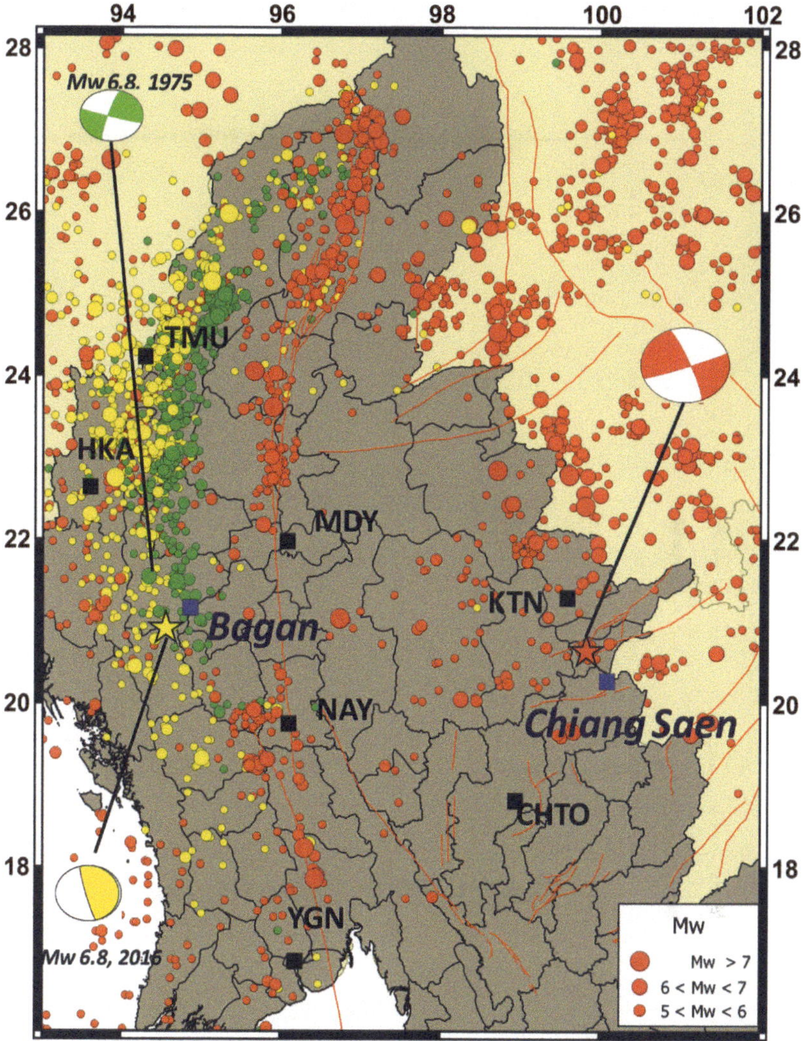

FIG. 8.2 Location of ancient cities discussed in this study (Bagan in *blue squares*). Shallow earthquakes (*red circles*, depth < 50 km), intermediate (*yellow circles*, 50 < depth < 100 km), and deep earthquakes (*green circles*, depth > 100 km). The *stars* represent damaging earthquakes discussed in this study. *Black rectangles* show recorded seismic stations, and the known active faults are displayed in *red* color lines.

time histories, and Fig. 8.4 shows their 5% damped elastic response spectra at Nyaugn-U station.

The acceleration time histories contain two separated parts that could be clearly identified. The initial part of the acceleration trace corresponds to high-frequency body-wave phases, and the long period waves begin to

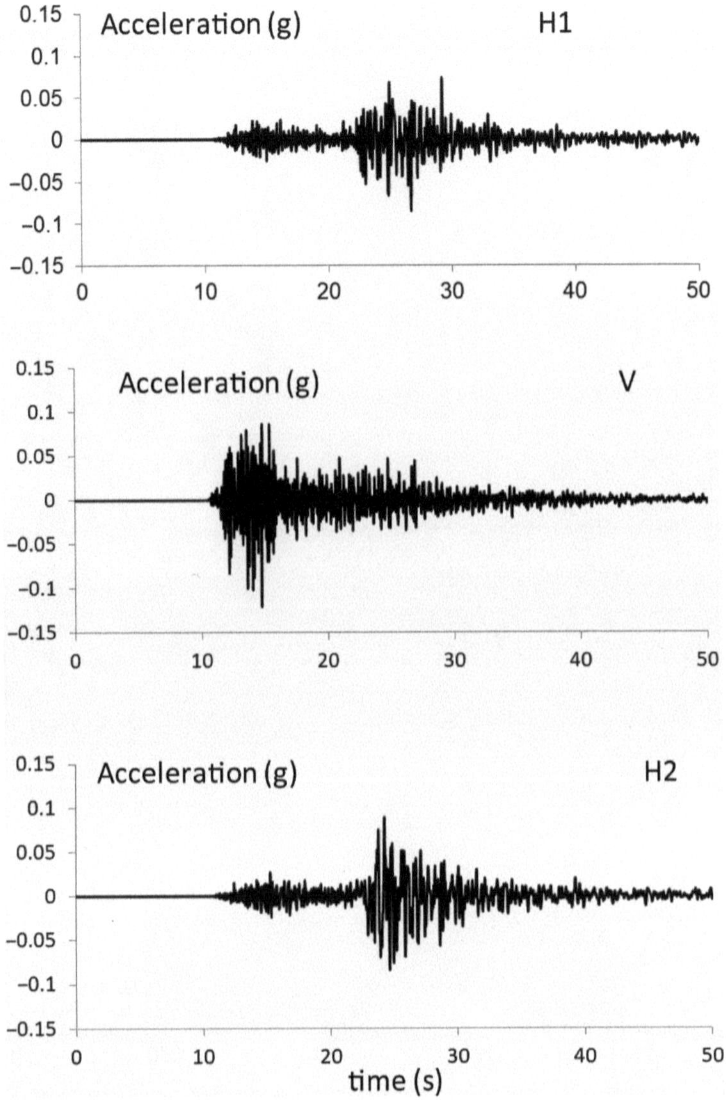

FIG. 8.3 Recorded accelerograms at Nyaung-U seismic station from Mw 6.8 August 24, 2016 event with an epicentral distance of 48 km.

arrive. At short vibration periods, there is pronounced difference between vertical and horizontal accelerations due to the effect of body-wave. For moderate periods ($0.5 < T < 1.0$ s), the horizontal spectra contain large energy, which could explain severe damage to moderate to large ancient monuments in Bagan. In addition, based on intensity data, the August

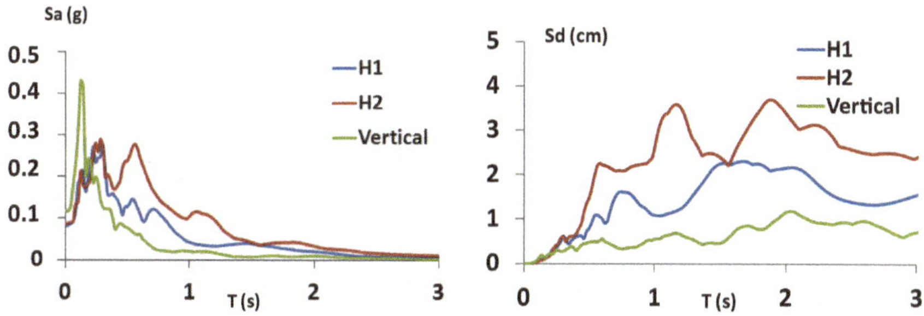

FIG. 8.4 The 5%-damping horizontal and vertical acceleration and displacement spectra at Nyaung-U seismic station from Mw 6.8 August 24, 2016 event with epicentral distance of 48 km.

24, 2016 event could widely be observed in Rangoon, Calcutta, and Bangkok with hypocentral distance of 493, 672, and 1000 km, respectively, as shown in Table 8.1.

Although large, vertical peak ground acceleration reported at the Nyaung-U station is a typical characteristic of intraslab ground motion recorded within epicentral region due to P-wave. At longer distance, the horizontal PGA values recorded at Nay Pyi Taw and Mandalay stations become larger than those recorded in vertical direction, as shown in Fig. 8.5. The Mandalay station has a lower PGA than Nay Pyi Taw station, given similar distances, as shown in Table 8.1. This is possibly because Mandalay station is located at the basement of Mandalay hill. Because of the availability of the modified Mercalli intensity (MMI) data, the reported felt intensity data from public are also provided based on

TABLE 8.1 List of strong motion stations recorded mainshock and aftershock of Mw 6.8 Chauk earthquake on August 24, 2016.

Station	Lat	Long	Focal distance (km)	Vertical PGA (g)	Geometric mean PGA (g)	MMI
Nyaung-U	21.20	94.91	97	0.12	0.08	5
Mandalay	22.02	96.11	195	0.002	0.003	3
Naypyitaw	19.75	96.11	217	0.02	0.03	5
Haka	22.64	93.60	230	0.002	0.003	3
Tamu	24.23	94.30	365	0.0005	0.001	3
Yangon	16.86	96.15	487	0.001	0.003	3
Keng Tung	21.29	99.59	529	0.001	0.001	2
CHTO	18.81	98.94	461	0.004	0.005	3

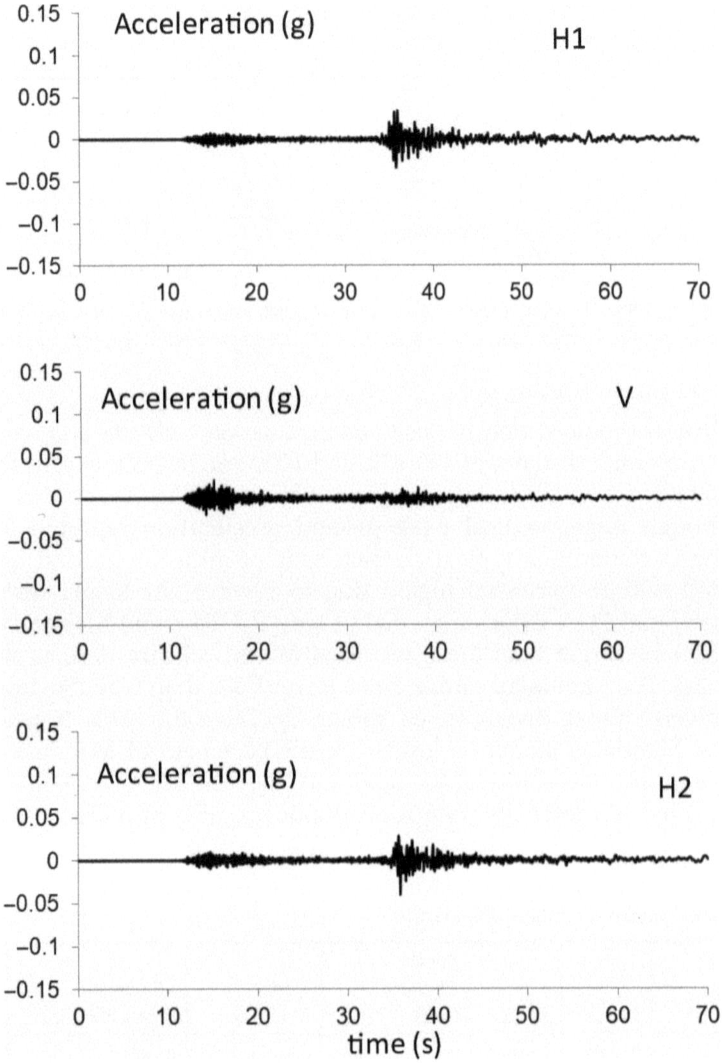

FIG. 8.5 Recorded accelerograms at Nay Pyi Taw seismic station from Mw 6.8 August 24, 2016 event with epicentral distance of 200 km.

local media, which are summarized in Table 8.1. Moreover, following improving seismic networks inside Myanmar since 2016, the recorded ground motion and MMI data from Mw 6.9 April 13, 2016 Kani intraslab earthquake ground motion is also reported in Table 8.2. In Fig. 8.6, the comparison between MMI intensity and PGA based on two intraslab earthquake ground motions is shown, with the relationship between MMI or EMS-98 with PGA developed by Worden, Gerstenberger,

TABLE 8.2 List of strong motion stations recorded mainshock and aftershock of Mw 6.9 Kani earthquake on April 13, 2016.

Station	Lat	Long	Focal distance (km)	Vertical PGA (g)	Geometric mean PGA (g)	MMI
Nyaung-U	21.20	94.91	97	0.015	0.01	2.5
Mandalay	22.02	96.11	195	0.001	0.001	3
Naypyitaw	19.75	96.11	217	0.001	0.001	2
Haka	22.64	93.60	230	0.020	0.05	2.5
Tamu	24.23	94.30	365	0.014	0.03	3
Yangon	16.86	96.15	487	0.004	0.002	3
Keng Tung	21.29	99.59	529	0.0003	0.0006	–

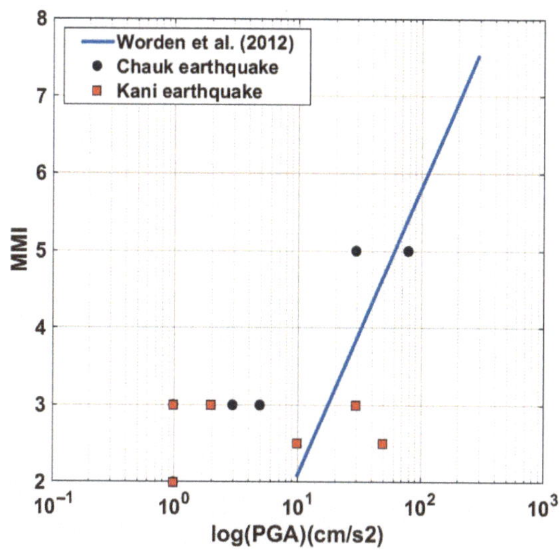

FIG. 8.6 Observed PGA and MMI Intensity plot for Mw 6.8 August 24, 2016 Chauk and Mw 6.9 April 13, 2016 Kani earthquake events.

Rhoades, and Wald (2012). It could be observed that Worden et al. (2012) model also provide a relatively good correlation for intraslab events.

8.4 Observed damage of heritage structures

The earthquake caused extensive damages to Bagan historical monuments, but the effects were seen to vary for each structure. An overall description of monuments in Bagan can be classified generally into two types. The first one is the solid pagoda (or stupas) structures, such as Shwezigon Pagoda, which is a dome-shaped structure. In contrast to the stupas, the second type is the masonry arch structures, which transfer their own gravity load through the arch. The example of these ancient monuments is Thatbyinnyu Temple. The construction materials of these heritage monuments are made with bricks and earthen plaster, and they have limited earthquake-resistant designs built from the 9th to 13th centuries. Severe crack or collapse of top part is the usual damage pattern for both types of these structures. This upper portion of these heritage structures can be significantly damaged even for low-intensity ground shaking due to the dynamic amplification of the ground motion, depending upon different structural characteristics. The collapse of top bell can also be the source of additional damage to other structural parts. This situation was observed in Sularmani temple. Overall, the majority of these heritage structures had been repaired following seismic damage due to the 1975 earthquake, although it is quite common to observe the same pattern of damage, implying the lack of proper detailing, which increases seismic vulnerability to some ancient heritages.

The field survey had been carried out with five groups; within each group, there is at least one local civil engineer to inspect ancient monuments and damaged buildings in Bagan Nyaung U and Chauk areas. The Department of Archeology (DOA) also assigned an official staff to each team to coordinate the works and identify the monuments and buildings in the affected area. To acquire the damage data properly, the rapid visual survey form has been used. During field survey, more specific damage patterns are acquired, including observed damage at the foundations, walls, below top bell, or above top bell. The survey form is originally developed based on FEMA P-154 with some modification to reflect real application for ancient structures, as shown in Table 8.3. Baseline data for each monument are collected from information provided by DOA staff, including building height, construction period, previous retrofitting method, etc.

Based on initial damaged monument list, the survey groups assessed 270 monuments during the period from 3rd to 6th September 2016. The assessment is based on damage level of primary and secondary structural systems, and the results of each monument were identified as follows: red tag for severely damaged primary and secondary structural system monuments, yellow tag for moderately damaged monuments, and green tag for lightly damaged monuments. Consultations and discussions for immediate emergency measures had also been carried out occasionally to prevent consequential damages due to aftershocks and rainwater penetration, together with engineers from Ministry of Construction and

TABLE 8.3 Rapid ancient monument damage survey form adopted from FEMA P-154.

RVS for Seismic Damage Assessment Adapted FEMA P-154 Data Form	High Seismicity	Group:
	Name:	Date:

Photo, Sketch

Identity	
Pagoda ID:	....
Name:	....
Latitude:	....
Longitude:	....

	Form					
F1	Type:	1- စေတီ	2- ဂူဘုရား	3- အခြား		
F2	Built:	1- 10 CE	2- 12 CE	3- 14 CE	4- >20 CE	
F3	Last Reno:	1- 5 yr	2- 10 yr	3- 20 yr	4- >50 yr	
F4	Dimension:	1- <10'	2- 10' to 20'	3- 20' to 50'	4- >50'	
F5	Plan:	1- Circle	2- Square	3- Rectglr	4- Other	
F6	no. of Story:	1- 1	2- 2	3- ≥3		
F7	Center Piller:	1- Yes	2- No	3- DNK		
F8	Wdth/Heigt:	1- ≤1:1	2- 1:1	3- ≥ 1:1		
	Hazard					
H1	Ground Level	1- မြေမညီ	2- မြေညီ	3- မြေအောက်	4- မသိ	
H2	Soil	1- ကျောက်	2- မြေမာ	3- မြေပျော့	4- မသိ	
	Damage					
D1	Foundation	1- မထိခိုက်	2- နဲနဲထိ	3- အထိများ	4- အလွန်ထိ	5- ပြိုကျ
D2	Walls	1- မထိခိုက်	2- နဲနဲထိ	3- အထိများ	4- အလွန်ထိ	5- ပြိုကျ
D3	Below Bell	1- မထိခိုက်	2- နဲနဲထိ	3- အထိများ	4- အလွန်ထိ	5- ပြိုကျ
D4	Above Bell	1- မထိခိုက်	2- နဲနဲထိ	3- အထိများ	4- အလွန်ထိ	5- ပြိုကျ
P1	Primary Str	1- မထိခိုက်	2- နဲနဲထိ	3- အထိများ	4- အလွန်ထိ	5- ပြိုကျ
P2	Secondry Str	1- မထိခိုက်	2- နဲနဲထိ	3- အထိများ	4- အလွန်ထိ	5- ပြိုကျ
P3	Posting	1- Green	2- Yellow	3- Red		

COMMENT

BUILDING TYPE	W1	W2	C1	C2	C3	URM	RM1	BN1
Basic Score	3.6	2.9	1.5	2.0	1.2	1.0	1.7	0.8
Severe Vertical Irregularity, V_{L1}	-1.2	-1.2	-0.9	-1.0	-0.7	-0.7	-0.9	-0.6
Moderate Vertical Irregularity, V_{L1}	-0.7	-0.7	-0.5	-0.6	-0.4	-0.4	-0.5	-0.3
Plan Irregularity, P_{L1}	-1.1	-1.0	-0.6	-0.8	-0.5	-0.4	-0.7	0.0
Pre-Code	-1.1	-0.9	-0.4	-0.7	-0.1	0.0	-0.5	NA
Post-Benchmark	1.6	2.2	1.9	2.1	NA	NA	2.1	NA
Hard Soil	0.1	0.5	0.4	0.5	0.3	0.3	0.5	0.0
Soft Soil (1-3 stories)	0.2	0.1	0.0	0.0	-0.2	-0.2	-0.1	0.0
Soft Soil (>3 stories)	-0.3	-0.9	-0.5	-0.7	-0.3	-0.2	-0.5	NA
Minimum Score, S_{MIN}	1.1	0.7	0.3	0.3	0.3	0.2	0.3	0.2
FINAL LEVEL 1 SCORE								

Department of Archeology. The monuments with high cultural values and well-known structures were given priority and emphasized to minimize the damages. According to survey results, there were 50 monuments in red damage, 96 monuments in yellow damage, and 124 monuments in green damage.

From Table 8.4, it is quite clear that the heaviest damaged historical monuments are, as expected, large and medium monuments (base width greater than 12 m). These 36 ancient temples (1.5% of around 2000 pagodas in Bagan) are classified as heavy and moderate damage (red

TABLE 8.4 Damage statistic reported from field survey due to 2016 Chauk earthquake.

	Red	Yellow	Green	Total
Large	18	31	21	70
Medium	18	43	64	125
Small	14	22	39	75
Total	50	96	124	270

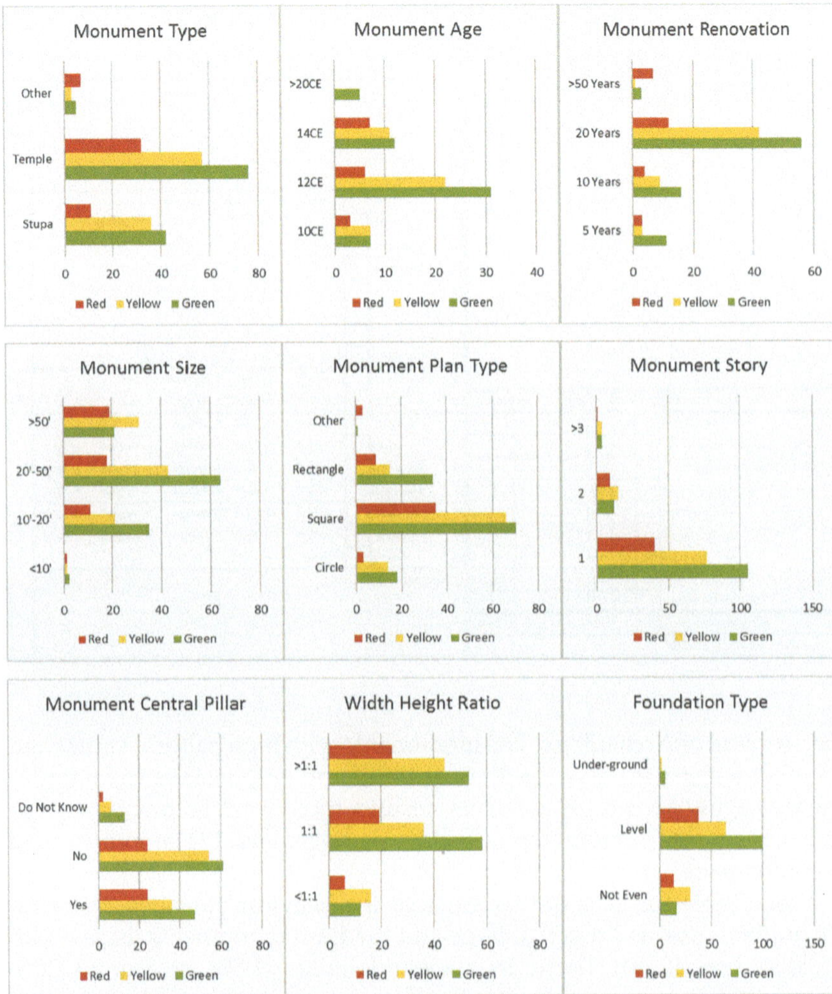

FIG. 8.7 Detail ancient monument damage based on different structural characteristics.

and yellow classes). In addition, it could be clearly observed that the larger the height per width ratio, the higher numbers of red and yellow damage ancient monuments could be expected, as shown in Fig. 8.7. In contrast to the 1975 event, there is no report of collapsed pagoda from 2016 Chauk earthquake. In addition, Fig. 8.8 displays spatial distribution of damaged pagoda (red, yellow, and green tags), where it could be noticed that the distribution of red and yellow tag pagodas in Old Bagan area uniformly distributed and extended entirely across the city, from Irrawaddy River to the north into the flat plain in the south.

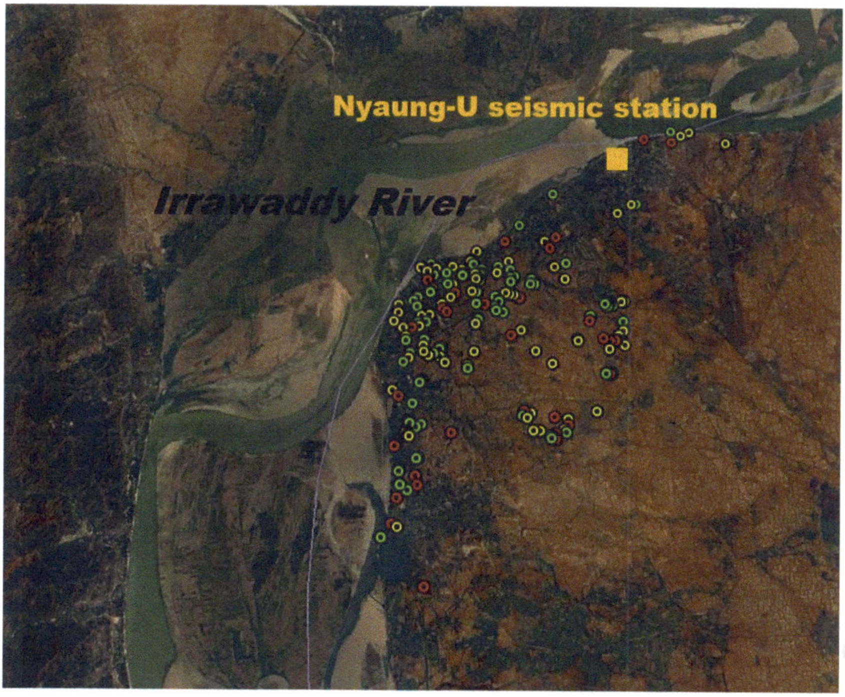

FIG. 8.8 Different damage states of ancient monuments in Bagan (*green*, *yellow*, and *red* tags) from Mw 6.8 earthquake on August 24, 2016, following structural assessment by MEC.

In addition to damage statistics of ancient monuments, damage description and basic information of six important monuments, visited during the earthquake reconnaissance, are provided in Table 8.5. The information included name, construction period, structural type, height, number of stories, etc. To complement the data presented in Table 8.4, additional details are provided in the following section with respect to the damages that were observed in some of these monuments.

The first moderate damage ancient monument is Thubyinnyu temple, as shown in Fig. 8.9. It is a four-storied temple with a large solid core constructed in AD 1144. The height of this temple is 61 m, considered as the highest historical monument in Bagan. The solid core, approximately 44 by 44 m, is designed to bear the loads from top structural part. All four sides of the walls at the base of the "Kundaung," the third story, were slanted outward with large cracks. The arched halls below those walls were also damaged with peeling of plasters and stuccos and with large cracks at the crown of the arched roof. In some part, a small portion of bricks on the roof was even pushed down about 5–7 cm. A similar damage pattern was also found in Dhammayangyi and Htilominlo temples.

TABLE 8.5 Baseline data for survey ancient monuments.

Name	Thatbyinnyu	Htilominlo	Sularmani
Damage state	Yellow	Yellow	Red
Structural type	Arch and masonry solid core load-bearing system	Arch and masonry solid core load-bearing system	Arch and masonry solid core load-bearing system
Construction period	12th century	13th century	12th century
Height (m)	61	46	55
Ground floor solid core (m)	44	37	33
Height/ width ratio	>1.0	1.0	>1.0
Total base width	55	65	54
Number of stories	4	2	2

Mostly, the structure at ground floor was basically unharmed. The intervention works of reinforced concrete corner ties installed after 1975 earthquake were intact and had worked very well noticeably.

Among the visited ancient monuments, Sulamani is the most severed damaged one due to the collapse of the top bell and extensive damage to the roof in the second story, as shown in Fig. 8.10. Because the construction system of this structure was seen to be similar to that of Thatbyinyu and Htilominlo Temples, the level of observed damage was surprisingly high. The difference between observed damage of these three temples might be because the vertical irregularity of Sulamani temple, which differs from the more regular shape of Thatbyinyu temple. Additional factor that might be able to explain the differences in observed damage is the slenderness ratio of Sulamani temple (height per width of solid core) is about 1.6 (55/34), which differs from the slenderness ratio of other monuments (e.g., 1.4 for Thatbyinyu temple and 1.3 for Htilominlo temple). The collapse of top part seems to be similar to damage patterns that have been observed in the previous Mw 6.8 in 1975. Based on local interview, the upper part "Kundaung" was reinstalled in the 1990s with reinforced concrete, and that upper part was fallen down on the roof of the upper terrace, making a large hole at the top. The viewing stair and railing walls

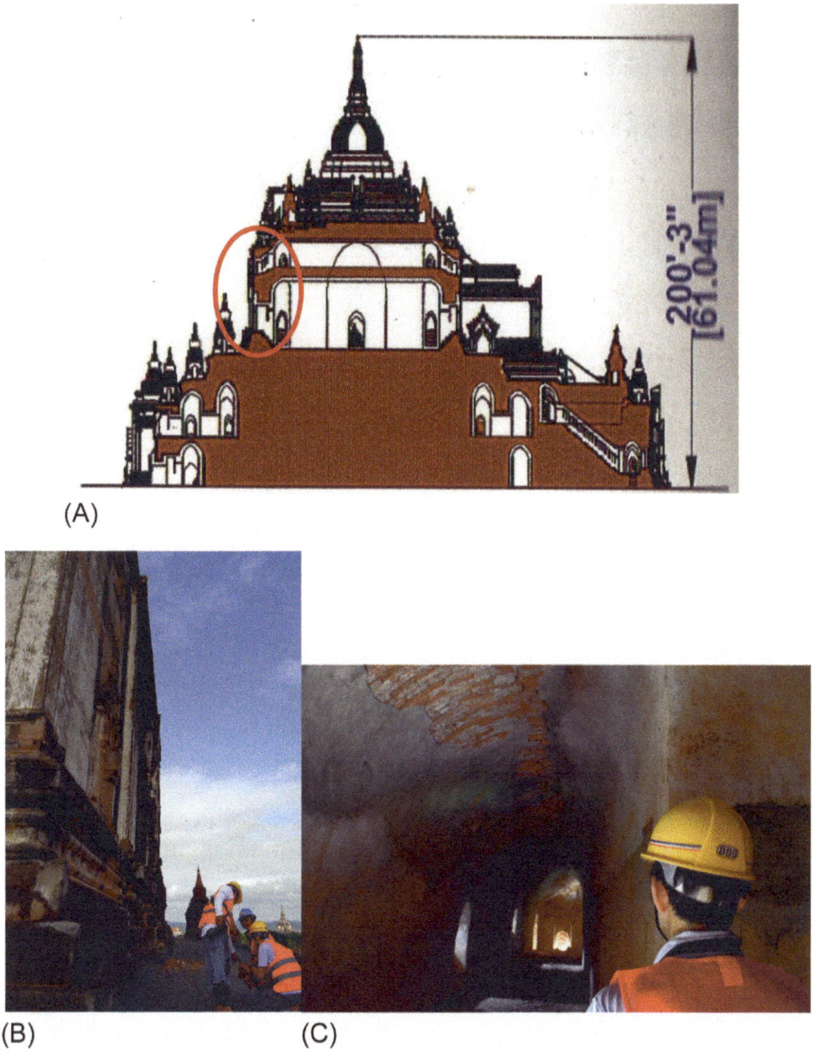

FIG. 8.9 (A) Cross-section of the Thatbyinnyu temple, (B) observed wall damage in the third story due to high acceleration on top part *(red circle)*, and (C) concentration of cracking at the central part of masonry arch in the fourth story.

were also damaged significantly as consequences. However, the base of the Pagoda and structure below the upper terrace were intact and strong despite the large impact of falling upper part. At the time of the reconnaissance survey, the team was not able to get inside only this temple due to the ongoing restoration operation.

FIG. 8.10 Damages to Sula Mani pagoda, including collapse of "Kundaung" and severe damage to the eastern side roof at the second story due to fall of top part.

8.5 Conclusion

The 1975 earthquake and Chauk event in 2016 were both subduction intraslab earthquakes with hypocentral depths at 100 and 85 km, respectively. Based on available records, the 2016 Chauk earthquake seems to show considerably less damage compared to the 1975 event, which might be because some retrofitting have been implemented from local authorities and foreign experts. However, same damage patterns reported after the 1975 event to several importance heritage monuments could still be observed after the 2016 earthquake, which could indicate improper retrofitting and strengthening methods. The renovation and rebuilding efforts made during the 1990s have several weaknesses, such as adding extra loads without strengthening the supporting existing structure below, no proper bond between new addition and old structure, and so on. Therefore it is obvious that these ancient masonry monuments are much vulnerable than normal buildings even at low ground shaking intensity due to several weaknesses pointed out in this study.

References

Bilham, R. (2004). Earthquakes in India and the Himalaya: Tectonics, geodesy and history. *Annals of Geophysics*, *47*(2), 839–858.

Fitch, T. J. (1970). Earthquake mechanisms in the Himalayan, Burmese, and Andaman regions and continental tectonics in Central Asia. *Journal of Geophysical Research*, *75*, 2699–2709.

Shedlock, K. M., Giardini, D., Grünthal, G., & Zhang, P. (2000). The GSHAP global seismic hazard map. *Seismological Research Letters, 71*(6), 679–689.

Simons, W., Socquet, A., Vigny, C., Ambrosius, B., Haji Abu, S., Promthong, C. , et al. (2007). A decade of GPS in SE Asia: Resolving Sundaland motion and boundaries. *Journal of Geophysical Research. 112*, B06420 https://doi.org/10.1029/2005JB003868R.

Socquet, A., Vigny, C., Chamot-Rooke, N., Simons, W., Rangin, C., & Ambrosius, B. (2006). India and Sunda plates motion and deformation along their boundary in Myanmar determined by GPS. *Journal of Geophysical Research. 111*, B05406 https://doi.org/10.1029/2005JB003877.

Storchak, D. A., Di Giacomo, D., Bondár, I., Engdahl, E. R., Harris, J., Lee, W. H. K. , et al. (2013). Public release of the ISC-GEM global instrumental earthquake catalogue (1900–2009). *Seismological Research Letters, 84*(5), 810–815. https://doi.org/10.1785/0220130034.

Swe, T. L. (2015). *Historical earthquakes of Myanmar 12th annual meeting Asia Ocean Geosciences Society, 2nd to 7th August 2015, Singapore.* .

Tha, S. G. (1976). *Historical earthquakes and the Bagan earthquake.* Yangon, Myanmar: Sanay Literature Group. in Burmese.

Worden, C. B., Gerstenberger, M. C., Rhoades, A. D., & Wald, D. J. (2012). Probabilistic relationships between ground-motion parameters and modified Mercalli intensity in California. *Bulletin of the Seismological Society of America, 102*, 204–221. https://doi.org/10.1785/0120110156.

CHAPTER

9

Structural characteristics of Nepalese pagoda temples

Manjip Shakya[a] *and Jeevan Gwachha*[b]

[a]Khwopa Engineering College, Bhaktapur, Nepal,[b]Lecturer Kathford International College of Engineering and Management, Lalitpur, Nepal

9.1 Introduction

Nepalese pagoda temples are often referred to as tiered temples. Built as structures reserved for religious or spiritual activities, they began to appear extensively around the middle of the 14th century during the Malla Dynasty. The uniqueness of such temple structures in comparison to other traditional masonry structures is due to their extensive wall thickness, multitier roof, regular configuration, and high plinth section. These temples were constructed using brick masonry and timber members with tiles or metal roof coverings.

Most of the cultural heritage site monument buildings and structures are constructed using an unreinforced brick masonry wall structural system. These structures were supposed to be built using the best materials available and with the best workmanship and knowledge of that era (Ranjitkar, 2006). However, the basic construction materials used, such as bricks and mud mortar as the binding material, are known to have limited strength, making these structures vulnerable to earthquake excitation.

On 25 April 2015, the Gorkha earthquake of magnitude 7.8 severely damaged a number of Nepalese cultural heritage sites. In the Kathmandu Valley and other districts, out of 750 damaged monuments, including temples, chaityas, stupas, sattal, shrines, palace complexes, etc., 133 have completely collapsed, 95 partially collapsed, and 522 were slightly damaged. In particular, the seven monument zones of the Kathmandu Valley UNESCO World Heritage Site suffered extensive damage, placing

https://doi.org/10.1016/B978-0-12-821087-1.00010-7

the "Outstanding Universal Value" criteria of the property at risk. Out of 195 surveyed monuments, 38 completely collapsed and 157 were partially damaged. The Hanumandhoka Durbar Square Monument Zone has been most affected: out of 39 damaged monuments, 12 completely collapsed and 27 were partially damaged. The Swayambhu Monument Zone also suffered major damage to smaller temples around the stupa including surrounding buildings: out of 13 damaged monuments, 4 completely collapsed and 9 were partially damaged (DoA Report, 2015).

Typical damage observed in the heritage structures included collapse of gable walls, collapse of the top floor roof, heavy tensile shear cracks to masonry walls, out-of-plane collapse of masonry walls, collapse of the outer layer of masonry walls, failure of the timber connection, and brittle failure of the structures leading to total collapse. It has been reported that lack of quality monitoring, periodic repair, and maintenance/restoration, as well as some mistakes made during previous restoration projects, were the primary causes of the extensive damage to the historic structures (Shrestha, Shrestha, Shakya, & Maskey, 2017).

This chapter presents the research that has been carried out on the Bhairavnath Temple, located in the Bhaktapur district, to determine the vulnerability of the temple by evaluating its seismic capacity. For this, three-dimensional models were developed using shell elements to investigate the (a) responses, displacements, and stresses of the structure; (b) different dynamic properties of the masonry temple; (c) different failure modes of the masonry temple; and (d) weak elements of the temple under earthquake excitation. Seismic capacity evaluation was performed using the seismic coefficient method and response spectrum method. The results show that the most critical parts are the piers between the openings and most of the failure modes are associated with tensile and shear stresses.

9.1.1 Pagoda structural system

The historic temples are tiered with multiple overhanging roofs in the pagoda style. Pagoda temples were constructed to express religious and cultural beliefs. Although there are different construction styles of pagoda temples, similar structural concepts maintain the overall shape, construction technologies, and materials. Pagoda style temples have distinct characteristics compared with other Nepalese temples, which can be summarized as (Shakya, Varum, Vicente, & Costa, 2014):

- Square large plinth to limit shear forces at foundation,
- Large wall thickness composed of mud-masonry brick,
- Walls constructed of three layers with decreasing story heights,
- Heavy timber frame elements embedded in wall to support the roof deadweight,

- Multitiered roof levels with timber overhangs to protect the masonry walls from heavy rain and sun,
- Inclined timber struts to support the roof and transfer the load from purlins to the masonry walls, and
- Timber roofs covered by tiles or metal for protection.

Pagoda-style temples consist of four essential components: masonry walls, roof system, timber members, and foundation.

9.1.1.1 Foundation

The foundation of the tiered temple is as wide as the plinth platform itself, appearing as a masonry mat. If existing old temples are observed, it can be seen that the plinth mat is built directly off the ground level or on a thin brick soling (Tiwari, 2009). However, the temples with shallow plinths rise from some depth below ground. The temples with a massive plinth in a pyramid shape yield more complexity. These huge plinth-based massive foundations will benefit from eliminating the earthquake risks associated with soft soils (Nienhuys, 2003).

9.1.1.2 Masonry wall

The masonry wall is the most important structural member, as it has the crucial job of resisting vertical gravity loads and lateral loads, such as an earthquake load. Masonry walls used in the Nepalese temple have their own thickness characteristic, which varies from 50 cm to 1.5 m, depending on the number of stories. The thickness of the wall decreases with increase in the floor level of the temple. (Shakya et al., 2014). The Nepalese temple is mainly constructed in three layers of sun-dried bricks and fire-burnt bricks with mud mortar. The outer layer is constructed with fire-burnt bricks to protect it from exposure to the sun and other climatic conditions. The inner portion of the wall is constructed using sun-dried bricks, as they have the most limited exposure to environmental impacts (Thapa, 2011). The middle layer is constructed with random rubble, brick pieces, and mud mortar; thus it is the weakest part and the most vulnerable to earthquake loads.

9.1.1.3 Roof system

Nepali temples are distinguished by the unique architectural design of the roof system, known as a pagoda. The pitched roof originates radially from the center of the system and slants in all directions, supported by the overhanging eaves. Pagoda temples have a single-, double-, triple-, or five-tiered roof. The main reason for this design of the pagoda roofs is to reduce the effects of climate, including both monsoon rains and extreme sun exposure. The symmetrical pitches of the temple roofs are made up of small rafters stemming from each corner in a radial configuration to the central point of the structure (Bonapace & Sestini, 2003). The inner rafters are connected to the primary structural masonry walls via beams, while the outer rafters are banded with purlins supported by wooden struts. The inclined timber struts play an essential role in distributing the roof loads to the wall. All of the wooden elements are held together using wooden wedges (Shakya et al., 2014; Shakya, Varum, Vicente, & Costa, 2012).

9.2 Structural analysis: A case study of Bhairavnath Temple

9.2.1 History

The Bhairavnath Temple is situated at the east of Taumadhi Square in Bhaktapur. The site of the Bhairavnath Temple dates back more than 1343 years, according to the inscriptions found in different periods. A Lichchhavi inscription is found on the temple premises that states that in 88 B.S. (Bikram Sambat) on the day of Poush Shukla Tritiya, a Tamrakalash[a] was offered in the name of Lalitmaheshwor (Raj, 2061 B.S.). The inscriptions and their details are published in different issues of a monthly magazine called *Pasuka*. This shows the temple may have existed from the Lichchhavi period. The origin of the temple is not clearly known, but some available inscriptions mention later additions and reconstruction and renovation work.

9.2.2 Geometrical characteristics

The Bhairavnath Temple, located at the center of Bhaktapur Municipality in Taumadhi Square, has become a major attraction for visitors and is listed by UNESCO. This is a pagoda temple of Lord Bhairab, the dreadful aspect of Lord Shiva. Table 9.1 tabulates the construction details of the Bhairavnath Temple.

9.2.3 Material characteristics

The materials used in the structural analysis are assumed to be homogeneous, isotropic, and linearly elastic. The properties of the materials are taken from Jaishi, Ren, Zong, and Maskey (2003) and are given as:

9.2.3.1 Mud mortared brick masonry

Density $= 20\ \text{kN/m}^3$
Young's modulus of elasticity $= 800$ MPa
Poisson's ratio $= 0.1$

a Tamrakalash - A kalash made up of copper

TABLE 9.1 Description of Bhairavnath Temple.

Temple name	Bhairavnath (Fig. 9.1, Fig. 9.2)
Location	Taumadhi Square, Bhaktapur
Terrain type	Plain
Type of structure	Brick in mud mortar masonry
Plan configuration	Rectangular (Fig. 9.3)
No. of stories	Seven
Vertical configuration	Receding walls and roofs typical of tiered architectural style
Position of building block	Isolated building
Building dimensions	Refer attached architectural drawing
Total plinth area	172.15 m^2
Story height	2.35 m in ground floor, 2.97 m in first floor, 2.72 m in second floor, 2.71 m in third floor, 2.48 m in fourth floor, 2.67 m in fifth floor, 1.78 m in top floor
Wall thickness	1.18 m in ground floor, 0.93 m in first floor, 0.84 m in second floor, 0.76 m in third floor, 0.73 m in fourth floor, 0.74 m in fifth floor, 0.69 m in sixth floor
Floor structure	Joists with planking
Roof structure	Lower roofs of tiles and topmost roof of metal

Basic compressive stress in axial compression = 0.606 MPa
Permissible bending and tensile stress = 0.05 MPa
Permissible shear stress = 0.201 MPa

9.2.3.2 *Timber members*

Density = 8 kN/m^3
Young's modulus of elasticity = 1250 MPa
Poisson's ratio = 0.12

9.2.4 Assumptions for numeric modeling

9.2.4.1 *Participating components*

In general, any structure is made up of structural and nonstructural components. Only those components that provide the stiffness and significant loading to the structure are considered. The main structural elements, masonry wall, and timber elements that contribute the loads are considered. The roof is not considered as a main structural component, as the failure of the roof does not mean the failure of the whole structure (Jaishi et al., 2003). The struts, roof cladding, roof joists, and small beams connecting to the heavy masonry wall are not considered in the modeling, but the contribution of those elements to the mass is considered.

The base at the level of the plinth is considered rigid. The walls below the plinth are assumed to have considerable stiffness, allowing full seismic load to be transferred to the structure from the ground.

The floors are modeled as a flexible diaphragm. The joists as per site are modeled in software.

9.2.4.2 *Load cases*

Dead load

The dead load of the model consists of the self-weight of all primary structural elements, which includes the walls, columns, and beams of the structural frame. These values were calculated within SAP2000 based on the cross-sectional geometry and material density of each element. The total self-weight of Bhairavnath Temple is 9761 kN.

Roof load

The roof load of the model includes the self-weight of nonstructural components including the timber, roof beams, purlins, roof tile, and mud mortar. The roof load is applied as the concentrated load on each node that corresponds to a floor level as a lumped mass. The roof's weight calculation was based on the measured roof dimensions from the measured drawing. The total weight of the roof is 1389 kN. Some assumptions were considered for the roof's weight calculations as recommended in previous literature (Jaishi et al., 2003). The thicknesses of the timber roof, inclined timber, and clay roof tiles were assumed to be 30, 20, and 15 cm, respectively.

9.2.4.3 *Load combination*

According to NBC 105 (1994), the load combinations for design of a structure using the working stress method are as follows:

(a) DL + LL
(b) DL + EQx
(c) DL − EQx
(d) DL + EQy
(e) DL − EQy
(f) DL + LL + EQx

FIG. 9.1 Bhairavnath Temple description.

(g) DL + LL + EQx
(h) DL + LL + EQx
(i) L + LL + EQx

9.2.4.4 Structural model

The shell element has been adopted for macro finite element (FE) modeling of the Bhairavnath Temple (Fig. 9.4). Axisymmetric solid and plate bending behavior can be analyzed using four noded shell elements. Gravity and uniform loads may be introduced in any direction. The plate-bending behavior includes plate rotational stiffness and a translational stiffness component in the direction normal to the plane of the element, excluding the effect of shear deformation. However, line elements are used for modeling the timber elements.

FIG. 9.2 Elevation and section of Bhairavnath Temple.

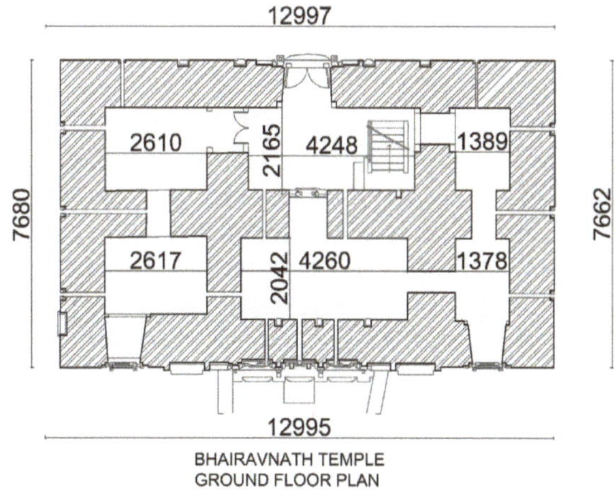

FIG. 9.3 Ground floor plan of Bhairavnath Temple.

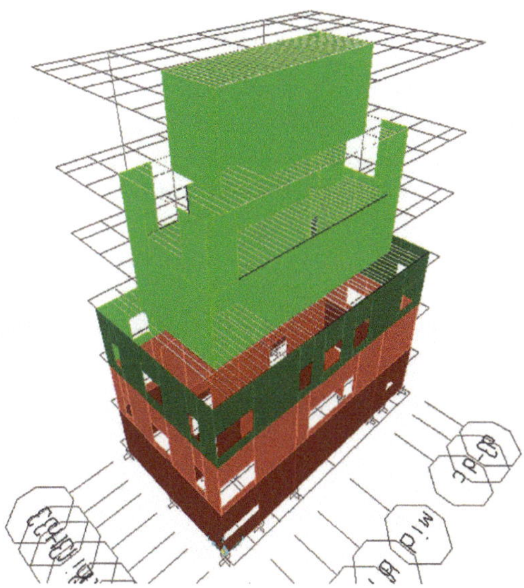

FIG. 9.4 FE shell model of Bhairavnath temple.

9.2.4.5 *Limitations on modeling*

(a) The material is considered to act only linearly, excluding the nonlinearity.
(b) Geometrical nonlinearity, P-Δ effect, temperature effect, age effect, and shrinkage effect are not considered.
(c) Soil-structure interaction is not considered.
(d) In finite element calibration, among various influencing parameters, only certain parameters such as Young's modulus of elasticity of mud mortar brick and boundary conditions are considered. Other influencing parameters are not considered.

9.2.5 Modal analysis

Dynamic characteristics of any structure without influence of any external excitation can be found using modal analysis. Dynamic characteristics include the natural frequency and different mode shape of the structure. The natural frequency is the fundamental or normal mode of vibration in which a structure oscillates due to ambient excitation. The deformed shape under the natural frequency of the structure is the mode shape. Modal analysis is also known as eigenvalue analysis, in which the eigenvalue represents natural frequencies and the eigenvector represents the mode shape of the structure. Dynamic characteristics depend upon the structural properties such as mass and stiffness and also the boundary conditions of the structure. Determination of natural mode shapes is of primary importance, because if the natural frequency of a structure matches the frequency of the earthquake expected, the response of the structure during the earthquake will be amplified and greater damage will be experienced.

According to Newton's second law of motion, the generalized equation of motion is expressed as Eq. (9.1).

$$[M]\{\ddot{U}\} + [C]\{\dot{U}\} + [K]\{U\} = \{F\} \tag{9.1}$$

where

$[M]$ is a mass matrix,
$[C]$ is a damping matrix,
$[K]$ is stiffness matrix,
$\{U\}$ is a displacement,
$\{\dot{U}\}$ is a velocity,
$\{\ddot{U}\}$ is an acceleration, and
$\{F\}$ is external force vector.

For modal analysis, discounting the external force and damping, the equation of motion changes as expressed in Eq. (9.2).

$$[M]\{\ddot{U}\} + [K]\{U\} = 0 \tag{9.2}$$

Displacement response can be solved in harmonic solution as Eq. (9.3).

$$\{U\} = \{\phi\}(\omega t) \tag{9.3}$$

Harmonic motion is considered for free vibration systems in which only amplitude varies with the same mode shape in every degree of freedom. Substituting the harmonic solution into the preceding equation gives Eq. (9.4):

$$-\omega^2[M]\{\phi\}\sin\omega t + [K]\{\phi\}\sin\omega t = 0 \tag{9.4}$$

Further simplifying this equation, we get Eq. (9.5):

$$([K] - \omega^2[M])\{\phi\} = 0 \tag{9.5}$$

This equation is called an eigenvalue equation; $\lambda = \omega^2$ is the eigenvalue and $\{\phi\}$ takes the form of the eigenvector (eigenmode). Each eigenvalue and eigenvector defines a free mode of vibration. For n degrees of freedom, there are the same number of eigenvalues.

9.2.6 Seismic analysis

Several methods have been adopted for seismic analysis or determination of performance under seismic lateral force. Generally, seismic analysis depends upon the lateral forces to be applied to the structure, which are static load and dynamic load. In addition to the forces to be applied, the analysis methods also depend upon the consideration of material linearity and nonlinearity. Static analysis might be linear or nonlinear, depending upon the characteristics of the lateral load.

Linear analysis is a proportional analysis, where if the force is a known factor then the corresponding deformation can be derived, considering the principle of superposition within the material limit of proportionality and vice versa. In nonlinear analysis, progressive strength loss of the material, effects of crack propagation, plastic behavior of the material, and geometrical nonlinearity are considered. The static equivalent force method, displacement-based method, energy-based method, and capacity-based method are common linear static analysis methods, and pushover analysis is a nonlinear static seismic analysis method. The response spectrum method and linear time history method are linear dynamic methods and nonlinear time history analysis is a nonlinear dynamic analysis method. The equivalent static and response spectrum methods are considered in this chapter to analyze the seismic performance of the Bhairavnath Temple.

9.2.6.1 Equivalent static method

This linear static method is based on the assumption that all members undergo the same ductility demand and act more or less as a cantilever. Here the first mode of vibration is considered to be the fundamental response under lateral action. Base shear of the structure depends upon the approximate period, site specific acceleration, and soil type. This method is applicable for regular structures where no significant change in mass or stiffness occur.

According to IS, 1893 (Part I): 2002, clause 7.5.3, the design base shear (V_B) of the structure is obtained as Eq. (9.6):

$$V_B = A_H \times W \tag{9.6}$$

where

A_H = design horizontal seismic coefficient for a structure.
W = seismic weight

$$A_H = \frac{ZI\frac{Sa}{g}}{2R} \tag{9.7}$$

where

Z = zone factor (Table 2 of IS, 1893: 2002);

I = importance factor, which depends upon the functional use of the structure's characterized by hazardous consequences (Table 6 of IS, 1893: 2002);

R = response reduction factor, which depends upon the perceived seismic damage performance of the structure characterized by ductile or brittle deformations (Table 7 of IS, 1893: 2002);

Sa/g = average response acceleration coefficient for rock and soil sites as given by Fig. 9.2 of IS, 1893: 2002, depending upon the appropriate natural periods of the structure. The fundamental natural period for buildings is given in clause 7.6.

In equivalent lateral force analysis, the story force is distributed based on the fundamental period and shape of vibration of the structure, ignoring the mass distribution. IS, 1893 (Part 1): 2002, clause 7.7.1, adopts a parabolic distribution of lateral force along the height of the building as per the expression given in equation (9.8):

$$Q_i = V_B X W_i H_i^2 / \sum_{i=1}^{n} W_i H_i^2 \tag{9.8}$$

where

Q_i = design lateral force;
H = height of the floor from base.

9.2.6.2 Elastic response spectrum analysis

Response spectrum analysis is suitable for irregular buildings. In these cases, the torsional mode might be influential, and it is important to incorporate these behaviors. The response spectrum represents an envelope of all the responses of the single degree of freedom (SDOF) system obtained from an earthquake in the specific region with a specific return period. If a structure has infinite degrees of freedom, then the stiffness matrix and mass matrix are calculated and then eigenvalue analysis of the structure is performed. Each individual mode is considered to be an SDOF system. When the mode is known, the responses of the modes are picked from the code and, by combining all the responses of the modes, total response of the structure is obtained. But not all modes are important in properly reflecting the response of the building. In particular, those modes that are coincident with the main frequency content of the earthquake contribute more. Different codes state that 90% modal mass participation is enough to reflect the overall response of the building.

Modal mass can be calculated according to 7.8.4.5 (a) of IS, 1893 (Part I): 2002. It is given by Eq. (9.9):

$$M_k = \frac{\left[\sum_{i=1}^{n} W_i \phi_{ik}\right]^2}{\left[\sum_{i=1}^{n} W_i (\phi_{ik})^2\right] * g} \tag{9.9}$$

where

ϕ_{ik} = mode shape coefficient.

The modal participation factor should be calculated according to 7.8.4.5 (b) of IS, 1893 (Part I): 2002. It is given by equation (9.10):

$$P_k = \frac{\sum_{i=1}^{n} W_i \phi_{ik}}{\sum_{i=1}^{n} W_i (\phi_{ik})^2} \tag{9.10}$$

The design lateral force at each floor in each mode is calculated using clause no. 7.8.4.5 (c) of IS, 1893: 2002, which is given by equation (9.11):

$$Q_{ik} = A_k \phi_{ik} P_k W_i \tag{9.11}$$

9.2.7 Analysis of stresses

For the analysis, a four-node quadrilateral shell element is formulated. For the quadrilateral, the four joints need to be coplanar. Each shell element has its own element local coordinate system used to define material properties, loads, and output. The first two axes lie in the plane of the element with an orientation specified and the third axis is in the normal direction.

The shell element stress and internal forces are computed for the different types of analysis cases: loads, modes, specs, histories, and combos. Also, principal values and the associated principal directions are computed for the loads and modes. The internal forces such as reaction, moment, and shear are labeled F11, F22, M11, M22, M12, V13, and V23 (in SAP 2000). Similarly, the direct stresses are labeled S11 and S22, whereas shear stresses are labeled as S12, S13, and S23 (in SAP2000).

9.3 Calculations, results, and discussion

9.3.1 Lumped mass calculation

The lumped mass is calculated at floor level considering half mass from the lower and upper story, as tabulated in Table 9.2.

TABLE 9.2 Total lumped mass.

Story	Wall (kN)	Floor (kN)	Roof (kN)	Total (kN)
6	298.38852	11.1537	312.645	622.1872
5	850.05144	25.71912	200.9751	1076.746
4	962.350096	25.71912	264.6591	1252.728
3	1448.665696	162.2692	299.8261	1910.761
2	1877.64864	42.24	310.8261	2230.715
1	2626.69284	42.24		2668.933
			Total	9762.07

9.3.2 Base shear calculation

The seismic coefficient method from IS, 1893 (Part 1): 2002, as mentioned in an earlier section, is used to calculate the total base shear, as follows:

Base shear $= A_H \times W$

$$A_H = \frac{Z}{2} \times \frac{I}{R} \times \frac{Sa}{g}$$

$Z = 0.36$
$I = 1.5$
$R = 1.5$
$Sa/g = 2.5$

$$T_a = \frac{0.09\text{h}}{\sqrt{d}}\text{s}$$

$T_x = 0.412;\ h = 16.527;\ d = 13$
$T_y = 0.536;\ h = 16.527;\ d = 7.68$

Base shear coefficient = 0.45
Base shear = 4392.931 kN

9.3.3 Calculation of lateral forces

The distribution of forces along with the height of the building and lateral forces calculated are as shown in Table 9.3.

TABLE 9.3 Distribution of base shear.

Story	W_i (kN)	$h_{i(m)}$	h_i^2	$W_i h_i^2$	$Q_{i(kN)}$
6	622.1872	15.74	247.7476	154,145.4	1022.175
5	1076.746	12.84	164.8656	177,518.3	1177.167
4	1252.728	10.54	111.0916	139,167.6	922.8537
3	1910.761	7.74	59.9076	114,469.1	759.0721
2	2230.715	5.16	26.6256	59,394.12	393.8567
1	2668.933	2.58	6.6564	17,765.48	117.8072

FIG. 9.5 First mode (2.107 Hz).

9.3.4 Fundamental frequency and mode shape

The first three mode shapes and frequencies shown in Figs. 9.5–9.7 were obtained from modal analysis of the FE model in SAP 2000. The first and second mode correspond to bending into a mutually perpendicular direction. However, the third mode shows twisting behavior. The fundamental time period obtained from free vibration analysis is comparable with the time period obtained from the seismic coefficient method mentioned in the previous section.

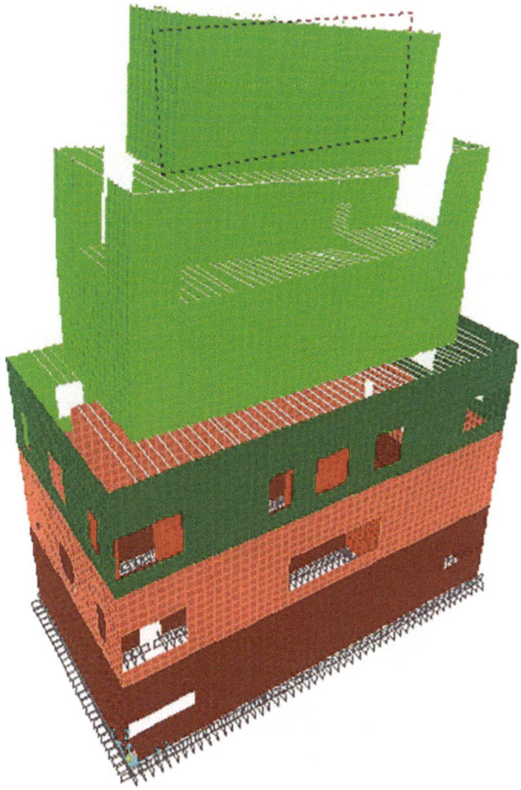

FIG. 9.6 Second mode (3.082 Hz).

9.3.5 Stress profiles in shell elements

The various stress contours shown in Figs. 9.8–9.13 compare with permissible values mentioned in the previous section. Bending stresses were found to be critical in both in-plane and out-of-plane walls. Bending stress is developed due to the wall's out-of-plane deflection. Hence, bending stress is critical at the location of maximum horizontal deflection. Shear stress at the base is not critical in this temple because of its considerable thickness, where the shear area is high. The results for shear show that only the top floor and fourth floor levels are critical in shear, while other levels are not likely to fail in shear because of very large wall thickness. The piers between the openings are more flexible than the portion of the wall below and above the openings, so practically all deformations take place in the pier that leads its failure. The masonry piers near the door and window openings fail in shear due to stress concentration.

FIG. 9.7 Third mode (4.078 Hz).

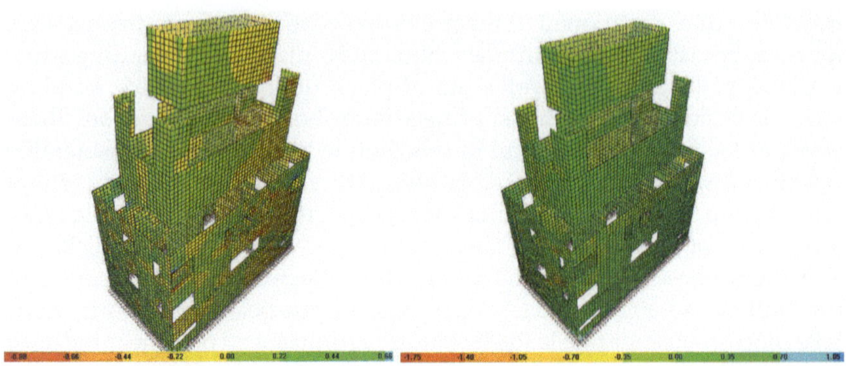

FIG. 9.8 Bending stress (Horizontal S11) contours under seismic action in *X* direction (left) and *Y* direction (right).

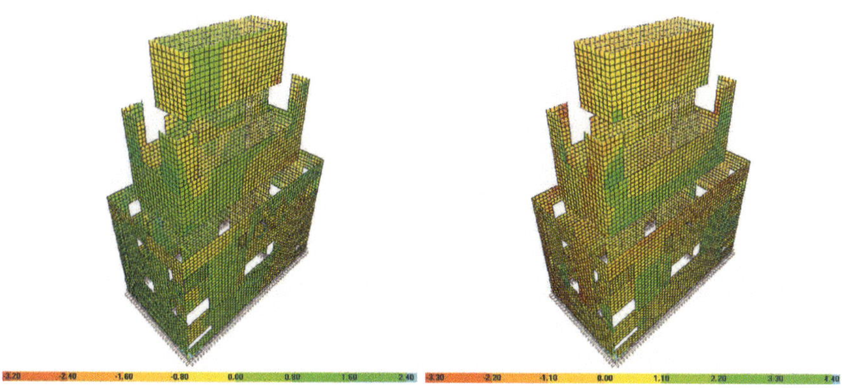

FIG. 9.9 Bending stress (Transverse Horizontal S11) contours under seismic action in X direction (left) and Y direction (right).

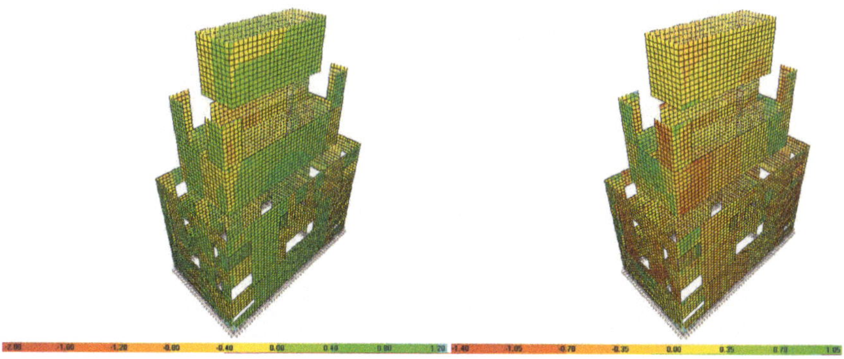

FIG. 9.10 Shear stress contours under seismic action in X direction (left) and Y direction (right).

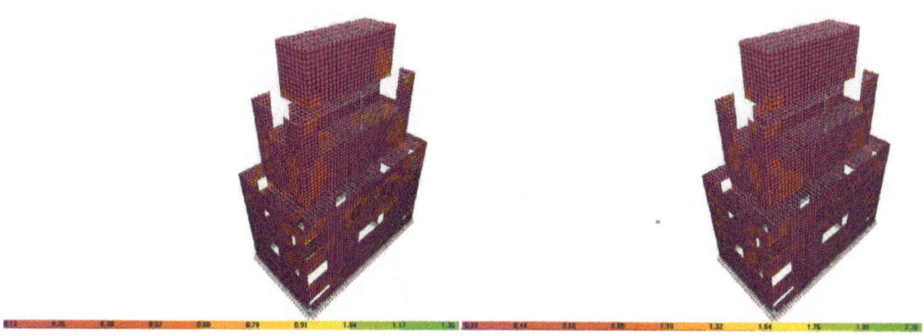

FIG. 9.11 Bending stress (Horizontal S11) contours under response spectrum analysis in X direction (left) and Y direction (right).

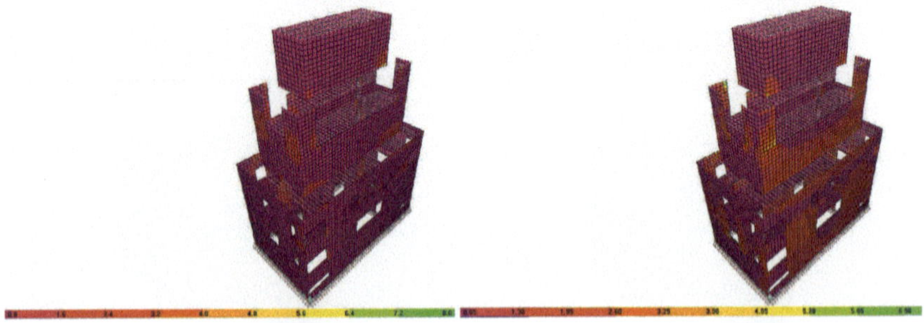

FIG. 9.12 Bending stress (Transverse Horizontal S11) contours under response spectrum analysis in X direction (left) and Y direction (right).

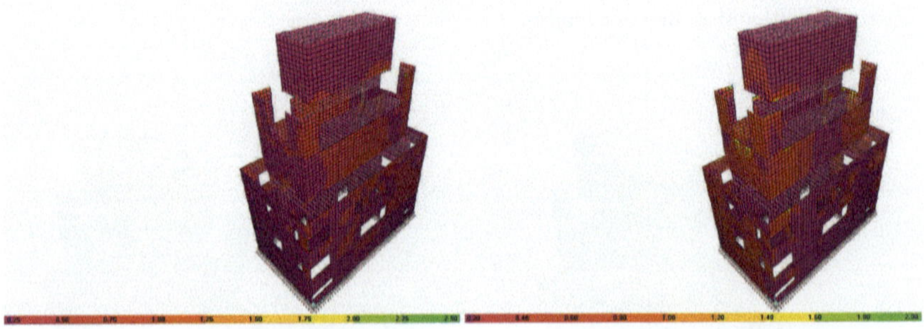

FIG. 9.13 Shear stress contours under seismic action in X direction (left) and Y direction (right).

9.4 Conclusion

The key structural elements of pagoda style temples are thick masonry walls of three layers with mud mortar, a multiple-tiered overhanging roof system, heavy timber members, and a fixed foundation. The unreinforced brick masonry walls have always been the structural system of traditional buildings and monuments from ancient and medieval periods in Kathmandu. Burnt clay bricks, mud mortar, stones, and wood used to be the only available materials for construction in the medieval period in Nepal. The seismic resistance of the unreinforced brickwork is limited, due to the density and brittleness, with low tensile strength and exhibiting low ductility. Seismic resistance of these buildings can only be achieved by basically planning the building with a regular configuration, and by proper use of the then-available materials with possible detailing.

Acknowledgements

The authors would like to express their sincere gratitude to the faculties and staff of Khwopa Engineering College, Khwopa College Engineering, and the Bhaktapur Municipality, Bhaktapur, Nepal for providing the drawings and sharing their expertise.

References

Bonapace, B., & Sestini, V. (2003). *Traditional materials and construction technologies used in the Kathmandu valley. Paragraphic for the United Nations Educational, Scientific and Cultural, 7 Place de Fontenoy, 75352 Paris 07 SP, France.* .

DoA Report (2015). *Preliminary list of affected monuments of Nepal by Gorkha earthquake 2015.* Department of Archaelogy.

IS 1893 (Part 1): 2002 (2002). Indian Standard Criteria for Earthquake Resistant Design of Structures Part 1 General Provisions and Buildings (Fifth Revision). *Indian Standard, 1*(5).

Jaishi, B., Ren, W. X., Zong, Z. H., & Maskey, P. N. (2003). Dynamic and seismic performance of old multi-tiered temples in Nepal. *Engineering Structures, 25*(14), 1827–1839. https://doi.org/10.1016/j.engstruct.2003.08.006.

NBC 105: 1994 (1994). *Nepal National Building Code, Seismic Design of Buildings in Nepal.* .

Nienhuys, S. (2003, September). *Options for reconstruction and retrofitting of historic pagoda temples, reconstruction of temples in Kathmandu.* HUYS ADVIES.(2003, September). http://www.nienhuys.info/mediapool/49/493498/data/Retrofitting_KVPT_HA_2003.pdf.

Raj, Y. (2061). *Bhaktapur Ta: Marihsthit Bhairavnath Mandirko Aitihasikta Bare Thap Jankari* (2061 B.S.) (2nd ed.). Bhaktapur, Nepal: Suntha Computers.

Ranjitkar, R. K. (2006). *Heritage homeowner's preservation manual. Kathmandu Valley World Heritage site, Nepal: Advice for maintenance of historic houses in the Kathmandu Valley.* Integrated Community Development and Cultural Heritage Site Preservation in Asia and the Pacific through Local Effort Programme (LEAP). UNESCO Bangkok, UNESCO Kathmandu. ISBN: 978-9937-2-0223-7.

Shakya, M., Varum, H., Vicente, R., & Costa, A. (2012). *Structural vulnerability of Nepalese pagoda temples. In 15th World Conference on Earthquake Engineering* (pp. 1–10). Retrieved from(2012). http://www.iitk.ac.in/nicee/wcee/article/WCEE2012_2919.pdf.

Shakya, M., Varum, H., Vicente, R., & Costa, A. (2014). Seismic sensitivity analysis of the common structural components of Nepalese pagoda temples. *Bulletin of Earthquake Engineering, 12*(4), 1679–1703. https://doi.org/10.1007/s10518-013-9569-6.

Shrestha, S., Shrestha, B., Shakya, M., & Maskey, P. N. (2017). Investigation and analysis of damages in the heritage structures: An aftermath of 2015 Gorkha (Nepal) earthquake. *Earthquake Spectra. 2017,* https://doi.org/10.1193/121616EQS241M.

Thapa, J. B. (2011). *Test and simulation of brick masonry wall of historic buildings.* Master thesis-Nepal: Tribhuvan University.

Tiwari, S. R. (2009). *Temples of the Nepal valley.* Kathmandu, Nepal: Himal Books.

CHAPTER

10

From *Tship Chim* to *Pa Chim*: Seismic vulnerability and strengthening of Bhutanese vernacular buildings

Nimesh Chettri[a], *Dipendra Gautam*[b], *and Rajesh Rupakhety*[c]

[a]Civil Engineering and Architecture Department, College of Science and Technology, Royal University of Bhutan, Phuentsholing, Bhutan, [b]Department of Architecture and Civil Engineering, City University of Hong Kong, Kowloon, Hong Kong, [c]Earthquake Engineering Research Center, Faculty of Civil and Environmental Engineering, School of Engineering and Natural Sciences, University of Iceland, Selfoss, Iceland

10.1 Introduction

Earthquakes are among the most destructive natural hazards in seismically active regions. Due to continuous convergence of the Indian plate beneath the Eurasian plate, the Himalaya is one of the most active seismic regions in the world. As Bhutan lies in the eastern fringe of the Himalaya, it faces the risk of moderate to large earthquakes. In recent years, several earthquakes struck Bhutan and neighboring areas, which prognoses that Bhutan is in high risk of major earthquake(s) in the near future (Bilham, 2019). The thrust fault that dips under Bhutan at a shallow angle extends beneath the entire kingdom. Bhutan lies in the seismic gap between the 1934 Bihar-Nepal earthquake and the 1950 Assam earthquake, and paleo seismic studies indicate that a large earthquake last occurred in this

https://doi.org/10.1016/B978-0-12-821087-1.00009-0

section of the plate boundary in ~1100 (JDWNRH, 2012). Researchers estimate that this earthquake was of moment magnitude 8 or greater. If a similar earthquake were to occur today, damages would likely be severe throughout Bhutan (JDWNRH, 2012). Still, the majority of Bhutanese people live in nonengineered vernacular buildings made up of rammed earth, stone masonry, wattle and daub, bamboo, and timbers (Chettri, Thinley, & Koirala, 2019). The study by Chettri, Gautam, and Rupakhety (2021) shows that such buildings perform poorly during earthquakes. Experience has shown that fatalities during earthquakes are mostly due to structural collapse, which is more common in weak masonry buildings like the ones used in Bhutan (Gautam, 2018; Gautam, Fabbrocino, & Santucci de Magistris, 2018; Sferrazza Papa & Silva, 2018; among others). The study on seismic capacity of these structures has become indispensable because many of them were heavily damaged during the 2009 Mongar (Mw 6.1) and the 2011 Sikkim-Nepal border (Mw 6.9) earthquakes.

The Bhutanese traditional houses are mostly nonengineered load-bearing constructions. The Bhutan Building Code does not provide design guidelines and postearthquake assessment measures for Bhutanese traditional houses. International codes may not be appropriate to the needs of the Bhutanese engineering sector in terms of format, content, and level of detail (Sethna, 2008). Having stated the wide range of problems from seismic hazard study to seismic vulnerability of existing structures, this chapter first presents the context of two recent earthquakes in Bhutan. The context and forensic interpretation are followed by construction and elucidation of taxonomy and vulnerability of existing buildings in Bhutan. Based on the field evidence from the two recent earthquakes, some retrofitting techniques are also proposed.

10.2 From history to hitherto: Timeline and tales of earthquakes in Bhutan

10.2.1 Seismic hazard in Bhutan

Records of earthquakes in Bhutan are available since 1713. Most of the records of earthquakes that occurred in the 18th century lack details regarding damages and losses. A summary of the occurrence and reported damage in Bhutan due to several earthquakes is presented in Table 10.1. As shown in Table 10.1, damage records of the earthquakes that occurred in 2003 are available. However, major damages were reported only during the 2009 and 2011 earthquakes. Details of both the earthquakes are presented in the following section.

TABLE 10.1 Historical earthquake in Bhutan.

Earthquake magnitude (Mw)	Date	Epicenter	Damages/loss incurred
7.0	1713	Border between Bhutan and Arunachal Pradesh (India)	No reports available
7.6	June 11, 1806	Eastern Bhutan	
8.0	June 12, 1897	60 km from south of Bhutan in Assam	
6.5	May 12, 1906	Border of Bhutan, India, and China	
5.7	August 13, 1910	India-Bhutan border	
8.0	January 15, 1934	India-Nepal border	
6.7	January 21, 1941	South East Bhutan	
7.3	July 29, 1947	India-China border	
8.7	August 15, 1950	India-China border	
6.4	February 23, 1954	Border of Bhutan, India, and China	
6.6	July 29, 1960	Southern Bhutan	
6.4	November 19, 1980	North East India	
6.8	August 20, 1988	Nepal	
5.5	March 26, 2003	Paro, Bhutan	– Few landslides along highways along Paro were recorded – Minor damages in building in Paro, Haa, and Thimphu were reported
5	February 11, 2006	Arunachal Pradesh, felt in Trashigang (Bhutan)	No reports available

Continued

TABLE 10.1 Historical earthquake in Bhutan—cont'd

Earthquake magnitude (Mw)	Date	Epicenter	Damages/loss incurred
5.7	February 14, 2006	Sikkim (India)	
5.8, 5.5	February 24, 2006	Dewathang, Bhutan	Around 126 houses sustained minor damages
6.1	September 21, 2009	Narang, Mongar, Bhutan	- 12 lives lost, 22 injured, damaged 4950 households - 45 BHUs, 117 schools, over 800 cultural heritage buildings, 29 RNR offices and 26 Gup's office - Total estimated loss of Nu. 2501million (US$ 42 million)
6.9	September 18, 2011	Sikkim-Nepal border	- Loss of one life due to landslides and 14 injuries. A total of 6977 rural home damaged, 36 schools, 22 hospitals, 286 heritage sites, monasteries, 27 RNR centers

10.2.2 The 2009 Mongar earthquake

On September 21, 2009, around 2:53 pm local time, an earthquake of moment magnitude 6.1 jolted Bhutan. Its impact was the worst in the eastern districts of Bhutan (World Bank, 2009). The epicenter was at 27.351°N and 91.425°E in Mongar district, with a focal depth of ~10 km, around 180 km away from capital city, Thimphu. Most of the damaged structures were stone masonry in mud mortar (Bhutan, 2009). The report highlights that seven aftershocks were recorded within half an hour of the main shock (Rodgers et al., 2014). The Indian state of Assam, West Bengal, Arunachal Pradesh, and adjoining areas of north-eastern India were also affected by this earthquake. The effects were also felt in parts of China, Myanmar, and Bangladesh. The United States Geological Survey (USGS) estimated an intensity of VI+ at the epicentral (World Bank, 2009) region. The intensity contours of the Mongar earthquake as mapped by the USGS are shown in Fig. 10.1. Damage statistics of the affected nine districts are presented in Table 10.2. As shown in Table 10.2, 9 out of 20 districts in Bhutan sustained considerable damage due to the earthquake. In total, 4614 buildings were affected; among them, 446 collapsed or were beyond repair, 1012 sustained major damage, and 3156 buildings sustained minor damage. According to (World Bank, 2009) official figures,

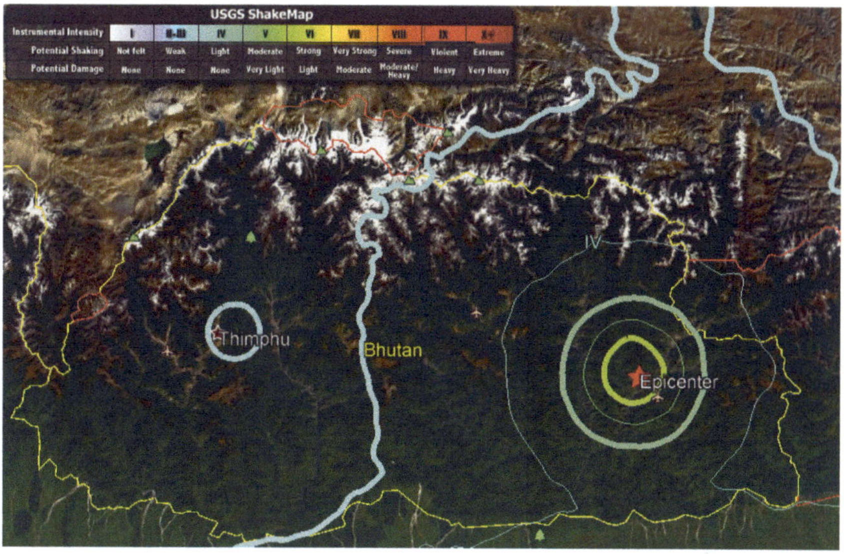

FIG. 10.1 Intensity contours of the Mongar earthquake in Bhutan and neighboring areas. *(Modified from United States Geological Survey; retrieved from: https://earthquake.usgs.gov/earthquakes/eventpage/usp000h1r4/executive.)*

TABLE 10.2 Summary of damage due to the 2009 Mongar earthquake in Bhutan.

Districts	Total collapse and beyond repair	Major damages (major repair)	Minor damages (minor repairs)	Total houses affected
Wangdue Phodrang	0	14	24	38
Trongsa	5	9	37	51
Samdrup Jongkhar	41	0	179	220
Trashiyangtse	49	52	417	518
Lhuentse	28	1	129	158
Mongar	136	438	694	1268
Trashigang	147	423	1011	1581
Pemagatshel	21	42	318	381
Zhemgang	19	33	347	399
Total	446	1012	3156	4614

12 fatalities, and 47 injuries occurred. Records from the media and official reports highlight that the aftershocks may have aggravated damage. Apart from residential buildings, 91 schools, 25 health centers and hospitals, 50 government offices, 281 monasteries, 485 stupas, and 7 district administration centers were reported to be damaged to various extents (Situation Report Earthquakes in Bhutan, 2009). According to official figures (World Bank, 2009), losses of approximately 23.3, 12.9, and 2.6 million US$ were sustained, respectively, by residential buildings, school buildings, and health facilities that include 25 basic health facilities and 3 hospitals. A loss of US$0.3 million was attributed to damage to the local administrative buildings, and ~US$13.5 million was attributed to damage to cultural heritage. The total economic loss was ~US$52 million.

10.2.3 The 2011 Sikkim-Nepal border earthquake

The 2011 Sikkim-Nepal border earthquake occurred on September 18, 2011 in the bordering areas of Nepal and Sikkim. The focal depth was estimated to be ~35 km by the USGS. The recorded peak ground accelerations (PGAs) at Gangtok and Kathmandu were about 0.15 and 0.05 g, respectively (Gautam, 2017). Spatial distribution of damage in Nepal, Sikkim, and Bhutan shows that the seismic rupture propagated to the east, causing more damage to the east of the epicenter. Further details regarding damage due to the 2011 earthquake can be found in Gautam et al. (2018) and Chaulagain, Gautam, and Rodrigues (2018). The Sikkim-Nepal border earthquake damaged 7965 buildings in Bhutan. The earthquake affected all 20 districts of Bhutan; however, the western districts were more affected than the eastern ones. For instance, Haa, Paro, Samtse, and Chhukha districts were the most affected ones (see Fig. 10.13). The earthquake caused 1 fatality, 14 injuries, and extensive damage in residential as well as public structures and infrastructures in Bhutan (PDNA, 2011). As reported in PDNA (2011), 345 residential buildings either collapsed or were damaged beyond repair; 1660 buildings suffered major damage; and 5960 suffered minor damage. Apart from residential buildings, 487 cultural heritage structures; 47 administrative buildings; 47 public buildings; 50 health facilities; 117 schools; and 31 Renewable Natural Resources (RNR) Centers were damaged by the earthquake (PDNA, 2011). The PDNA estimated losses of ~US$15.8 million due to damage in residential buildings. Similarly, loss attributed to government and public buildings was ~US$0.34 million; to cultural heritage structures was ~US$6.96 million; to school and educational buildings was ~US$0.22 million; and to agricultural infrastructure was ~US$0.12 million. The total loss in Bhutan was estimated to be ~US$24.46 million. Building damage in each district is summarized in Table 10.3.

TABLE 10.3 Damage statistics in Bhutan due to the 2011 Sikkim-Nepal border earthquake.

Districts	Total collapse and beyond repair	Major damages-repairable	Minor damages-repairable	Total houses affected
Haa	170	250	805	1225
Paro	90	706	1238	2034
Wangdue Phodrang	2	52	0	54
Samtse	34	263	590	887
Chhukha	9	83	712	804
Phuentsholing Thromde	0	16	69	85
Gasa	19	38	171	228
Trongsa	9	17	35	61
Bumthang	0	0	62	62
Sarpang	5	3	48	56
Samdrup Jongkhar	0	0	18	18
Trashiyangtse	7	52	324	383
Lhuentse	0	1	59	60
Mongar	0	26	434	460
Thimphu	0	0	131	131
Thimphu Thromde	0	10	44	54
Punakha	0	17	691	708
Tsirang	0	23	225	248
Dagana	0	0	117	117
Trashigang	0	100	115	215
Pemagatshel	0	0	32	32
Zhemgang	0	3	40	43
Total	345	1660	5960	7965

10.3 Learning from the past: Damages and lessons from historical earthquakes in Bhutan

Bhutan does not have structural guidelines of its own and relies on Indian design standards for reinforced cement concrete structures. All traditional buildings in Bhutan are built by skilled craftsmen without any technical plans. These knowledge and expertise are passed on to new generation through practice and experience. Most buildings damaged in 2009 and 2011 were nonengineered owner-built constructions. The losses of the past earthquake have sensitized Bhutanese Engineers to come up with measures to assess hazard and post damages caused by earthquake. The Royal Government of Bhutan also emphasized on creating awareness on earthquake and improve design of vernacular structures to stay prepare for the future disasters (Rodgers et al., 2014). The 2009 earthquake was Bhutan's first experience of a damaging earthquake after many years—ending a period of relative seismic inactivity. The eastern part of the country suffered major losses, mostly due to damage to stone masonry constructions and some rammed earth constructions. Damage analysis and interpretation of failure mechanism of Bhutanese vernacular structure during the 2009 and 2011 earthquake are presented in the following section.

10.3.1 Cracks

Various types of cracks (vertical, horizontal, lintel, putlog holes, dislocation or corner separation of wall lift and crack due to timber joist) were observed after the 2009 and 2011 earthquakes. Vertical cracks (sometime corner separation), as shown in Fig. 10.2A, formed at the corners of the wall, are due to poor connection between two orthogonal walls. Excess clay content in mortar caused the shrinkage of walls leading to irregular cracks. Vertical cracks were also noticed in near large openings encasing doors and windows that intercept the load path. Different blocks of rammed earth having vertical joints along the same line showed worse performance when exposed to lateral forces induced by earthquakes. Corner connector, stitch, or bands were lacking in the damaged buildings. Excessive cracking of this nature can lead to local collapse.

Horizontal cracks between lifts of rammed earth wall, like cold joints in concrete, were also common. This type of crack occurs due to poor ramming between adjacent lifts. Lack of sealing in the joints makes them more vulnerable to lateral forces. Such cracks were more prevalent near large openings as shown in Fig. 10.3.

Lintel cracks are typical vertical cracks along the edges of windows and doors. It occurs due to discontinuity of load path. Fig. 10.4 pictorially represents the cracks around lintel due to obstruction in the load path.

FIG. 10.2 Different forms of vertical cracks: (A) Vertical crack at the corner between orthogonal walls, (B) vertical crack near a door opening, (C) vertical corner crack due to rotation of stone masonry wall, (D) vertical wall crack, poor corner integrity.

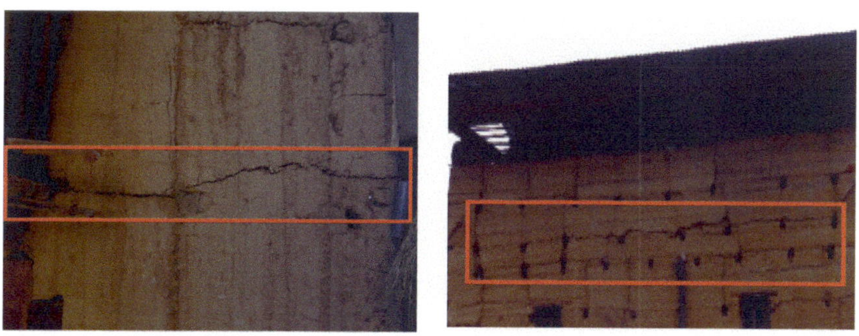

FIG. 10.3 Horizontal cracks.

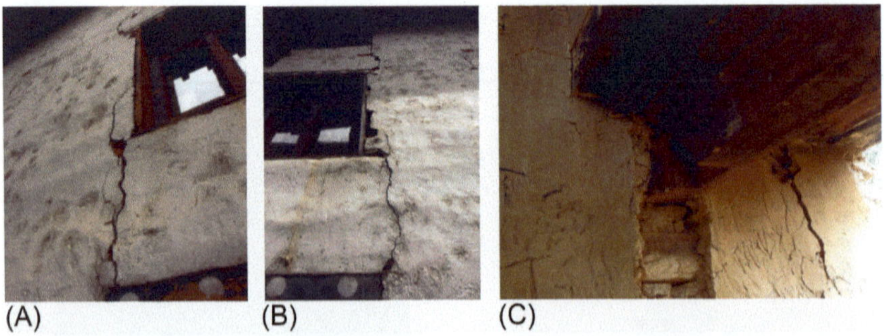

FIG. 10.4 (A) Cracks along window frame and lintel (B) vertical cracks passing through window frames and (C) lintel crack above a door.

Cracks due to putlog holes were common. Putlog holes are small holes made in the walls to support scaffolding. These putlog holes are kept unsealed after the construction is completed. Stresses tend to concentrate around these holes causing extensive cracking. It was observed that most of the cracks (both horizontal and vertical) were either originating on or along these putlog holes as shown in Fig. 10.5.

During field reconnaissance, several cracks were also observed on the wall immediately below the timber joist supporting the floor (Fig. 10.6). The cracks occur due to relative motion between the wall and the joists during horizontal shaking (Windstorm & Schmidt, 2013). As the moisture in the timber dries up, shrinkage occurs, loosening the contact between the joists and the walls. During lateral shaking, movement of the joints against the wall can initiate cracking.

FIG. 10.5 Cracks due to putlog holes: (A) vertical and horizontal crack, (B) vertical crack.

FIG. 10.6 Cracks due to floor joist.

10.3.2 Out-of-plane collapse of wall

Out-of-plane failures occur due to various reasons such as quality of construction materials, large span of wall without cross support, and large openings. Material quality is regarded as the prime factor as many failed samples were found to have poor material composition. Through stones and corner stones, which are important elements to bind adjacent walls together, were lacking in many stone masonry buildings. Similar cases were evident during the 2015 Gorkha earthquake in Nepal (Gautam, 2018; Gautam, Rodrigues, Bhetwal, Neupane, & Sanada, 2016). Lack of cross walls in rammed earth constructions led to many cases of out-of-plane failure as shown in Fig. 10.7. Buttress walls restraining the corners could be effective in avoiding such out-of-plane collapse.

10.3.3 Buckling of the walls

Structural integrity at wall corners is essential to generate box action and transmit dead load from higher floors. Lack of such action can result in buckling of walls at the corners as shown in Fig. 10.8A and B. Improper composition of the construction materials (mud mortar in case of stone masonry and rammed earth ingredient in case of rammed earth house) leads to weak bonding, which makes the wall corners more vulnerable to lateral forces. Presence of organic materials in soil and mortar can cause swelling of the walls and failure of the type shown in Fig. 10.8C.

10.3.4 Delamination of the wall

Multiwythe stone masonry walls can delaminate due to lack of integrity and proper binding. The outer layers usually fall apart and collapse as shown in Fig. 10.9. When the earth walls are exposed to moisture, they

(A) (B) (C) (D)

FIG. 10.7 (A) Wall collapse due to poor mortar quality, (B) collpase of rammed earth wall near a large opening, (C) and (D) complete destruction of stone masonry walls.

become weak and the outer plaster erodes from the walls. Delamination is common in all the rammed earth buildings and this mechanism primarily occurs on the outer side of the exterior walls. Such walls are more exposed to rainfall and other sources of moisture.

10.3.5 Gable failure

Gable walls are highly vulnerable to seismic actions especially when they are not connected to the roof and main structural systems. Such damages are essentially out-of-plane and usually affect a portion of the perimeter wall below the gable (Fig. 10.10).

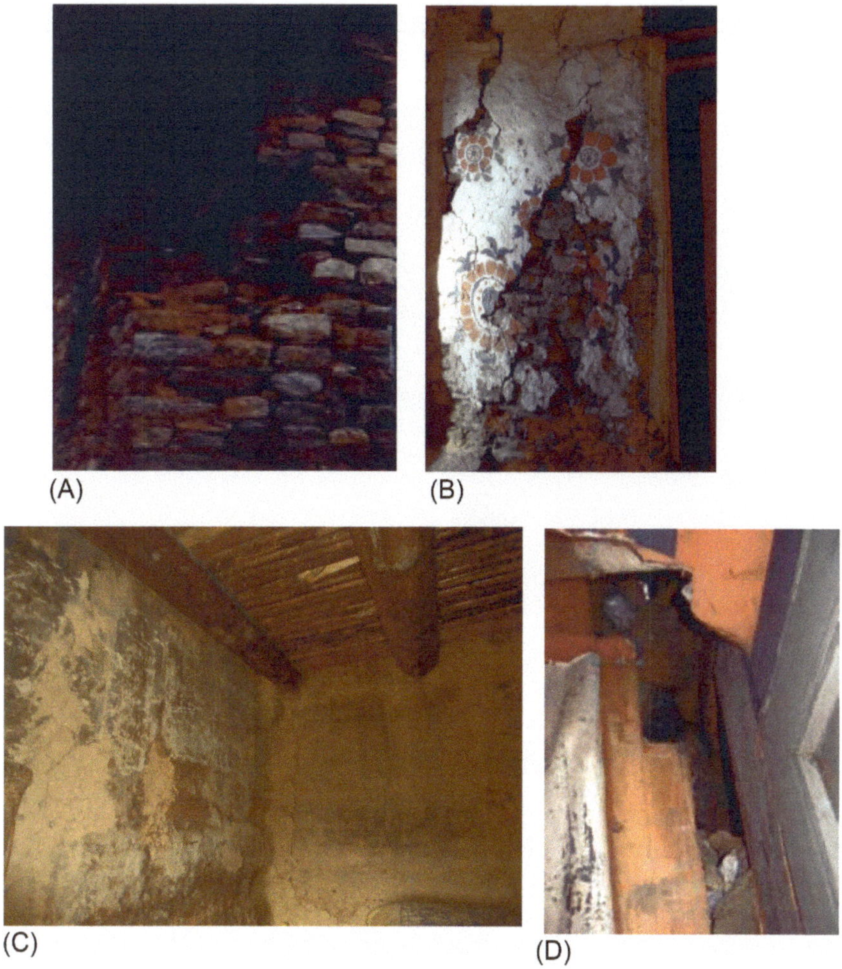

FIG. 10.8 (A) Poor wall integrity causes buckling and mass failure, (B) nonuniform stone shapes lead to improper masonry bonding, (C) swelling of corner, (D) corner collpase due to poor connection and unscreened material.

FIG. 10.9 Delamination of wall.

FIG. 10.10 Gable failure.

10.3.6 Tilting of Rabsel (Bhutanese decorative and ornamental window)

Rabsels are ornamental Bhutanese windows that are integrated to the main structure by fixing their timber members on walls. There is no additional anchorage to provide support against rotation of wall; and the overall stability of Rabsels depends on the thickness of the wall. Buildings with wall thickness of 600 mm or more were seen to have intact Rabsels during the earthquakes. Bonding failure between the Rabsel and the wall at the base of the opening causes the Rabsel to tilt away from the wall (Fig. 10.11). Such mechanisms were seen to be more common in thinner walls.

10.3.7 Dislocation of connection of roof members

Bhutanese traditional roofs lack structural integrity in their trusses. There are weak connections especially at the strut members such as purlin, rafter, and tie (Fig. 10.12A and B). It was observed that the strut supporting the trusses is not integrated to the roof floor or to any other rigid elements of the buildings. It was simply kept on the surface of the roof floor to bear the gravity load of the trusses and roof coverings. They performed poorly during lateral shaking caused by the earthquakes. Most critical connections in the roof were found to be weak and had no mechanism to support lateral forces (Fig. 10.12C and D).

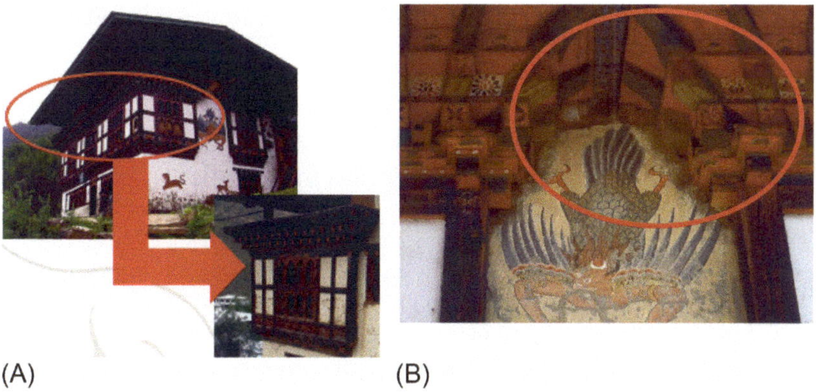

FIG. 10.11 (A) A typical Rabsel (Bhutanese ornamental window), (B) imbalanced Rabsel tilted away from the wall.

FIG. 10.12 (A) Weak member connection, (B) small prop size and simply supported truss, (C) and (D) collapse due to old and weak truss connection.

10.4 Taxonomy and vulnerability of Bhutanese buildings

According to the population and housing survey conducted by the Royal Government of Bhutan in 2017, Bhutanese buildings can be categorized into 13 types. However, generic classification can be made as: RC, masonry, and others. Number of buildings in each district per the 2017 database is mapped in Fig. 10.13. The south-western districts of Bhutan are more populated than the eastern and northern districts. Three categories are defined as per the number of buildings: districts having more than 10,000 buildings; districts having 5000–10,000 buildings, and districts having less than 5000 buildings. Fig. 10.14 shows the distribution of buildings in the 13 categories. Based on the construction description of each building type, a taxonomical classification system is developed as shown in Fig. 10.14. Three categories are defined: reinforced concrete, masonry, and other. The reinforced concrete construction alone comprises 33.37% of building stock, whereas the masonry construction as defined in Fig. 10.14 comprises 42.81%. The rest are classified as "other building." This

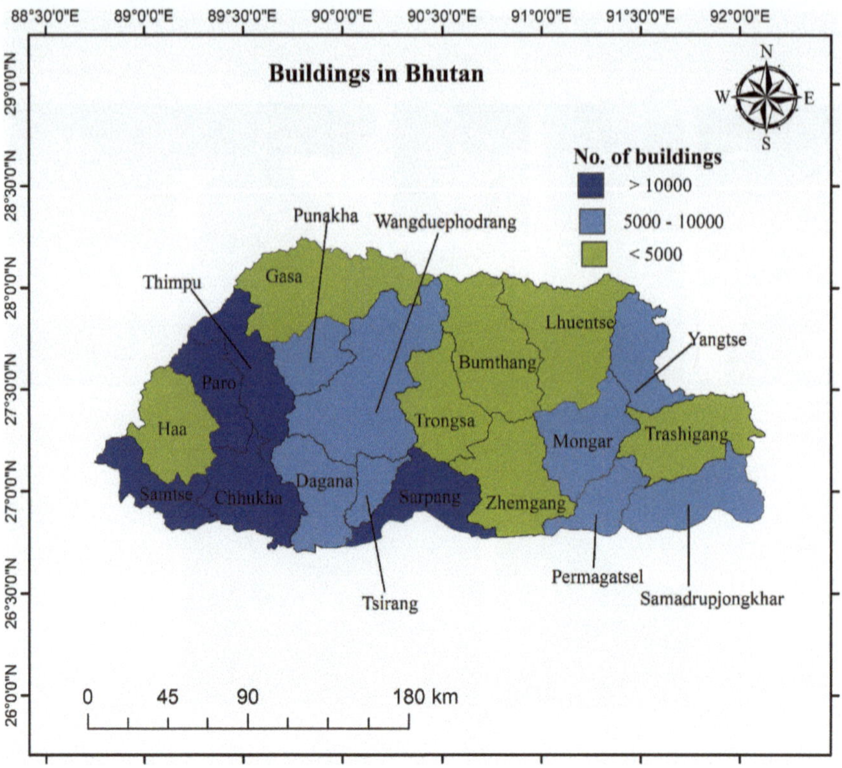

FIG. 10.13 Districtwide distribution of building population in Bhutan.

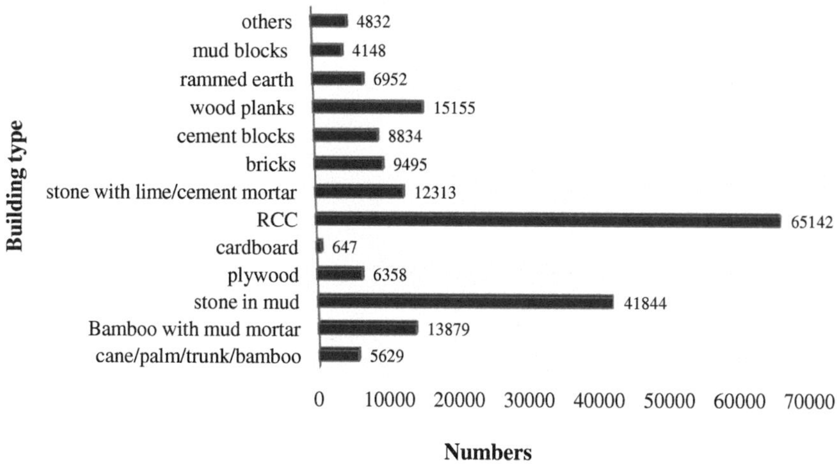

FIG. 10.14 Typological distribution of Bhutanese buildings.

category comprises wooden (7.76%), wattle and daub (7.11%), plywood (3.26%), cane/palm/trunk/bamboo (2.88%), other (2.48%), and cardboard (0.33%). To the best of the authors' knowledge, seismic fragility/vulnerability functions of Bhutanese buildings have not been reported in the literature. To have a preliminary understanding based on the European vulnerability conventions, we assigned each taxonomy three vulnerability categories: most likely, probable, and rare vulnerability considering A-F vulnerability classes as per the EMS-98. The vulnerability scenario of each building class in Bhutan is depicted in Fig. 10.15. Due to differences in workmanship and quality of construction materials, the European vulnerability classification may not be directly applicable to the Bhutanese building stock. However, preliminary understanding of vulnerability and prioritization in seismic strengthening could be achieved using Fig. 10.15. Figs. 10.14 and 10.15 together show that more than 40% of buildings in Bhutan can be classified under vulnerability class A and B, which means that even moderate earthquakes could cause significant damage to the masonry buildings, as was the case in the 2009 and 2011 earthquakes. For instance, during the 2009 earthquake, 22.57% stone masonry buildings in Trashigang district (~12 km from the epicenter) were damaged. In Zhemgang district, which was ~77 km from the epicenter, 25% of the buildings sustained damage. This shows that many masonry buildings in Bhutan can fail life safety performance level at a relatively low shaking. Similar evidence was observed during the Sikkim-Nepal border earthquake. The Samtse district, 135 km far from the epicenter, sustained damage to ~30% of its buildings during the earthquake.

Building type	Indication	Description	Vulnerability classes					
			A	B	C	D	E	F
Others	O1	Cane/palm/trunk/bamboo						
	O2	Wattle and daub						
	O3	Plywood						
	O4	Cardboard						
	O5	Wood planks						
Masonry	M1	Stone in mud						
	M2	Bricks						
	M3	Stone in lime/cement						
	M4	Cement blocks						
	M5	Rammed earth						
	M6	Mud blocks						
Reinforced concrete	RC1	RC frame without earthquake resistant design						
	RC2	RC frame with earthquake resistant						

Vulnerability: Most likely Probable Rare

FIG. 10.15 Taxonomy and vulnerability of Bhutanese buildings.

As the current Bhutan Building Regulation does not incorporate seismic provisions, all the existing RC buildings are classified under vulnerability class C (most likely). This signifies that Bhutan should improve the current construction practice to prepare for impending earthquakes. The other buildings fall under vulnerability class C-D. Construction systems

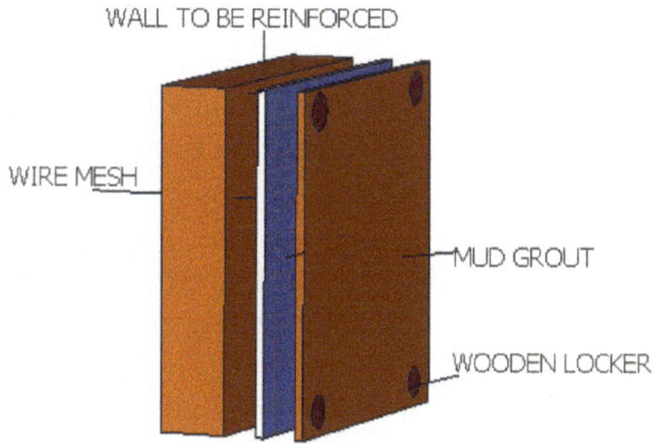

FIG. 10.16 Details of wire mesh, anchor, and mud plaster.

such as timber frames, buildings constructed with timber planks, bamboo, wattle and daub, among others may have better seismic performance if their connections and foundation system is well constructed and maintained. Furthermore, life safety assurance is superior in other building classes than the masonry construction.

10.5 Building regulations in Bhutan and future insights

The city corporation offices in the districts overlook the affairs of validating the technical drawings (architecture, structural, electrical, and sewage) in Bhutan. Such drawings need to comply to "Bhutan Building regulation 2018" and "Guidelines on Traditional Architecture." "Bhutan Building Rules 2002" was reframed and termed as "Bhutan Building regulation 2018" to facilitate and regulate safe building construction and to promote healthy living environment (BBR, 2018). "Traditional Architecture Guidelines" specifies height of building, room size, parking spaces, offset distance, paintings, and other cultural aspects. It also emphasizes fire safety provisions, ventilation, water supply, and access to differently abled people. The building height in rural areas should be maximum of three stories. Per the Table 3.0.2 of Thimphu Municipal Development Control Regulations (2004), maximum of five stories are permitted in urban villages (Thimphu Thromde, 2004). As there are no specific codes to cover masonry buildings in Bhutan, traditional knowledge and mason/carpenter knowledge-based constructions are widespread. Constructions are usually owner-built and mostly the shape, size, dimensions, and detailing are governed by the needs of the owners. Reinforced concrete construction was made obligatory to follow the Indian standard code of practice in 1997. Given its unique requirements in architecture, local culture, and construction materials and methods, Bhutan needs to work toward a seismic design code to suit its needs. Detailed study of seismic hazard is therefore necessary. In addition, consideration of sustainability approaches in construction practices is important. Prioritization of strengthening and retrofitting of the existing vulnerable building stocks as depicted in Fig. 10.15 is important to reduce the impact of future earthquakes.

10.6 Strengthening and beyond: Comparison of various retrofitting approaches for Bhutanese residential buildings

This section presents some strengthening solutions for the Bhutanese vernacular buildings. Some of the parameters used to model the structures using the finite element method are listed in Table 10.4. The parameters were adopted from various literature as no test results are available for Bhutanese buildings.

Equivalent lateral load analysis is performed considering a typical rammed earth house and vertical and lateral loads of 2165 and 242.532 kN, respectively. As there were no accelerograms recorded in Bhutan, the PGA of 0.15 g as recorded in Sikkim in the 2011 earthquake is considered for analysis. Finite element models with retrofitting solutions are prepared in ABAQUS, and a static analysis is carried out to estimate the response of the retrofitted walls.

10.6.1 Reinforcing with containment wire mesh and mortar

For cracks deeper than 200 mm, wire meshing (chicken wire mesh) in addition to grouting is recommended. Loose and damaged sections of both rammed earth and stone masonry are removed and replaced. A wire mesh is then anchored to the wall by timber pegs or dowels. (Sandbhor & Botre, 2015). This method improves tensile strength of the walls. The dimension of timber pegs should not be less than 15 mm diameter and must be inserted in the wall up to a minimum depth of 150 mm. The exposed surface of wire mesh is plastered with modified or unmodified mud grout (20 mm thick). The modified grout acts as a synthetic glue and enhances performance. This technique is most suitable for houses with rammed earth wall. It can be used in stone masonry walls if both

TABLE 10.4 Material properties used for finite element analysis.

Material	Density (KN/m^3)	Modulus of elasticity (N/mm^2)	Poisson's ratio
Mud	19.6	45	0.22
Timber	5.570	11,850	0.3
Steel	78	2.1×10^5	0.3

the faces are strengthened all the way to their base. After plastering, the surface has to be kept protected from external environment for 10 days (Silva, Olliveira, Schueremans, Miranda, & Machado, 2014). Pull-out test has to be done on the surface to make sure it does not fail at 1 MPa (Guidelines for Retrofitting of Buildings, 2006).

FE models of the wall panels (Fig. 10.17) are subjected to the gravity load as well as lateral load. The base of the wall is restrained against

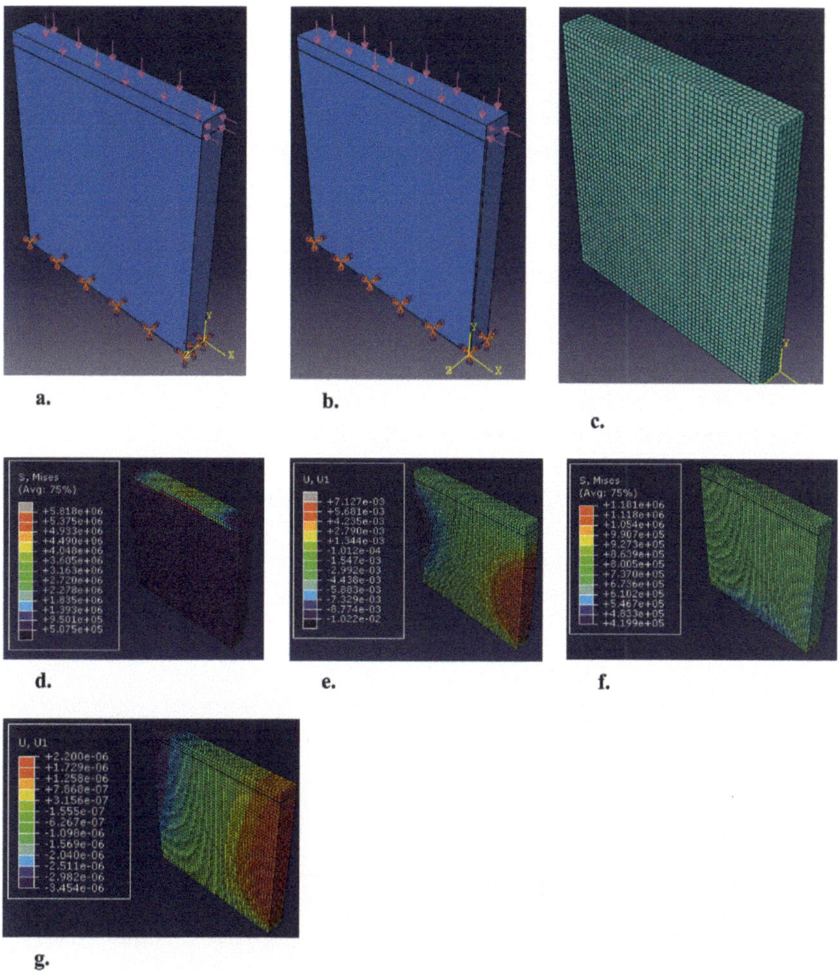

FIG. 10.17 Finite element analysis results of the wall shown in Fig. 10.16: (A) Model without wire mesh, (B) model with wire mesh, (C) meshed model, (D) stress distribution without wire mesh, N/m^2, (E) displacement along x direction without wire mesh, m, (F) stress distribution with wire mesh, N/m^2, (G) Displacement along x direction with wire mesh, m.

displacement along all directions. Wire mess is integrated into the wall by explicit tie surface. Fig. 10.17 clearly shows higher stress distribution on the wall panel when it is not reinforced. Lateral displacements of the wall are significantly reduced by this retrofitting method. This clearly shows the increased stiffness with the use of wire mesh.

10.6.2 Strengthening with cane, timber, and bamboo

Timber, cane, and bamboo can be used in various ways to reinforce failed segments of walls. Some examples are illustrated in Fig. 10.18. The loose material from the cracked wall should be removed or chipped, cleaned, grouted before timber battens or other measures are used to reinforce them. Use of timber battens is suitable for rammed earth houses. It helps reinforce the wall against tensile and buckling failures. The spacing between the consecutive batten must not be less than 2 times the thickness of the main wall or larger than 1/3 times the dimension of longer wall. The thickness of the batten can be in the range of 20–40 mm, whereas the width may vary from 80 to 100 mm. The batten must be of length equal to two times the thickness of the wall under consideration (Om, Gyeltshen, & Lama, 2015). Depending on the severity, batten can be provided on both faces of the wall and anchored or wire mesh can be provided as well.

The modeled wall has base explicitly fixed with $U1 = U2 = U3 = 0$. Loads as described above are applied along x direction, and the response is shown in Fig. 10.19. Comparison of stress and displacement in the unreinforced and reinforced walls as shown in Figs. 10.17D and 10.19 shows that this retrofitting technique increases the stiffness of the system, thereby decreasing its displacement and stresses.

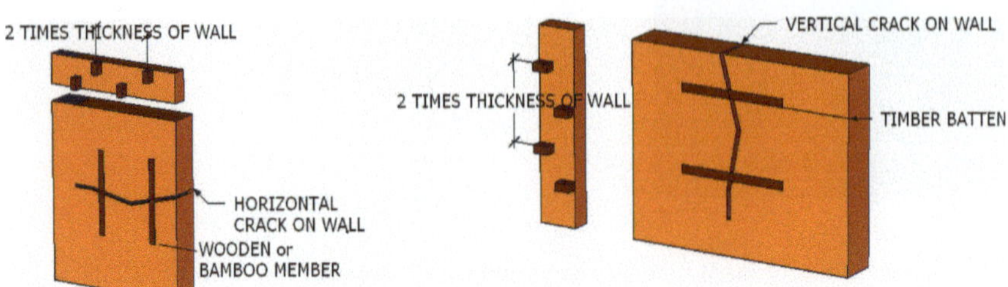

FIG. 10.18 Pictorial representation of retrofitting of horizontal and vertical crack, respectively, and strengthening with timber.

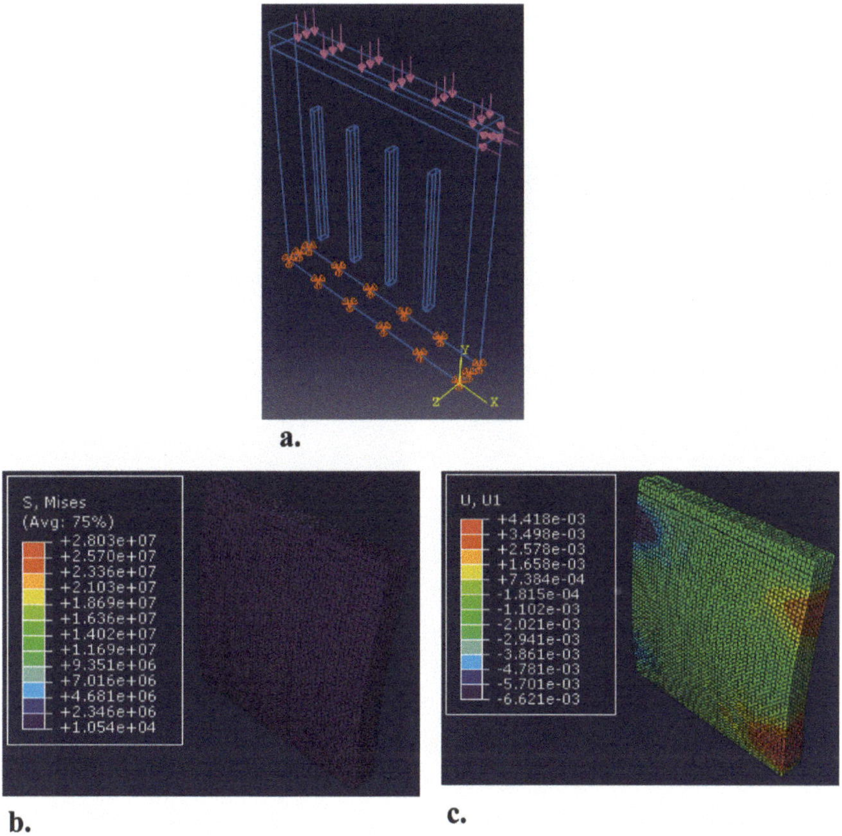

FIG. 10.19 FE model of timber-reinforced wall panel: (A) FE model with timber-reinforced wall panel, (B) stress distribution on timber-reinforced wall panel, N/m^2, (C) Displacement along x direction, m.

10.6.3 Corner reinforcement

As shown in Fig. 10.20, a typical corner failure occurs due to the poor corner interlocking and integrity between the orthogonal units. Collapse of corner, corner separation, cracks along the corner, rotation of the corner, and buckling were the common failure patterns observed during the past earthquakes.

Seismic performance of traditional houses in such context can be improved by various measures like providing seismic bands at plinth level, lintel level, and floor level (Murty, 2005). Retrofitting techniques like the ones shown in Fig. 10.21 can be adopted for such failure mechanisms. For rammed earth walls, the loose materials are removed and vertical timber dowel of 25–30 mm are erected at the corners and subsequently anchored to the foundation of the main wall. It is made sure that the

FIG. 10.20 Corner damage on rammed earth walls.

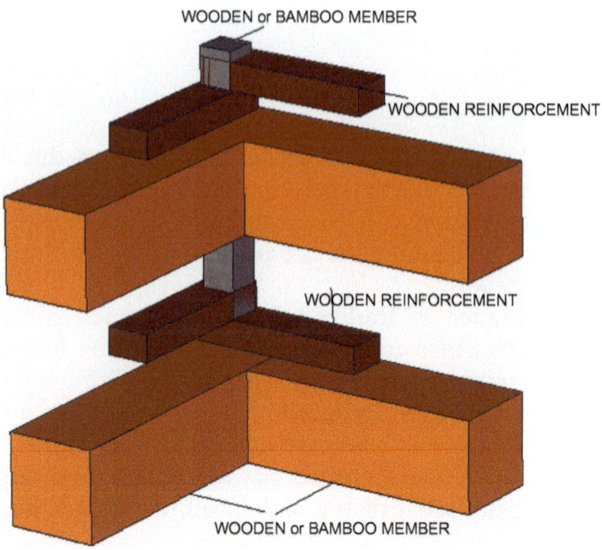

FIG. 10.21 Retrofitting at damaged corner.

successive rammed earth blocks of depth 500–600 mm are anchored by the timber member along the vertical direction. This will prevent development of vertical cracks, rotation, and overturning of walls. The minimum dimension of timber member depends on site condition and building layout.

In case of corner collapse of stone masonry wall, damaged portion should be carefully removed and reconstructed using rich cement mortar. Providing steel reinforcement anchored at foundation (jacketing) would serve as a better option. This system prevents development of wide cracks along bedding planes and helps re-cantering. The displacement and von Mises stresses induced by the loading conditions are shown in Fig. 10.22.

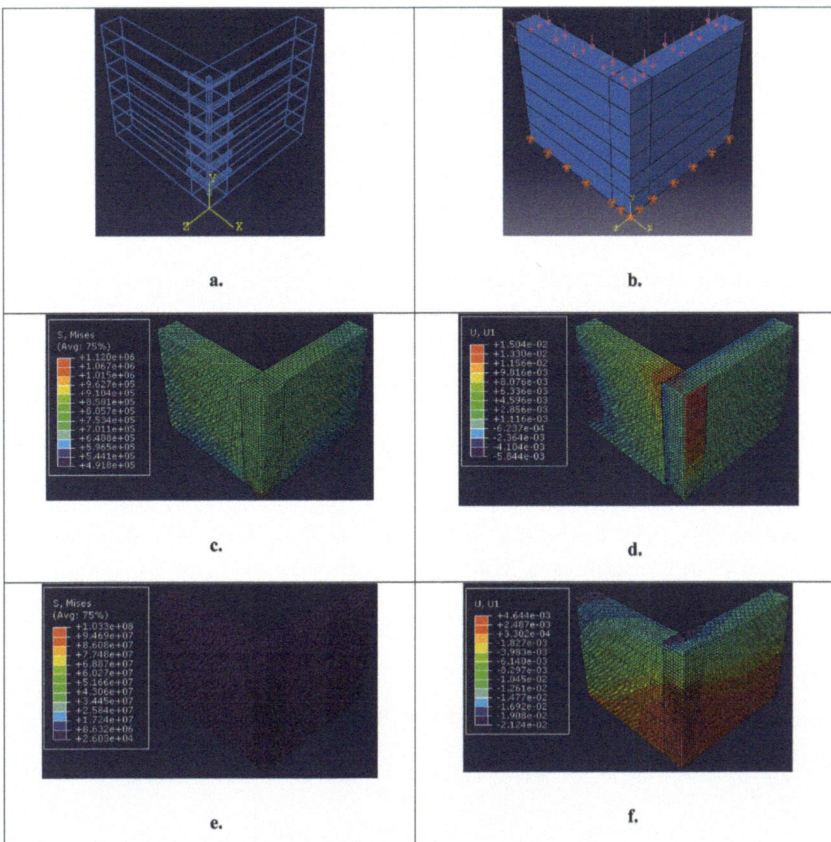

FIG. 10.22 FE analysis of orthogonal wall panel with proper corner connection: (A) FE model of corner reinforcement on rammed earth wall panel, (B) FE model of wall with loading condition, (C) stress distribution (without reinforcement in corner), N/m^2, (D) displacement under given loading scenario, m, (E) stress distribution (with reinforcement in corner), N/m^2, (F) displacement under given loading scenario, m.

As shown in Fig. 10.22, the corner reinforcement method can be effective in reducing displacement and stresses in the wall.

10.6.4 Timber reinforcement

Cracks of various patterns were observed in the periphery of doors, windows, and below timber floor joists. These are coined as lintel or joist crack depending on the location. As shown in Fig. 10.23, it also occurs when timber element (joist) are placed directly on soft element like mud wall and near windows where the opening has disturbed the load path.

The retrofitting process begins with removing the loose material and the damaged windows or door frames. The new frame must possess top and bottom flatter planks 38 mm thick so that the gravity load from above is spread over a larger area. The new frame must have vertical planks of at least 38 mm to stiffen the internal frame (Fig. 10.24). This technique is suitable for both rammed earth and stone masonry houses. Care should be taken in case of stone masonry by placing the frame over thick plaster (30 mm) of mud.

Stress and displacement distribution in a wall near an opening with and without retrofitting elements are shown in Fig. 10.25. The retrofitting elements significantly reduce stress concentration near the edges of the opening. Also, the region of severe displacement is less in case of openings where loads are distributed over the flat timber member (see Fig. 10.25C and D). It was found out that simply placing timber element around the periphery of wall opening would reduce the stresses on the wall; however, because it's not embedded inside the wall, it gives negative thrust on the corner, leading to the fracture at left corner. Therefore, it is advisable to anchor the timber elements toward the wall.

FIG. 10.23 Cracks in lintel area and joist area.

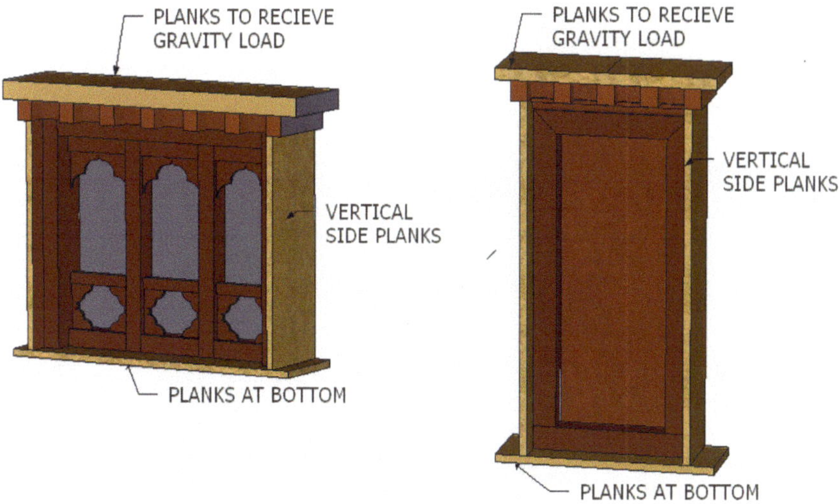

FIG. 10.24 Retrofitting against lintel and joist crack.

10.6.5 Timber band

Bhutanese architecture is unique, aesthetically soothing, and culturally important. However, structural integrity of its different elements is not robust. One peculiar structural weakness of Bhutanese architecture is the Bhutanese ornamental window (Fig. 10.11), which is a relatively large opening and heavier than conventional windows. When placed near the corners of a building, such large windows disrupt the load transmission path, making them more vulnerable. A potential strengthening method is illustrated in Fig. 10.26. A timber band (75 mm × 100 mm) is placed immediately below the Rabsel along with vertical supports of the same dimensions. Before placing the retrofitting band, the tilted Rabsel needs to be propped from the ground and supported by rope at the top. Finite element models with and without timber band is prepared and subject to gravity and lateral loads. The noggin timber members, which occur along vertical direction, were restrained along all directions ($U1 = U2 = U3 = 0$). The interaction of earth wall and timber members was considered to be non-cohesive. The lateral loads are applied along out-of-plane direction ($-z$ direction, indicated in Fig. 10.27). The results show that the retrofitting methods help to reduce stress concentration near the corner. It also reduces displacement of the wall.

High walls can undergo rotation and buckling during seismic action if they are not properly restrained. Timber bands can be an effective retrofitting solution in such cases. Tensile capacity of bands helps develop strut and tie mechanism. A 3 in. × 4 in. timber band and support system (timber noggin) can be embedded after chipping at the height of 2.8–3 m from

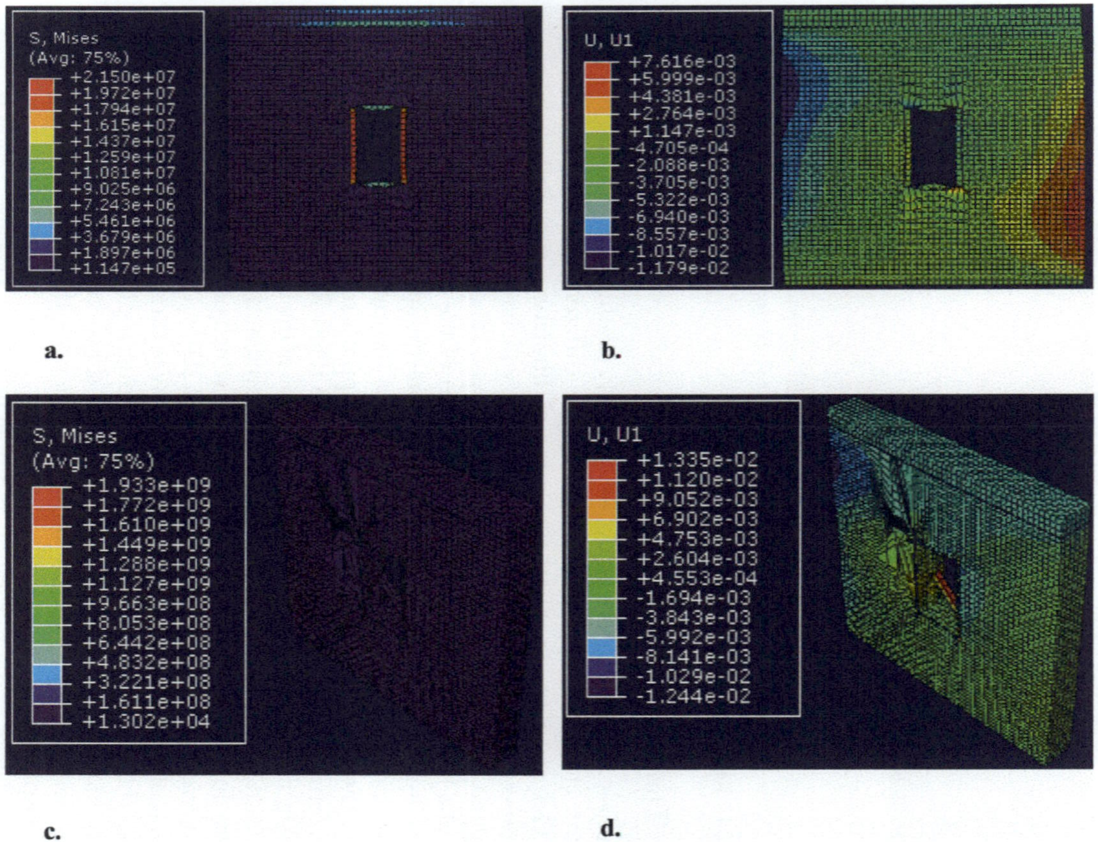

FIG. 10.25 FE analysis of wall panel with reinforced opening: (A) stress distribution (no retrofit), N/m^2, (B) displacement along in plane direction (without timber element) due to lateral load, m, (C) stress distribution (with retrofit), N/m^2, (D) displacement along in plane direction (with timber element).

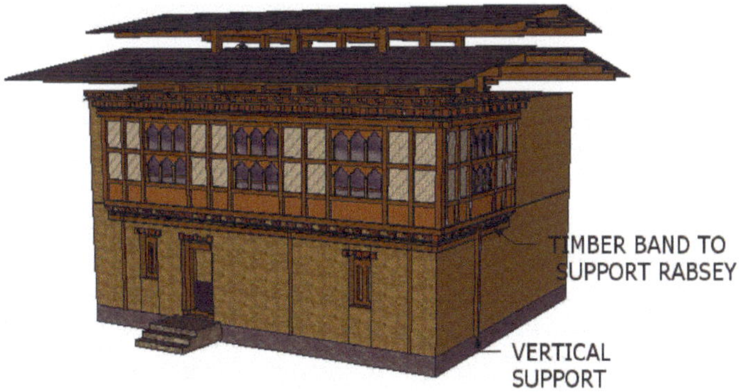

FIG. 10.26 Strengthening with timber band.

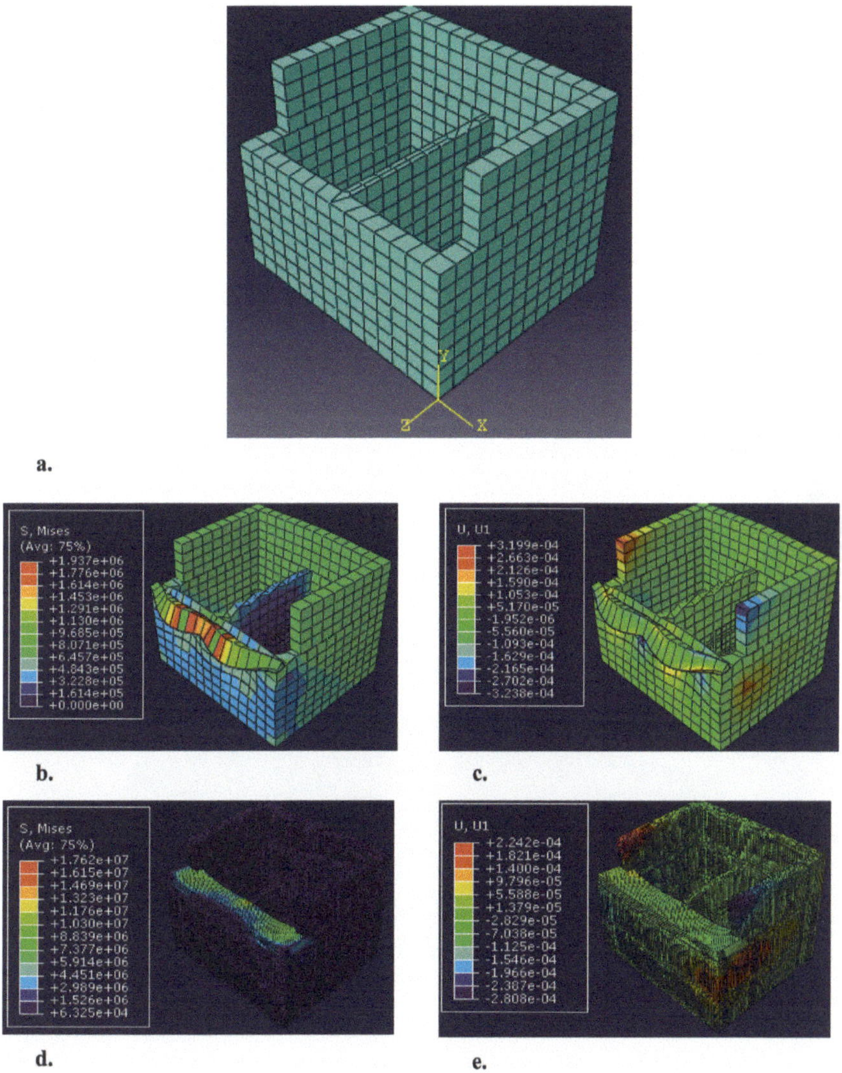

FIG. 10.27 Retrofitting with timber band FE analysis: (A) meshed model under consideration, (B) stress distribution without timber band, N/m^2, (C) displacement along +ve z direction, m, (D) stress distribution with timber band, N/m^2, (E) displacement along +ve z direction, m.

the floor as shown in Fig. 10.28 (Murty, 2005). These timber bands should be placed at plinth level, lintel level, and eave level to facilitate good box action.

FE model of a typical rammed earth house is prepared in Abaqus with timber band along the periphery of wall to support the Rabsel (Bhutanese ornamental windows). The results are shown in Fig. 10.27. It is clear that the timber band gives additional restraint against rotation of Rabsel and reduces stress intensity along the wall. It also reduces the displacement of the overhung window elements.

An alternative retrofitting method is the so-called polypropylene (PP) band. PP bands are commonly used in New Zealand (NZS 4297, 1998). Finite element models with and without timber band and noggin are prepared and subject to gravity and lateral loads, which are calculated above. The noggin timber members, which occur along vertical direction, were restrained along all directions ($U1 = U2 = U3 = 0$). The vertical timber noggin members are encasted to the ground and embedded within the wall. The interaction of earth wall and timber members was considered to be noncohesive. The lateral loads are applied along out-of-plane direction along positive x, and the entire periphery of the wall at top is subjected to the dead load. Because single floor building is considered for analysis, the results does not show significant variation in terms of stress. However, the displacement in unreinforced wall is 10^{-4} m, whereas after reinforcement, it is reduced to 10^{-2} m along $-x$ direction. The first wall from positive x direction is zoomed for better visibility of stresses and displacements as shown in Fig. 10.29.

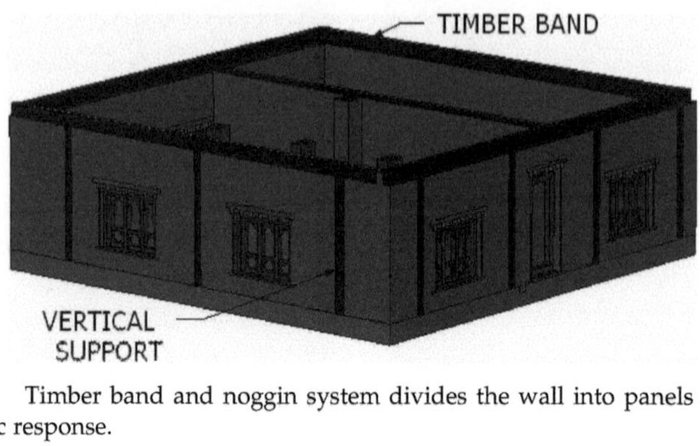

FIG. 10.28 Timber band and noggin system divides the wall into panels and facilates good seismic response.

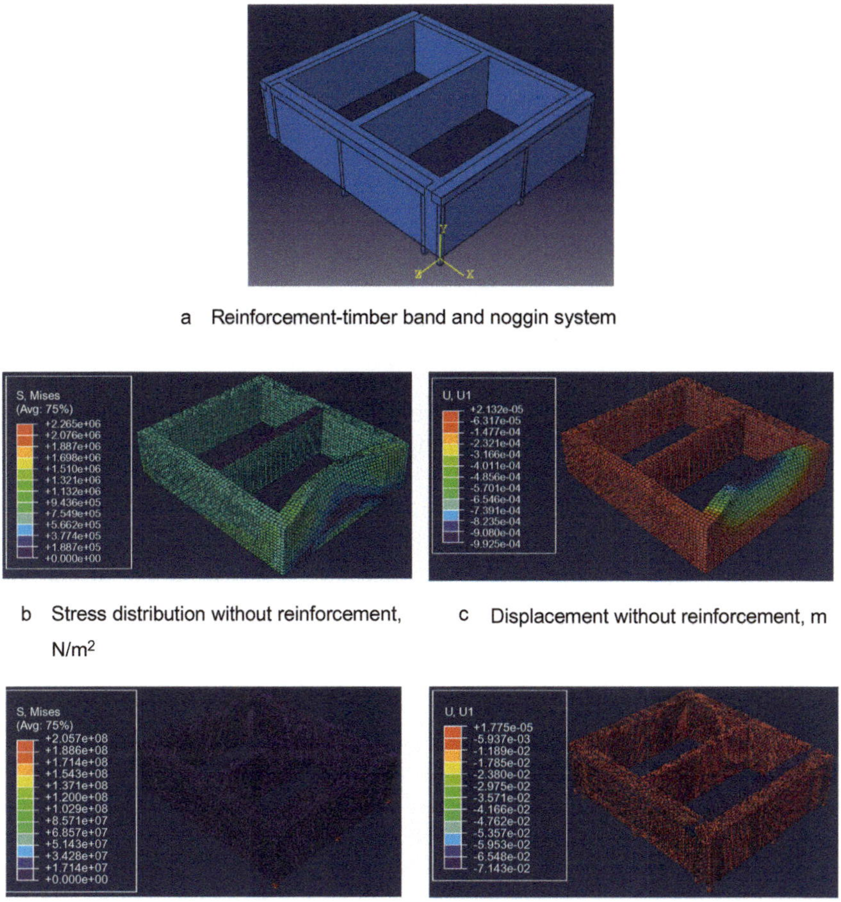

a Reinforcement-timber band and noggin system

b Stress distribution without reinforcement, N/m2

c Displacement without reinforcement, m

d Stress distribution with reinforcement, N/m2

e Displacement with reinforcement, m

FIG. 10.29 FE analysis of timber band and noggin system: (A) reinforcement timber band and noggin system, (B) stress distribution without reinforcement, N/m^2, (C) displacement without reinforcement, m, (D) stress distribution with reinforcement, N/m^2, (E) displacement with reinforcement, m.

10.6.6 Cross wall and masonry pier

Most walls of Bhutanese vernacular houses are slender because of their small thickness compared to their height and length. They fail to generate box action during seismic shaking and experience excessive bending near the mid span. Several out-of-plane collapse of such walls was observed during past earthquakes. Damaged walls can be strengthened by simple

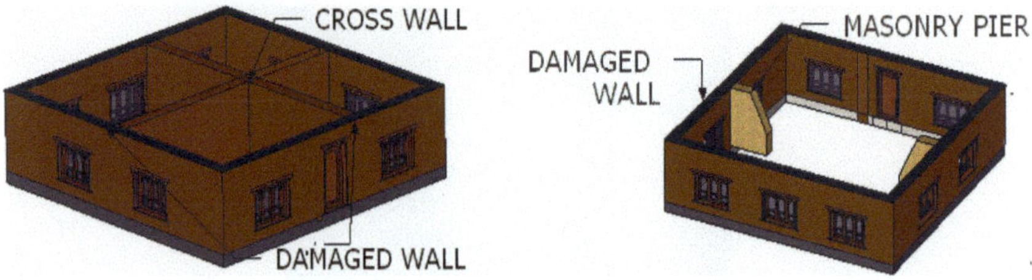

FIG. 10.30 Cross wall and earth pier strengthens the weaker direction wall.

cross walls and masonry piers. The thickness of the new cross wall or pier can be the same as that of the damaged one. An illustration of this method is shown in Fig. 10.30.

If the damage is not extensive, wire mesh and timber reinforcement can be used for repair and strengthening. For walls longer than 6 m, cross walls are recommended.

The largest stress in the model without pier and cross wall goes up to $10^5\ N/m^2$. These stresses are reduced about 10 times by the cross pier and cross walls. Fig. 10.31 shows that both cross wall and earth pier are good restraining elements, although cross wall performs better in terms of limiting the displacement.

10.6.7 Buttressing

During seismic action, buildings are subjected to combined inertial vibrations in its weaker and strong directions (Murty, 2005). In such circumstances, the load deposition to the foundation occurs through corners, where stress concentrations occur. Buttressing of 0.75 times the floor height can be used to reinforce corners as illustrated in Fig. 10.32 (Guide for the structural rehabilitation of heritage buildings, 2010). The main disadvantage of providing a buttress wall is that it affects the traditional aesthetics of the building. Note that buttress can also be provided in mid span of a long wall.

As shown in Fig. 10.33, buttresses are more effective in reducing largest stress but not as effective as cross walls in limiting the displacement along the direction of the lateral load.

10.7 Conclusions

We present the first comprehensive account of taxonomy, vulnerability, and strengthening of Bhutanese vernacular buildings considering the most dominant *Pa Chim* (rammed earth construction) and *Tship Chim* (stone masonry). Forensic interpretations of damage are reported considering

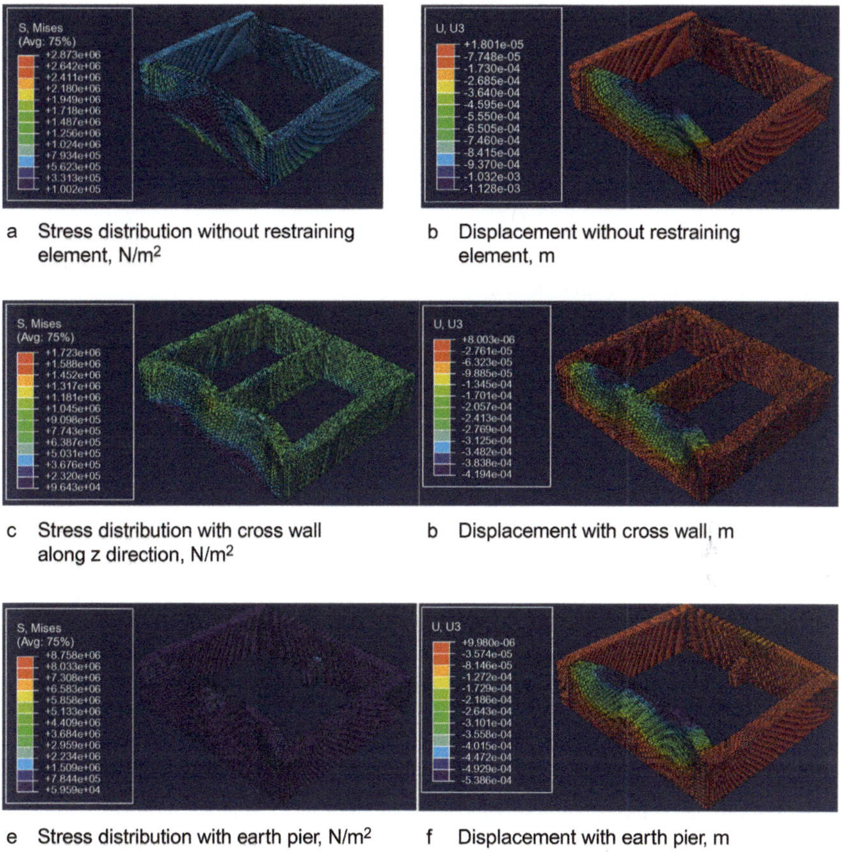

FIG. 10.31 FE analysis for cross wall and earth pier; (A) stress distribution without restraining element, N/m^2, (B) displacement without restraining element, m, (C) stress distribution with cross wall along z direction, N/m^2, (D) displacement with cross wall, m, (E) stress distribution with earth pier, N/m^2, (F) displacement with earth pier, m.

FIG. 10.32 Buttress in both x and z direction provides restrain against horizontal movement.

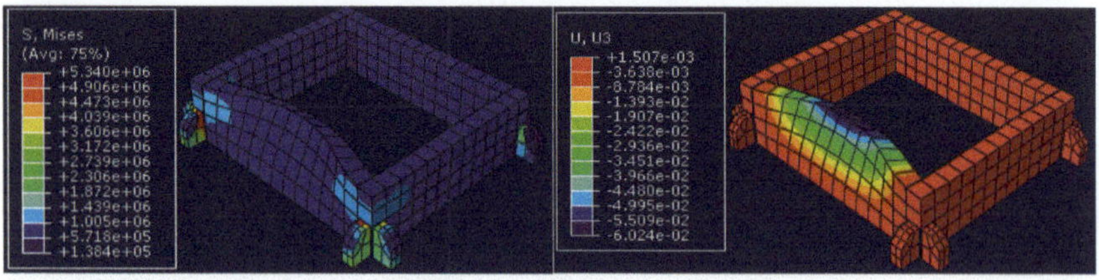

a. b.

FIG. 10.33 FE analysis of walls with buttresses: (A) stress distribution with buttress, N/m^2, (B) displacement with buttress, m.

the damage caused by the 2009 Mongar and the 2011 Sikkim-Nepal border earthquakes. We also present detailed accounts of construction systems in Bhutan. A broad taxonomy and vulnerability level for each building class per the newly formed taxonomy are presented. Existing construction practices are reviewed, and critical aspects of existing building regulations are discussed. Several methods of strengthening existing vulnerable and/or damaged buildings are demonstrated and described along with finite element analysis demonstrating their effectiveness.

The main finding is that existing buildings in Bhutan are highly vulnerable and may sustain severe damage during moderate to strong earthquakes. As moderate to strong earthquakes are likely to occur in Bhutan and neighboring areas, the existing building stocks in Bhutan need immediate strengthening to reduce loss of lives and properties during future earthquakes. There is also an immediate need of seismic hazard studies and formulation of new building codes compliant to seismic hazard and construction trends and techniques in Bhutan. As Bhutanese structures depicted intense damage even during relatively low shaking far away from the epicenter of the 2011 earthquake, it is likely that future earthquakes in the eastern part of Indian subduction region can be damaging. Proper understanding of seismic hazard and vulnerability, and at the same time, reducing building vulnerability through retrofitting and strengthening measures seem essential to alleviate the impending seismic risk.

Acknowledgments

The authors are grateful to the owners and inhabitants of the buildings investigated in this research for granting the permission to study their houses. We also acknowledge local masons and carpenters for answering our questions on construction techniques. Special thanks go to final-year civil students (2018 batch, College of Science and Technology Bhutan) and architect students (final year and third year from the same college) for assisting in the data collection process. RR acknowledges support from the University of Iceland research fund.

References

BBR (2018). *Bhutan building regulation (Issue 4).*

Bhutan (2009). Situation report earthquakes in Bhutan. In *India and Myanmar.* (issue September).

Bilham, R. (2019). *Himalayan earthquakes: A review of historical seismicity and early 21st century slip potential.* London: Geological Societyhttps://doi.org/10.1144/sp483.16. Special Publications, SP483.16.

Chaulagain, H., Gautam, D., & Rodrigues, H. (2018). Revisiting major historical earthquakes in Nepal: Overview of 1833, 1934, 1980, 1988, 2011, and 2015 seismic events. D. Gautam, & H. Rodrigues (Eds.), *Impacts and insights of Gorkha earthquake* (pp. 1–17). Vol. I(pp. 1–17). Elsevier. https://doi.org/10.1016/B978-0-12-812808-4.00001-8.

Chettri, N., Thinley, J., & Koirala, G. S. (2019). The comparative study on vernacular dwellings in Bhutan. *United International Journal for Research & Technology, 01*(02), 42–47. https://uijrt.com/volume-01-issue-04.

Chettri, N., Gautam, D., & Rupakhety, R. (2021). Seismic vulnerability of vernacular residential buildings in Bhutan. *Journal of Earthquake Engineering, 26*(1), 1–16. https://doi.org/10.1080/13632469.2020.1868362.

Gautam, D. (2017). Seismic performance of world heritage sites in Kathmandu Valley during Gorkha seismic sequence of April–May 2015. *Journal of Performance of Constructed Facilities. 31*(5)06017003https://doi.org/10.1061/(ASCE)CF.1943-5509.0001040.

Gautam, D. (2018). Observational fragility functions for residential stone masonry buildings in Nepal. *Bulletin of Earthquake Engineering.* https://doi.org/10.1007/s10518-018-0372-2.

Gautam, D., Fabbrocino, G., & Santucci de Magistris, F. (2018). Derive empirical fragility functions for Nepali residential buildings. *Engineering Structures.* https://doi.org/10.1016/j.engstruct.2018.06.018.

Gautam, D., Rodrigues, H., Bhetwal, K. K., Neupane, P., & Sanada, Y. (2016). Common structural and construction deficiencies of Nepalese buildings. *Innovative Infrastructure Solutions, 1*(1), 1–18. https://doi.org/10.1007/s41062-016-0001-3.

Guide for the structural rehabilitation of heritage buildings, C. C. W. S (2010). *Guide for the structural rehabilitation of heritage buildings, CIB commission W023—Wall structures.*

Guidelines for Retrofitting of Buildings (2006). *Guidelines for retrofitting of buildings.*

JDWNRH (2012). *Initial seismic vulnerability assessment of Jigme Dorji Wangchuck National Referral Hospital.*

Murty, C. V. R. (2005). Learning earthquake design and construction. *Resonance, 10*(11), 89–92. https://doi.org/10.1007/bf02837649.

NZS 4297 (1998). *New Zealand standards.* (Vol. 4297).

Om, N., Gyeltshen, K., & Lama, U. (2015). *Retrofitting of traditional rammed earth buildings. Zorig Melong – Technical Journal.*

PDNA (2011). *Joint rapid assessment for recovery, reconstruction and risk reduction. Royal Government of Bhutan.* The W. B. and U. Nation.

Rodgers, J. E., Lizundia, B. J., Yangdhen, S., Lotay, Y., Hortaçsu, A., & Tshering, K. D. (2014). Postearthquake safety evaluation procedures for Bhutan buildings. In: *NCEE 2014 – 10th U.S. National Conference on Earthquake Engineering: Frontiers of Earthquake Engineering*https://doi.org/10.4231/D3TX35694.

Sandbhor, S., & Botre, R. (2015). A systematic approach towards restoration of heritage buildings- a case study. *International Journal of Research in Engineering and Technology, 02* (03), 229–238. https://doi.org/10.15623/ijret.2013.0203004.

Sethna, Z. (2008). *A sustainability approach to standards for rammed earth constructions in Bhutan:* (p. 1). University of Cambridge.

Sferrazza Papa, G., & Silva, B. (2018). Assessment of post-earthquake damage: St. Salvatore church in Acquapagana, Central Italy. *Buildings*. *8*(3)https://doi.org/10.3390/buildings8030045.

Silva, R. A., Olliveira, D. ., Schueremans, L., Miranda, T., & Machado, J. (2014). Shear behaviour of rammed earth walls repaired by means of grouting. In *9th international masonry conference* (pp. 1–12).

Thimphu Thromde (2004). *Thimphu municipal development control regulations*.

Windstorm, B., & Schmidt, A. (2013). *A report of contemporary rammed earth construction and research in North America*: (pp. 400–416). https://doi.org/10.3390/su5020400.

World Bank (2009). *Joint rapid assessment for recovery, reconstruction and risk reduction*. October.

CHAPTER

11

Local out-of-plane failure modes in traditional block-masonry buildings

Claudia Casapulla

Department of Structures for Engineering and Architecture, University of Naples Federico II, Naples, Italy

11.1 Introduction

In masonry buildings without a box-like behavior but with good quality masonry, as most existing unreinforced masonry buildings in the historic centers, local out-of-plane failures can take place, even under low intensities of ground motion and especially for peripheral walls. Recurrent vulnerabilities in these buildings are weak connections between orthogonal walls, absence of connecting ties, insufficiently rigid floor diaphragms, low strength and deterioration of materials, and irregular patterns of openings in the walls.

It has been recognized, in particular, that the most recurrent failure modes caused by seismic forces acting orthogonally to the building walls involve simple or complex rocking mechanisms of portions of masonry walls (Casapulla, Giresini, & Lourenço, 2017). In the case of simple rocking, mainly regarding façades, poor wall-to-wall and wall-to-horizontal structure connections cause the separation of the façade from the orthogonal walls and its autonomous rocking motion under seismic excitations (Fig. 11.1A). The complex rocking, instead, can be caused by similar vulnerability conditions but with better connections between interlocked walls, so that the rocking mechanism may involve the façade together with parts of orthogonal walls (Fig. 11.1B) or portions of the orthogonal walls at the corners (Fig. 11.1C).

https://doi.org/10.1016/B978-0-12-821087-1.00001-6

(A) (B) (C)

FIG. 11.1 Simple (A) and complex (B) rocking walls and rocking corner (C).

These kinds of local failures generally may occur to walls characterized by proper transversal connections within their cross-section, capable of assuring a monolithic behavior. Rather than rubble masonry, this is the case of traditional masonry block walls with cut stones (ashlar) or bricks, where the portions of walls involved in the mechanism can be modeled as rigid blocks. For this kind of masonry, it has widely been recognized since the pioneering work of Heyman (1966) that the limit equilibrium method, with a standard or nonstandard model, is one of the most suitable methods of analysis to assess the onset and evolution of out-of-plane failure modes.

In fact, according to the kinematic theorem of standard limit analysis (Heyman, 1966), local mechanisms in masonry buildings subjected to seismic loading can be schematized as kinematic chains of masonry portions, regarded as rigid macro- or micro-block assemblages, interacting through contact interfaces. Horizontal seismic actions can be represented as static forces, applied to the center of mass of each moving block, in addition to other loadings. These forces are related to the gravitational loads through a load factor, which can be defined by means of the limit equilibrium condition of any virtual mechanism to represent an upper bound multiplier. On the other hand, the application of the static theorem implies that a stable equilibrium state of a masonry structure can be guaranteed by the existence of an internal force system equilibrating the external loads (safe theorem). This approach provides a lower-bound or safe solution of the collapse load factor, whereas the solution that satisfies the hypotheses of both theorems, equilibrium, compatibility, and material conditions, is the correct solution and provides the collapse load multiplier for the specific problem. The material condition is based on the simple assumption that the blocks have infinite strength in compression and no tensile strength, whereas sliding is not allowed at block interfaces.

Despite its simplicity, the standard limit analysis presents some limitations: (i) the Heyman's safe theorem can fail, that is, collapse may occur despite the existence of an equilibrated force system (Bagi, 2014); (ii) some particular shapes of masonry structures, for example, flat arches, would never collapse unless sliding occurred (Heyman, 1966); (iii) the Heyman's results for simple rocking may be too conservative if the more realistic assumptions of the presence of sliding is not considered (Casapulla & Argiento, 2016).

Although limitation (i) can be prevented by using the kinematic approach, the other two limitations can only be overcome by adopting a nonstandard model, based on the possibility of sliding at block interfaces according to the Coulomb's law (Drucker, 1953). This model is even more suitable to ancient unreinforced block masonry buildings, where the quality of the contact surfaces in dry-joints or the mortar joints themselves might be so deteriorated as to substantially reduce the original friction coefficient and cohesion.

The use of a nonstandard model requires nonstandard limit analysis to get the "exact" solution (nonassociative solution) within a range of both statically and kinematically admissible solutions, because the unique solution is no longer guaranteed by the static and kinematic theorems in their classic forms (Drucker, 1953). Several computational micro-modeling approaches have recently been developed within the dry rigid block limit analysis (Baggio & Trovalusci, 1998; Ferris & Tin-Loi, 2001; Gilbert, 2001; Gilbert, Casapulla, & Ahmed, 2006; Livesley, 1978; Orduña & Lourenço, 2005; Portioli, Casapulla, Gilbert, & Cascini, 2014), or the discrete element modeling (Bui, Limam, Sarhosis, & Hjiaj, 2017; Caliò, Marletta, & Pantò, 2012; Pantò, Giresini, Sassu, & Caliò, 2017) to get such solutions, but a considerable computational effort is still required by them, especially when nonlinear analyses are developed. Moreover, interactions among rocking, sliding, and twisting of connected rigid microblocks are still not well taken into account to define the most likely in-plane and out-of-plane failure mechanisms (Casapulla & Maione, 2018).

An alternative method capable of tackling larger engineering problems is sought by the macroblock modeling approach based on the kinematic method, which is suitable to find the solution when nonassociative friction is assumed. In fact, under this assumption, the minimization of the load factor leads to a value that bounds from below the majority of the many solutions that are both statically and kinematically admissible and it can, therefore, represent a safe solution (Gilbert et al., 2006).

In addition, this modeling approach is also suitable to the application of the displacement-based assessment method, aimed at the construction of pushover curves from the static multipliers (nonlinear kinematic analysis). These are obtained by the application of the principle of virtual work, considering kinematic varied configurations of the examined mechanism, at large displacements. Along the incremental kinematic analysis, the contribution of links can be taken into account, till the ultimate equilibrium condition (Lagomarsino & Cattari, 2015). Pushover curves referred to ideal rigid blocks are represented by force-displacement laws with negative stiffness, whereas a trilinear simplified force-displacement law can be assumed for blocks with geometrical imperfections, namely "semi-rigid" (Casapulla et al., 2017; Casapulla, Maione, & Argiento, 2019; Griffith, Magenes, Melis, & Picchi, 2003; Lagomarsino, 2015). Although neglecting the dynamic effects, nonlinear kinematic analysis still represents a valid alternative to a more complex dynamic analysis to evaluate the displacement capacity of rocking rigid blocks (Abrams, AlShawa, Lourenço, & Sorrentino, 2017; Casapulla et al., 2017; Giresini, Fragiacomo, & Lourenço, 2015; Lagomarsino, 2015).

Within this context, an advanced macroblock model has recently been developed (Casapulla & Argiento, 2016; Casapulla & Argiento, 2018) with the purpose of describing the role of frictional resistances both at the onset and during the evolution of local failure mechanisms in masonry buildings. It has been applied to the analysis of simple and complex rocking mechanisms and validated on the basis of the comparison with experimental findings and other models based on limit analysis (Casapulla & Argiento, 2018) and discrete element analysis (Bui et al., 2017), with a considerably lower computational effort in obtaining optimized solutions. A further validation has recently been derived from some comparisons with the results of dynamic analyses (Casapulla, Giresini, Argiento, & Maione, 2019), confirming that the proposed macroblock model and solution procedure are a robust and reliable tool to provide a good assessment of both the geometry of the most likely mechanism and its capacity in terms of loads and displacements. The great potential of this model is mainly related to the strategy adopted to define the actual frictional resistances while using the kinematic approach to obtain optimized conservative results. It can be applied to any kind of local failure mechanisms in masonry buildings, provided that the external and internal loadings are clearly defined. In this chapter, however, it is presented with reference to the most recurrent simple and complex rocking mechanisms.

In the following, Section 11.2 deals with the modeling approach with reference to the simple and complex rocking masonry walls, with the application of the displacement-based method to construct pushover curves from the static multiplier (nonlinear kinematic analysis). In Sections 11.3 and 11.4, the proposed method of analysis is applied to real case studies, that is, the failures of a masonry façade of a historic structure and of a corner in a masonry school building, both buildings hit by the 2016–17 Central Italy seismic sequence. The selected case studies represent the two most recurrent rocking mechanisms in masonry buildings vulnerable to earthquakes. Concluding remarks are provided in Section 11.5.

11.2 Nonlinear static analysis of rocking masonry block structures using a macroblock modeling approach with frictional resistances

11.2.1 Method of analysis and modeling approach

The method of analysis adopted in this chapter is based on the plasticity theory for masonry structures, using the nonlinear kinematic approach to address the nonstandard material problem. According to this approach, as anticipated in the Introduction, it is possible to identify the onset of the most likely failure mode among all the feasible mechanisms, to which the lowest value of the load factor corresponds; this load factor represents the maximum strength capacity of the system to resist earthquakes. The evolution of the mechanism is then analyzed in terms of displacement capacity by considering kinematic varied configurations, at large displacements, that is, using a displacement-based approach.

The proposed macroblock model is based on the assumption that the general failure involves a number of cracks, which separate the structure into a few macroblocks, and all the possible relative motions among masonry units are concentrated along these cracks. Masonry block walls with regular units and staggering are concerned (single-leaf walls arranged in a running bond pattern), whereas the cracking and crushing of blocks are ignored. In fact, the macroblocks and the constituent units have infinite strength in compression, tension, and shear, whereas no tension and frictional behavior are assumed at their contact interfaces along the cracks (Coulomb failure criterion). The adoption of a single-leaf stone/brick wall is a simplified schematization based on the condition that the actual disposition of the units, defining the structure of the transversal section of the wall, can guarantee a monolithic behavior. Different bond patterns can easily be implemented, but out of the scope of this chapter.

The inclinations of the main crack lines, considered as average inclinations of the discontinuous lines that tend to separate the walls into macroblocks, represent the variables of the minimization problem, that is, they are not assigned a priori but depending on geometrical and mechanical parameters. Using an optimization routine, these can be found by imposing a minimum condition on the expression of the onset load factor (kinematic approach), accounting for the corresponding frictional resistances.

According to what previously proposed (Casapulla & Argiento, 2018), the actual value of the resultant frictional resistance F_w is assumed as a function of the crack inclination angle α_c (blue dashed Line 2 in Fig. 11.2A), to the extent of a weighted value of the maximum one F_{max}, that is:

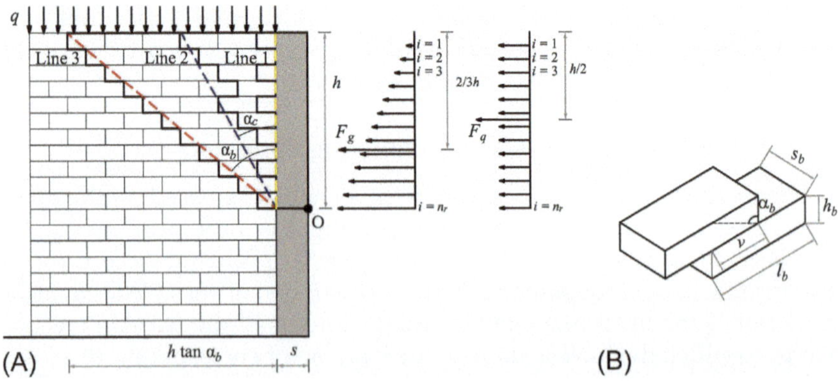

FIG. 11.2 Possible crack line and maximum frictional resistances (A); masonry unit dimensions (B).

$$F_w = \left(1 - \frac{\alpha_c}{\alpha_b}\right) F_{max} = \left(1 - \frac{\alpha_c}{\alpha_b}\right)(F_g + F_q) \tag{11.1}$$

where F_g and F_q are the maximum contributions due to the self-weight and the overloading, respectively, α_b is the maximum admissible crack angle in compliance with the unit staggering ratio ($\alpha_b = \tan^{-1}(v/h_b)$ in Fig. 11.2B), $v = l_b/2$ is the overlapping length between two units, and F_{max} is the maximum resultant frictional resistance allowed in the wall (corresponding to $\alpha_c = 0$, yellow dashed Line 1 in Fig. 11.2A). In particular, the two contributions are:

$$F_g = \gamma s_b^2 h_b \frac{n_r(n_r+1)}{2} f; \quad F_q = n_r q v f \tag{11.2}$$

where γ is the specific weight of the material, s_b is the width of the masonry unit corresponding to the wall thickness s, n_r is the number of the involved rows ($n_r = h/h_b$), f is the friction coefficient, and q is the uniformly distributed overload per unit length of wall. The application points of these stabilizing forces were found to be at about $2/3$ of the height h from the top of the panel for F_g and at about $h/2$ for F_q (Fig. 11.2A).

Eq. (11.1) means that the more the crack slope α_c is close to the maximum angle α_b ($\alpha_c = \alpha_b$ corresponds to pure rotation, red dashed Line 3 in Fig. 11.2A), the lower friction forces contribute to the weighted value ($F_w \rightarrow 0$), and, on the contrary, the more α_c is close to zero (Line 1 in Fig. 11.2A), the larger weight the maximum frictional resistance has on F_w, i.e., $F_w \rightarrow F_{max}$.

On the other hand, during the evolution of the rocking mechanism around the hinge O (Figs. 11.2A and 11.3), the frictional contributions gradually decrease after a certain displacement, due to the progressive

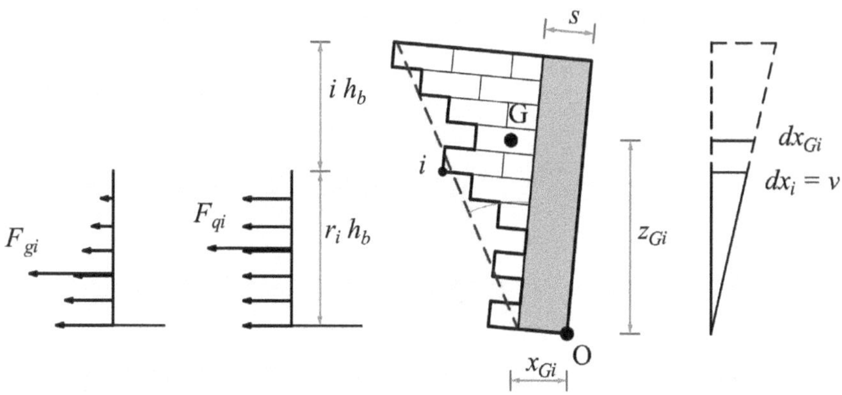

FIG. 11.3 Variation of frictional resistances at increasing wall rotation.

loss of contact in the units along the crack line. Thus, the variation of these forces can be represented by nonlinear functions of the decreasing number of involved rows, as described in the following.

The effectiveness of these forces calculated by Eq. (11.1) on the whole height h of the rocking wall is guaranteed as long as the first two rows at its top lose their contact with the fixed portion of the wall along the crack (first threshold displacement). As the wall rotation increases, the subsequent threshold displacements are assumed to be attained at every two underlying rows of the rocking wall that lose the contact.

This means that, considering the generic detached row i (at the top $i=0$) and the center of mass G of the moving portion of the wall (Fig. 11.3), the threshold displacement d_{xGi} is reached when $d_{xi}=v$. In fact, being r_i the number of rows still involved in frictional contact, that is:

$$r_i = n_r - i \quad \text{for} \quad i = 0,2,4,\ldots,n_r \tag{11.3}$$

overlapping of units is lost at the vertical distance from the hinge equal to $(r_i\ h_b)$.

Thus, it will be:

$$dx_{Gi} = \frac{z_{Gi}}{r_i h_b} v \tag{11.4}$$

while the corresponding reduced resultants F_{gi} and F_{qi} can easily be calculated by Eq. (11.2), by replacing n_r with r_i.

In the case of simple rocking walls, the crack line is vertical ($\alpha_c = 0$, yellow dashed Line 1 in Fig. 11.2A), and the frictional resistances on the side walls due to the overloading can be neglected (Casapulla & Argiento, 2016). As an example of this case, Fig. 11.4A shows a nonlinear

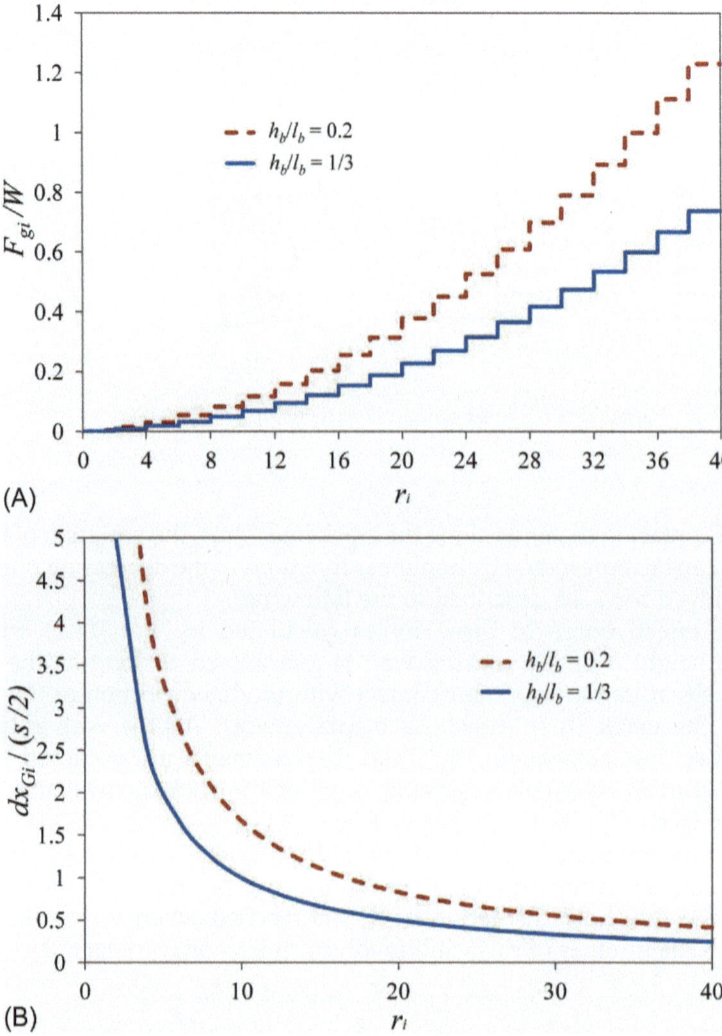

FIG. 11.4 Dimensionless frictional resistance (A) and displacement of the control point (B) vs. the number of rows in frictional contact.

relationship (step chart) between the dimensionless frictional resistance (W is the self-weight of the rocking wall) and the number of involved rows. This figure highlights that this stabilizing effect plays a major role at smaller displacements (higher r_i) and for smaller unit aspect ratios h_b/l_b (slender units). Instead, the displacement of point G sharply increases when the connection between the orthogonal walls tends to reduce in the lower part (smaller r_i) and this effect is greater for slender units (smaller h_b/l_b), as shown in Fig. 11.4B.

11.2.2 Nonlinear kinematic analysis of simple rocking walls

The most likely out-of-plane mechanisms of a façade wall connected with two side walls mainly depend on the geometry and boundary conditions. In particular, if flexural effects are prevented, simple or complex rocking can be considered depending on the efficacy of the walls connection at the corners.

In the case of weak connections, the simple rocking of the façade wall can take place without involving the sidewalls, and the crack pattern can actually be identified a priori as characterized by two main vertical-stepped crack lines along the corners with height H, generally coincident with the floor height (Fig. 11.5A). These separate the moving macroblock of the façade from the resting parts of the system, whereas frictional resistances develop on the interfaces of the involved units. This means that the façade can rotate around its external bottom edge OO′, while the units along the two vertical crack lines exhibit a rocking–sliding motion with a clear prevalence of sliding.

As the main crack lines are vertical, the frictional contribution F_q due to the overloading can be neglected and the resultant frictional force

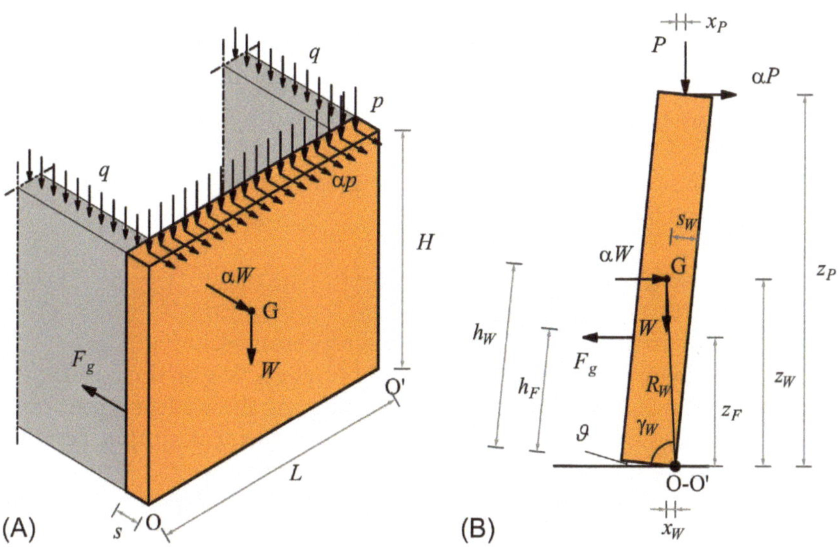

FIG. 11.5 Simple rocking of a façade connected with two sidewalls (A) and its varied kinematic configuration (B).

transmitted to the rocking façade by each sidewall is derived by Eqs. (11.1) and (11.2) as the maximum value of F_g referred to whole height $H = n_r\, h_b$, that is:

$$F_w = F_g = \left[\gamma s_b^2 h_b \frac{n_r(n_r+1)}{2}\right] f = F_{max} \tag{11.5}$$

It is worth highlighting that such an expression has been adopted by the Commentary to the Italian technical standard for construction CNTC19 (eq. C8.7.1.2 of Ministero delle Infrastrutture e dei Trasporti, 2019), with a reduction of 20% in favor of safety.

The onset and the evolution of this mechanism till collapse are then investigated by means of nonlinear kinematic analysis. This allows the construction of a pushover curve, which relates the load factor to the horizontal displacement of the façade center of gravity, assumed as the control point (Olivito & Porzio, 2019). The curve can be obtained by applying the principle of virtual work, considering varied kinematic configurations of the examined mechanism, at large displacements. Along this incremental kinematic analysis, the contribution of different links can be taken into account, till the ultimate equilibrium condition (Casapulla & Argiento, 2016; Lagomarsino, 2015).

In particular, considering the generic kinematic configuration of the rocking wall, represented by the finite rotation ϑ in Fig. 11.5B, the virtual work performed, for example, by the gravitational force W and the related inertial force $\alpha(\vartheta)W$, both applied on the center of mass G, can be expressed by the relation:

$$L_W = \alpha(\vartheta) W z_W \phi - W x_W \phi \tag{11.6}$$

where ϕ is the virtual rotation and x_W and z_W are the horizontal and vertical distances of the application point of the two forces from the hinge, respectively.

These distances, together with the angular position γ_W of the same point, have the following expressions:

$$x_W = R_W \cos(\gamma_W + \vartheta);\;\; z_W = R_W \sin(\gamma_W + \vartheta);\;\; \gamma_W = \tan^{-1}\left(\frac{h_W}{s_W}\right) \tag{11.7}$$

where the parameters R_W, h_W, and s_W are represented in Fig. 11.5B.

The virtual work performed by the other forces acting on the façade can be expressed in a similar way by only considering their vertical and horizontal distances from the hinge. Thus, the load factor is simply derived from the application of the virtual work principle as:

$$\alpha(\vartheta) = \frac{W x_W \pm P x_P + 2 F_g z_F}{W z_W + P z_P} \tag{11.8}$$

where P is the resultant overloading on the façade wall, giving positive or negative moment depending on its stabilizing or overturning effect, respectively.

It is noteworthy that the onset condition is obtained by Eq. (11.8) by putting $\vartheta = 0$ and the conservative constraint $\alpha \leq f$ could also be considered to prevent pure sliding of the wall, rather hard to take place (Casapulla & Argiento, 2016). Moreover, with reference to the generic kinematic configuration identified by the angle ϑ, the horizontal displacement dx_G of the center of gravity G, assumed as the control point, is:

$$dx_G(\vartheta) = s_W - x_W \tag{11.9}$$

The incremental analysis is extended till the ultimate value d_0 of the control displacement, for example, $d_0 = s_W = 0.5\ s$ in the presence of the sole self-weight. If different overloads are taken into account (e.g., plane and vaulted floors), different phases of motion, depending on the loss of effectiveness of their connections, can be considered till the ultimate equilibrium condition before overturning, which is characterized by the null value of the load factor ($\alpha(\vartheta) = 0$). It should be noticed, however, that this displacement capacity corresponds to the loss of static equilibrium but can be overcome in dynamic conditions, keeping the stability (Giresini, Fragiacomo, & Lourenço, 2015).

As mentioned above and described in detail in (Casapulla & Argiento, 2016), it is noteworthy that although the weights are constant during the wall rotation, the frictional resistances along the vertical crack lines keep constant till the façade starts to detach from the sidewalls. Then, they gradually decrease according to the nonlinear (step) function of the decreasing number of involved rows sketched in Fig. 11.4.

11.2.3 Nonlinear kinematic analysis of complex rocking walls

In the case of stronger connections at the corners of the façade with the two sidewalls, the most likely failure mode is the complex or compound mechanism (Fig. 11.6A), characterized by the rocking motion of part of the façade wall together with a portion of the sidewalls. In this case, the inclination of the crack lines and the height of the rocking macroblock are not known a priori and require the optimization routine introduced above. The frictional resistances exerted by the sidewalls depend of the optimized geometry as well, according to Eq. (11.1).

With reference to the onset configuration (Fig. 11.6A) and using parameters similar to those for the simple rocking, the optimization problem for the complex rocking is formulated as follows:

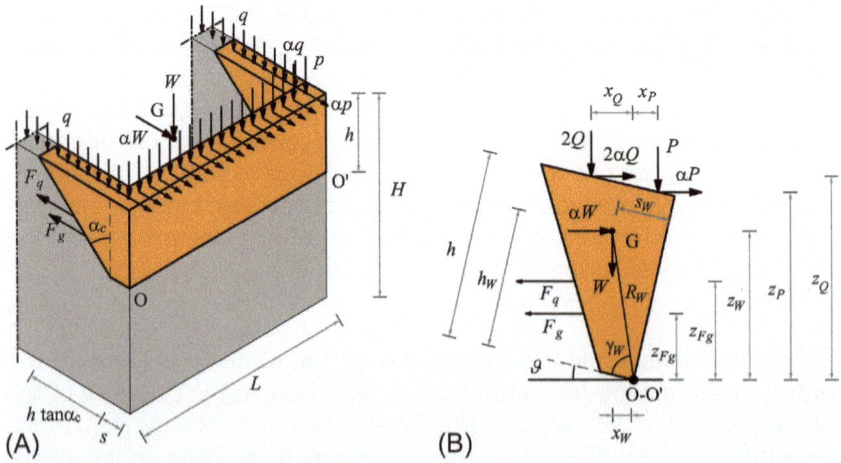

FIG. 11.6 Complex rocking of a façade connected with two sidewalls (A) and its varied kinematic configuration (B).

$$\textit{objective function}: \ \min \alpha_0 = \frac{Ws_W + P\dfrac{s}{2} + 2Q\left(\dfrac{hh\tan\alpha_c}{2} + s\right) + 2\left(F_g\dfrac{h}{3} + F_q\dfrac{h}{2}\right)\left(1 - \dfrac{\alpha_c}{\alpha_b}\right)}{Wh_W + (P+2Q)h} \tag{11.10}$$

$$\textit{variables}: \ \alpha_c, h \tag{11.11}$$

$$\textit{constraints}: \ 0 \leq h \leq H; \quad 0 \leq \alpha_c \leq \alpha_b \tag{11.12}$$

where F_g and F_q are the maximum contributions of the resultant frictional resistances according to Eq. (11.2), W is the self-weight of the moving portion applied on the center of masses G, and P and Q are the resultants of the overloading on the façade and each moving part of the sidewalls, respectively. It is important to highlight that W, Q, s_W, and h_W strictly depend on the geometry of the moving portion, that is, on the variables of the mathematical problem.

Hence, in the case of mixed rocking–sliding failure mode, Eqs. (11.10) and (11.11) will represent the effective load factor and the actual geometry of the moving wedge, respectively. The pure sliding mechanism for the same portion of wall is governed by:

$$\alpha_{0,\text{sliding}} = \frac{(W+P+2Q)f + 2F_{max}}{W+P+2Q} \tag{11.13}$$

which is easy to verify that is always greater than f, as occurs for simple rocking. This means that the conservative constraint $\alpha_0 \leq f$ could also be added to Eq. (11.12), although the pure sliding failure mode is rather hard to take place.

The evolution of this mechanism till collapse is still investigated by means of nonlinear kinematic analysis. In particular, taking into account Eqs. (11.6) and (11.7) and with reference to the generic kinematic configuration of the portion of the rocking wall, represented by the finite rotation ϑ in Fig. 11.6B, the application of the virtual work principle will provide the load factor:

$$\alpha(\vartheta) = \frac{Wx_W \pm Px_P + 2Qx_Q + 2\left(F_g z_{Fg} + F_q z_{Fq}\right)\left(1 - \dfrac{\alpha_c}{\alpha_b}\right)}{Wz_W + Pz_P + 2Qz_Q} \tag{11.14}$$

whose parameters are still governed by the relations in Eqs. (11.7) and (11.9), aimed to construct pushover curves, similarly.

Like the simple rocking mechanisms, these curves can be characterized by different phases of motion, depending on the loss of effectiveness of the connections among the different elements of the system.

11.2.4 The complex rocking of the corner

The corner failure in a masonry building generally involves well-connected masonry walls under orthogonal forces, whose geometry is not known a priori. The possible mechanism can be simplified as a combination of two failure modes: the bidirectional rocking–sliding failure, with the formation of two main cracks that break the continuity of the interlocked walls (Casapulla, Giresini, et al., 2019b), and the horizontal flexure failure of the two interlocked walls at the corner, with the formation per each wall of two macroblocks rotating around three cylindrical hinges (Casapulla & Maione, 2020). The geometrical parameters strongly affect the prevalence of one over the other, and in this chapter, only the former is considered, as belonging to the family of rocking mechanisms. In this case, the failure mode is characterized by a prismatic wedge rotating around a hinge placed at its external vertex, involving rocking–sliding motions along the crack lines. Mostly, the corner failure involves the top level of a masonry building for the whole floor height because it is mainly activated in the presence of roofs inclined in both directions with thrusting elements and openings close to the edges. Fig. 11.7 presents a recurrent geometry of the corner failure with the presence of openings within the interlocked walls.

These are denoted Wall 1 and Wall 2, respectively, oriented along the Cartesian Y and X axes and are assumed to have the same thickness s, for

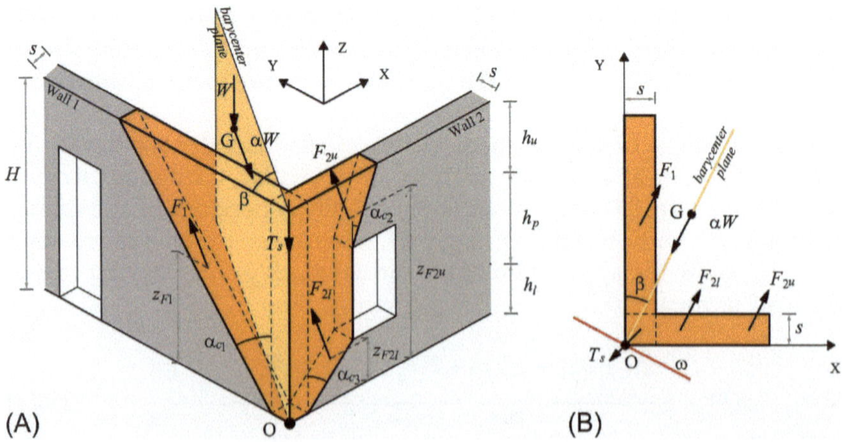

FIG. 11.7 Corner failure (A) and projection of the rocking wedge on the horizontal XY plane (B).

the sake of simplicity. The vertical overloads of a hipped roof are transferred to the intersection of the walls by a hip rafter, disposed along the bisector plane of the corner, as commonly occurs. The external horizontal loads are the inertial forces and the static thrust T_s due to the hip rafter, inclined of 45° with respect to the X and Y axes. Point G is the center of all the involved masses, that is, the mass of the rocking macroblock plus the masses of the roof.

Due to nonsymmetric conditions in geometry and loading of the 3D failure mechanism, the vertical plane of rotation does not coincide with the corner bisector plane, but with a plane whose trace on the horizontal XY plane is represented by the joining line OG. Thus, the inclination β of this plane with respect to the YZ plane is dependent on the geometry of the unknown crack pattern (Fig. 11.7B) and on the coordinates of point G. On the other hand, it is easy to recognize that the direction of the horizontal forces and the frictional resistances is parallel to the line OG and the rotation axis ω is orthogonal to this line on the horizontal plane.

The internal and external actions displayed in Fig. 11.7 are: F in different positions as the weighted frictional resistances, W as the sum of weights and overloads, and T_s as the horizontal thrust transferred by the hip rafter. In particular, it is worth highlighting that the frictional resistances are evaluated by applying the weighting criterion described above for the simple and complex rocking, even if, unlike the latter mechanisms, these resistances are not contained in the wall plane but differently oriented.

Taking into account the parameters introduced above, which are now referred to the corner failure represented in Fig. 11.8, the onset configuration is formulated by the following optimization problem:

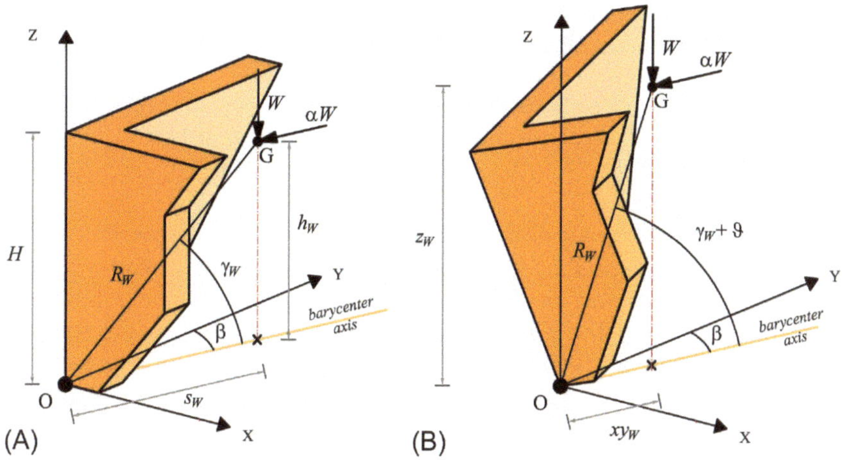

FIG. 11.8 Initial (A) and varied (B) kinematic configurations of the corner.

$$\textit{objective function}: \ \min \alpha_0 = \frac{Ws_W + F_1 h_{F1} + F_{2l} h_{F2l} + F_{2u} h_{F2u} - \frac{\sqrt{2}}{2} T_s (a_1 + a_2) h_T}{W h_W} \tag{11.15}$$

$$\textit{variables}: \ \alpha_{c1}, \alpha_{c2} \tag{11.16}$$

$$\textit{constraints}: \ 0 \leq \alpha_{c1}, \alpha_{c2} \leq \alpha_b \tag{11.17}$$

where $a_1 = \sin\beta$, $a_2 = \cos\beta$. The conservative constraint $\alpha_0 \leq f$ could be considered for this case as well.

Similarly, the pushover analysis can be made by considering geometric nonlinearity, that is, by evaluating the load factor for varied kinematic configurations, as a function of the displacement of a control point. In this case, still considering Eqs. (11.6), (11.7), and (11.9), the virtual work principle will provide the load factor:

$$\alpha(\vartheta) = \frac{Wxy_W + F_1 z_{F1} + F_{2l} z_{F2l} + F_{2u} z_{F2u} - \frac{\sqrt{2}}{2} T_s (a_1 + a_2) z_T}{W z_W} \tag{11.18}$$

which coincides with Eq. (11.15) for $\vartheta = 0$.

On the other hand, even for the more complex corner mechanism, the pushover curves can be characterized by different phases of motion, depending on the active contributions of the involved links, till the ultimate equilibrium condition. The evolution of these phases in terms of load factor and displacement is governed by the same Eq. (11.18) with updated coordinates of the application points, according to Eq. (11.7) with x_W replaced by xy_W. The control point can be assumed as the same point for all phases, for example, the center of mass of the masonry portion only. Its displacement is affected by the variation of xy_W, that is:

$$d_{xyG}(\vartheta) = s_W - xy_W \tag{11.19}$$

The incremental analysis is extended till the control displacement for which $\alpha(\vartheta) = 0$. The pushover curves so obtained are constructed with reference to the forces and displacements belonging to the plane of rotation. Nevertheless, when the seismic capacity is compared with the seismic demand reported in given directions, the curves can also be represented as their projections in different planes, as shown for the case study in Casapulla, Giresini, et al. (2019).

11.3 The case study of the rocking façade of a masonry church hit by the 2016–17 Central Italy earthquakes

The first heritage structure herein examined by using the nonlinear kinematic approach presented in the previous section is the façade of the San Michele Arcangelo Church in Lisciano (a municipality of Ascoli Piceno), struck by the 2016–17 Central Italy seismic sequence. This church is representative of the typical typology with a single nave of the 13th century and exhibits a clear incipient simple rocking of the main façade detaching from the sidewalls (Fig. 11.9). Previous analyses of this case study were carried out by the author in collaboration (Casapulla,

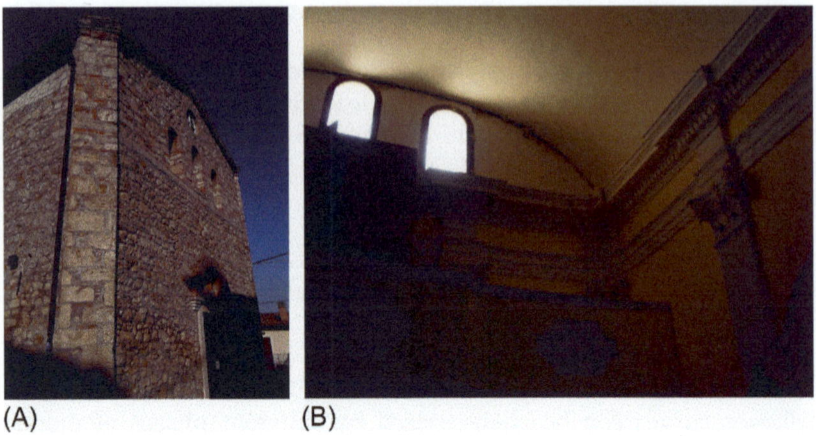

FIG. 11.9 San Michele Arcangelo Church (Ascoli Piceno): main façade (A) and internal view of its detachment (B).

Maione, et al., 2019; Giresini, Casapulla, Denysiuk, Matos, & Sassu, 2018), mainly to investigate the structural response to real accelerograms. Conversely, this case study is herein summarized to emphasize the reliability and suitability of the modeling approach with reference to the prediction of the seismic safety of the structure based on the observed damage and on the current Italian seismic code (Ministero delle Infrastrutture e dei Trasporti, 2018; Ministero delle Infrastrutture e dei Trasporti, 2019).

The masonry typology of the walls can be considered as square rubble stone masonry with good texture and variable dimensions. To calculate the frictional resistances with Eq. (11.5), it is assumed, with some approximation, that the restrained height of the façade is composed of 54 rows and the height h_b of the single stone is about 15 cm (Fig. 11.2B). The so computed resultant frictional resistance is also reduced by 20% according to the CNTC19 (Ministero delle Infrastrutture e dei Trasporti, 2019). Moreover, being that the length l_b of the stone blocks is rather variable, two reference limit values are considered for the parameter $\rho = l_b/h_b$, that is, $\rho = 1$ and $\rho = 2$. As shown later within the capacity curves, this parameter strongly affects the frictional resistances and the capacity of the façade wall with respect to the onset of the mechanism, whereas its effect decreases during the evolution of the mechanism.

Finally, the thickness of the walls is assumed coincident with that related to the stone blocks, the specific weight $w = 21$ kN/m^3 is considered for this masonry typology, according to table C8.5.1 of the CNTC19

TABLE 11.1 Geometrical and mechanical features of the rocking façade.

Parameter	Variable	[−]	Free rocking façade	Restrained façade ($\rho = 1$)	Restrained façade ($\rho = 2$)
Restrained height	H	[m]	–	8.1	8.1
Thickness	s	[m]	0.65	0.65	0.65
Width	L	[m]	9.30	9.30	9.30
Number of rows	n_r		–	54	54
Friction coefficient	f		–	0.6	0.6
Masonry specific weight	w	[kN/m^3]	21	21	21
Position radius of the self-weight	R_W	[m]	4.57	4.57	4.57

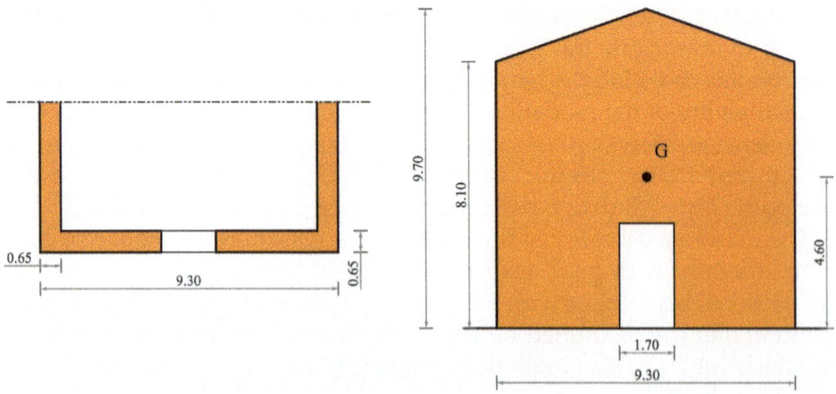

FIG. 11.10 Representation of the geometrical dimensions (in meters) of the façade wall of San Michele Arcangelo Church (Ascoli Piceno, Italy).

(Ministero delle Infrastrutture e dei Trasporti, 2019), and the friction coefficient is taken as $f = 0.6$, according to the literature (Casapulla & Maione, 2018; Lee, Park, Cho, & You, 2001; Lourenço & Ramos, 2004; Vasconcelos & Lourenço, 2009). Table 11.1 reports the geometrical and mechanical features of the rocking façade, whereas its geometrical dimensions are also reported in Fig. 11.10.

11.3.1 The capacity curve of the rocking façade

The pushover curve $\alpha(d_G)$ is obtained from the application of Eqs. (11.8) and (11.9) to the façade under study, neglecting the overloading (the façade is not a load-bearing wall) and accounting for different restraining conditions of the orthogonal walls, that is, the absence of restraints (free rocking) and the presence of frictional resistances related to two values of parameter ρ ($\rho = 1$ and $\rho = 2$). The capacity curve $a(d)$ of the equivalent single degree of freedom system is obtained from the pushover curve, by simply assuming $a(d) = \alpha(d_G)g/F_C$ and $d = d_G$, being g the gravitational acceleration and F_C a confidence factor that takes into account the available level of knowledge of structural properties. These conditions derive from the code formulations (Ministero delle Infrastrutture e dei Trasporti, 2019):

$$a(d) = \frac{\alpha(d_G)g}{e^* F_C}; \quad d = \Gamma d_G \tag{11.20}$$

where e^* is the rate of the total mass participating in the rocking mechanism and Γ is the transformation factor of the displacements. Being $\delta_{x,j}$ the virtual horizontal displacement of the j-th lamping mass, the two introduced factors are:

$$e^* = \frac{\left(\sum_j W_j \delta_{x,j}\right)^2}{\sum_j W_j \sum_j W_j \delta_{x,j}^2}; \quad \Gamma = \frac{\sum_j W_j \delta_{x,j}^2}{\delta_{x,G} \sum_j W_j \delta_{x,j}} \tag{11.21}$$

which are both equal to 1 for the case of the rocking façade under study. A confidence factor $F_C = 1$ is also assumed for this analysis, because some uncertainties of the structural properties are taken into account with the variability of the most influencing parameter ρ.

Fig. 11.11A and B represent the capacity curves and the relations between the frictional resistances and displacements, respectively, with reference to the restrained façade with $\rho = 1$ and $\rho = 2$ and the free rocking condition. It clearly emerges that the frictional resistances strongly increase the acceleration activating the mechanism, and their stabilizing contribution increases with the aspect ratio ρ of the units. Moreover, the trend of the capacity curves along the evolution of the mechanism is strictly related to the variation of the frictional resistances, as is evident by comparing Fig. 11.11A and B. In fact, till displacement d_B or $d_{B'}$, the first descending branch of both capacity curves is linear (Fig. 11.11A), because the frictional resistances are active along the whole height of the walls (constant branch in Fig. 11.11B). After such a displacement, both capacity curves and the frictional resistances have a step-wise linear decreasing trend, due to the progressive loss of contact, starting from the top of the interlocked walls; the frictional resistances become null at the displacement d_C or $d_{C'}$, depending on the aspect ratio ρ of the units. In general, longer units would imply enlarged interlocking between them and, consequently, greater displacement capacity, after which the capacity curves become linear and coincide with those corresponding to the free rocking.

11.3.2 Seismic safety assessment of the rocking façade according to the nonlinear kinematic approach of the Commentary to the Italian technical standard (CNTC19)

In this section, the seismic safety of the San Michele Arcangelo Church façade is assessed according to the nonlinear kinematic approach of the Commentary to the Italian technical standard CNTC19 (Ministero delle Infrastrutture e dei Trasporti, 2019), which is based on the capacity spectrum method (CSM) using over-damped elastic response spectra (Freeman, 1998; Lagomarsino, 2015). This commentary represents one of the most advanced in Europe on the assessment of the local out-of-plane mechanisms in existing masonry buildings. In fact, its earlier version,

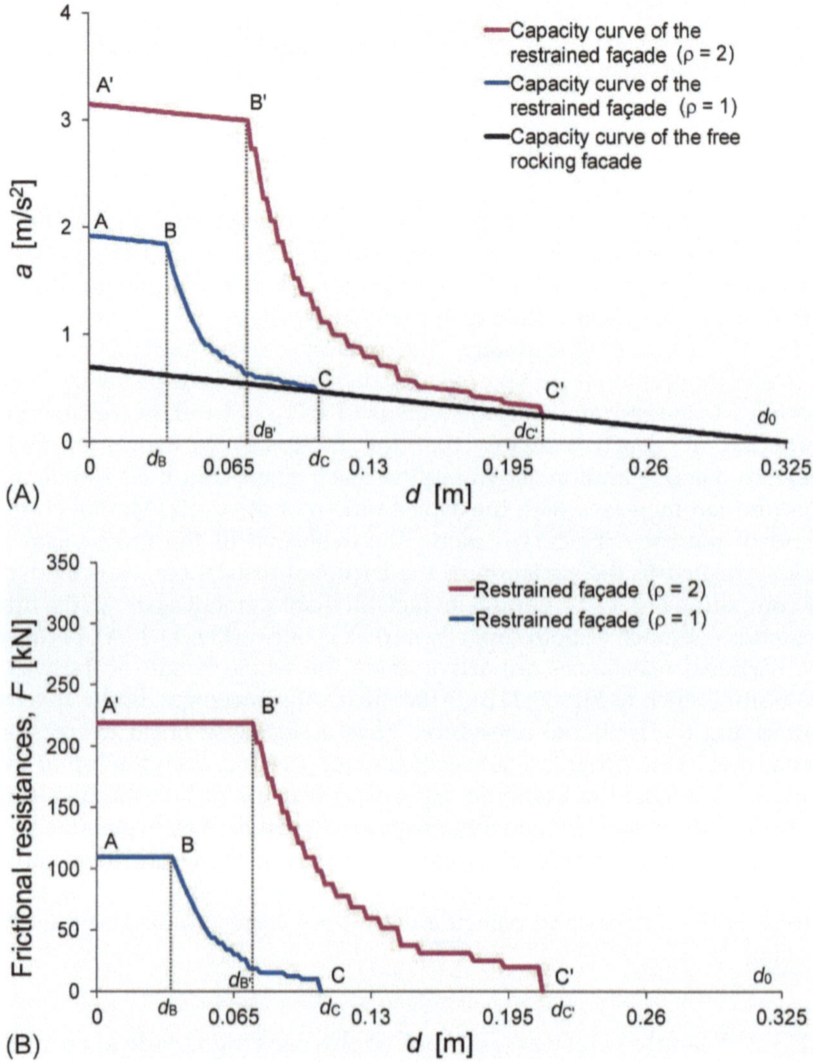

FIG. 11.11 Capacity curves (A) and frictional resistances (B) related to the free rocking façade and different restraining conditions with square ($\rho = 1$) and slender units ($\rho = 2$).

called CNTC09 (Ministero delle Infrastrutture e dei Trasporti, 2009), already transposed the most recent scientific advancements in modeling and verifying these structures, becoming a reference on this topic more updated than Eurocode 8 (Godio & Beyer, 2019; Sorrentino et al., 2016).

The seismic safety is herein evaluated by the ζ index summarizing the comparison between the expected seismic capacity and demand, with

TABLE 11.2 Reference values of the capacity parameters for the two conditions of the façade under study, related to the LS0, LS1, and LS2 limit states.

Limit state	Capacity parameter	[−]	Free rocking façade	Restrained façade ($\rho = 1$)	Restrained façade ($\rho = 2$)
LS0	$a_{LS0} = \alpha_0 g$	[m/s²]	0.69	1.92	3.14
LS1	$d_{LS1} = 0.4\ d_0$	[m]	0.13	0.13	0.13
LS2	$d_{LS2} = 0.6\ d_0$	[m]	0.195	0.195	0.195

reference to three meaningful limit states (Giresini et al., 2018), that are rocking initiation (LS0), moderate rocking (LS1), and near collapse (LS2) conditions. LS0 is representative of serviceability limit states, whereas LS1 and LS2 are considered as life-safety and collapse limit states, respectively. These reference values of the capacity parameters for the façade under study are reported in Table 11.2.

Once defined the reference displacement capacities for the LS1 and LS2 limit states, the application of the CSM according to the provisions of the Italian Commentary CNTC19 (Abrams et al., 2017) requires deriving the corresponding values of the equivalent periods through the following relations:

$$T_{LS1} = 1.68\pi\sqrt{\frac{d_{LS1}}{a(d_{LS1})}}; \qquad T_{LS2} = 1.56\pi\sqrt{\frac{d_{LS2}}{a(d_{LS2})}} \tag{11.22}$$

where $a(d_{LS1})$ and $a(d_{LS2})$ are the accelerations corresponding to the reference displacements on the capacity curve. The equivalent periods allow identifying the corresponding seismic demands from the over-damped acceleration–displacement response spectra (ADRS) curves. In fact, the displacement demand for a specific limit state can simply be obtained from the intersection between the ADRS curve and the line starting from the origin of the axes, which inclination depends on the equivalent period (Lagomarsino, 2015).

For the case under study, a soil type B (deposits of dense sand, gravel, or stiff clay) is assumed, which implies the soil factor $S = 1.19$ accounting for topographic and stratigraphic conditions. Following the indications of the CNTC19 (Ministero delle Infrastrutture e dei Trasporti, 2019), the assumed values of the damping ratio of the over-damped ADRS are 8% and 10% for the LS1 and LS2 limit states, respectively, whereas the predicted peak ground accelerations (PGAs) corresponding to the three limit states are $a_{g,\ LS0} = 0.072g$, $a_{g,\ LS1} = 0.178g$, $a_{g,\ LS2} = 0.23g$. The spectral acceleration for LS0 will be $a_{g,\ LS0}$ multiplied by S.

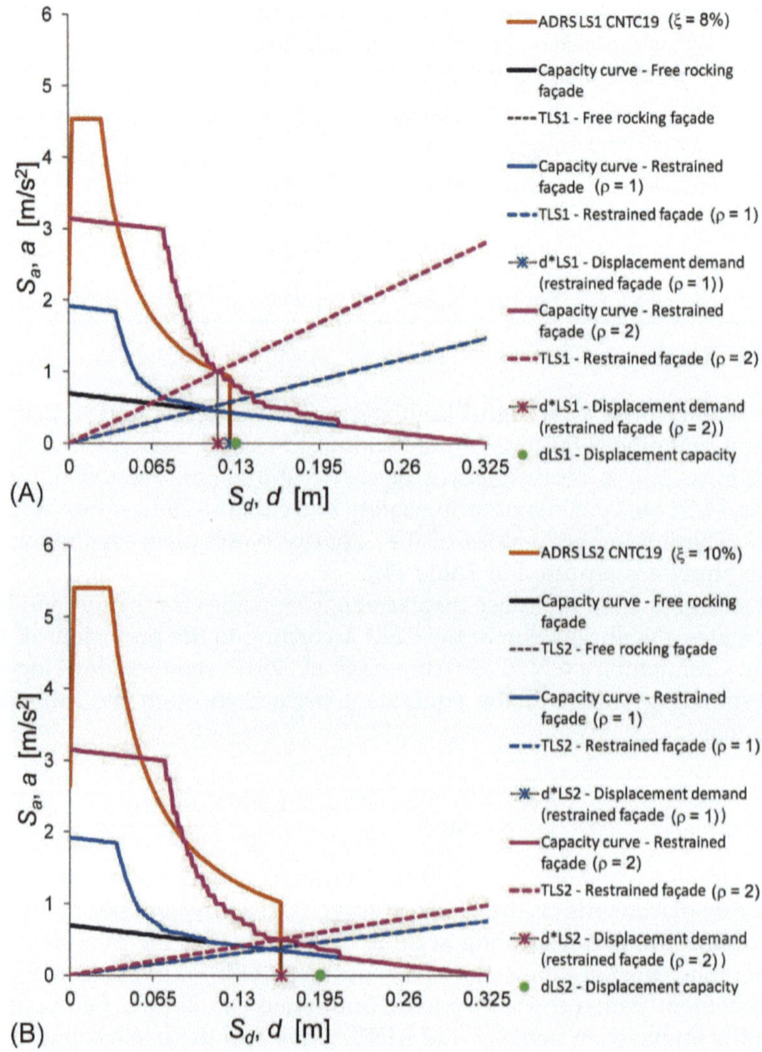

FIG. 11.12 Superposition of the capacity curves related to San Michele Arcangelo Church façade and ADRS provided by the CNTC19 (Ministero delle Infrastrutture e dei Trasporti, 2019), for the LS1 (A) and LS2 (B) limit states.

Fig. 11.12 shows the superposition of the capacity curves (in terms of a and d) related to the San Michele Arcangelo Church façade and the over-damped ADRS (in terms of S_a and S_d). The displacement demand related to each restrained conditions is highlighted by cross-type points together with the capacity displacements (green circles), related to the LS1 (Fig. 11.12A) and LS2 (Fig. 11.12B) limit states.

TABLE 11.3 Seismic safety indexes for LS0, LS1, and LS2 and different boundary conditions.

	Limit state	Seismic capacity	Seismic demand	Seismic safety index ζ
Free rocking façade	LS0 [g]	0.071	0.086	0.825
	LS1 [m]	0.130	0.125	1.039
	LS2 [m]	0.195	0.164	1.186
Restrained façade ($\rho = 1$)	LS0 [g]	0.196	0.086	2.283
	LS1 [m]	0.130	0.125	1.039
	LS2 [m]	0.195	0.164	1.186
Restrained façade ($\rho = 2$)	LS0 [g]	0.321	0.086	3.742
	LS1 [m]	0.130	0.116	1.125
	LS2 [m]	0.195	0.164	1.186

In Table 11.3, the evaluation of the seismic safety of the façade of San Michele Arcangelo Church is synthesized through the index ζ, representing the capacity/demand ratio for each limit state and boundary condition. From this table, it is first evident that the activation of motion (LS0) can only be predicted for the free rocking condition ($\zeta < 1$), whereas the capacity is higher than the demand in the other conditions. Actually, it was verified in (Casapulla, Maione, et al., 2019) that the seismic demands provided by the design spectra were underestimated with respect to the seismic inputs really occurred and affecting the case study. In fact, the activation of motion that could really be observed after the 2016–17 Central Italy earthquakes was determined by values of PGA higher than those assigned by the Italian seismic code (Ministero delle Infrastrutture e dei Trasporti, 2018) to all the limit states. This means that the seismic safety is expected to be not assured ($\zeta < 1$) even under the condition $\rho = 1$ of the restrained façade, if the real accelerograms are considered.

Instead, the safety indexes for the LS1 and LS2 limit states are greater than unity for all conditions and seismic inputs, meaning that the moderate and severe rocking motions are prevented, as actually occurred. It is also worth noting from (Casapulla, Maione, et al., 2019) that for LS1 and LS2, the seismic demand derived by the code is in good agreement with those recorded in situ. However, as a general trend, the predictive values obtained from the CSM are in rather good agreement with the real behavior of the façade under the 2016–2017 Central Italy earthquakes.

11.4 The case study of the corner failure of a masonry school building hit by the 2016–17 Central Italy earthquakes

As a further case study, the complex rocking of the corner mechanism is analyzed in this section with reference to the primary school P. Capuzi in the municipality of Visso (Macerata, Central Italy), hit by the seismic events of the 2016–17 Central Italy earthquake (Fig. 11.13). The main damage occurred in the building after the seismic shocks of August and October 2016 is the total collapse of one of the upper corners due to inappropriate connections between walls and timber roof.

Previous analyses with a detailed description of this case study and comparisons between different modeling approaches, in static and dynamic fields, were already carried out (Casapulla, Giresini, et al., 2019b; Ferrero, Lourenço, & Calderini, 2018; Giresini, Solarino, Paganelli, Oliveira, & Froli, 2019). Some of these results are herein summarized and the suitability of the proposed modeling approach is herein stressed with reference to the seismic safety of the structure, based on the observed damage and on the current Italian seismic code (Ministero delle Infrastrutture e dei Trasporti, 2018; Ministero delle Infrastrutture e dei Trasporti, 2019).

To analyze the corner failure, involving the upper level of the building (first floor and attic), it is important to highlight that the thrusting elements of the roof are the hip rafters sitting on the corners of the building along their bisector planes and rising to the central ridges (Fig. 11.14A and B); to each side of these hip rafters, a number of purlins, with variable length, are framed parallel to the two intersecting walls. This means that the weight of the roof around the corner is transferred to the walls in major part through the hip rafter as a concentrated inclined load and in minor part (just the bottom strips of the roof) along the walls as uniformly distributed loads. The calculation of the load transferred by the hip rafter to

(A) (B)

FIG. 11.13 The P. Capuzi primary school in Visso (Macerata) before (A) and after (B) the 2016–17 Central Italy earthquakes.

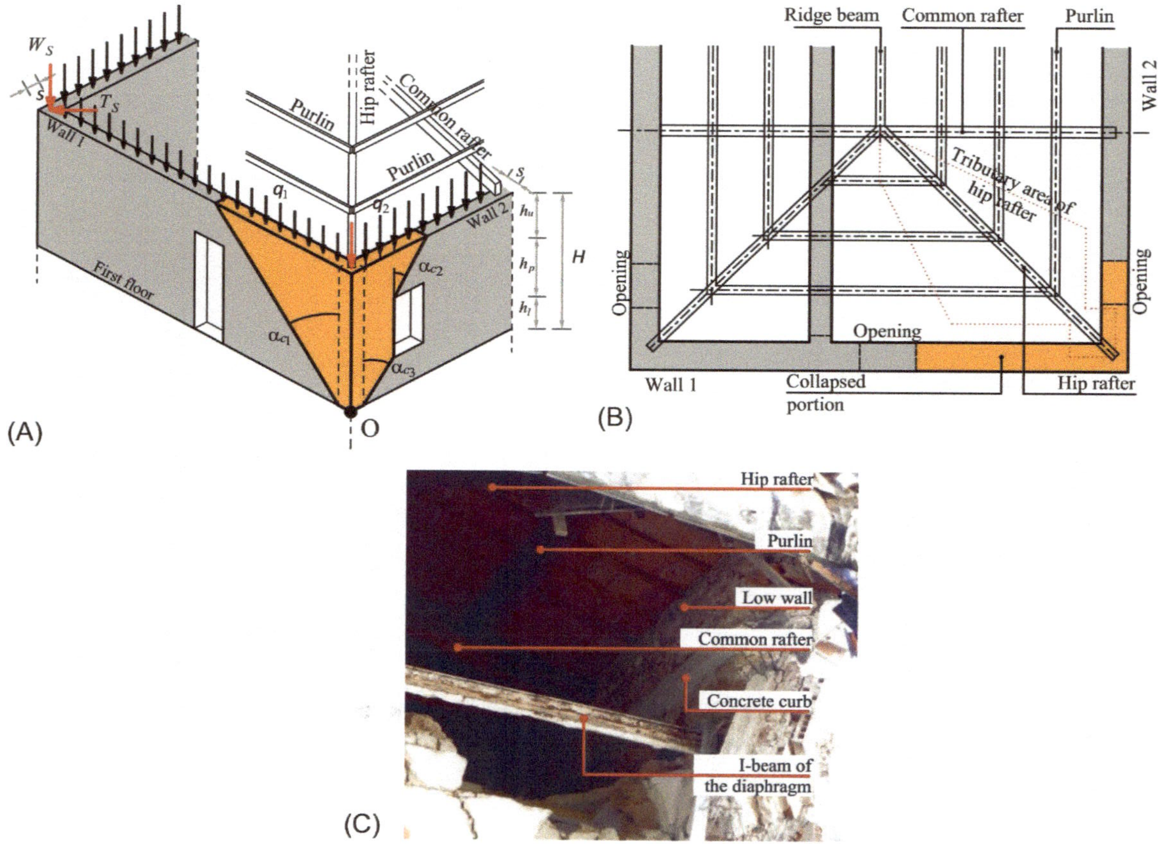

FIG. 11.14 3D (A) and 2D (B) layouts of the school wooden roof and masonry corner and details of other overloads on the walls (C).

the corner in the case study building is described in detail in (Argiento, Maione, & Giresini, 2019).

The masonry of the corner under study is composed of two outer layers of split stones and a rubble inner core, defining a cross-section with the same thickness s of about 65 cm for the two connected walls. The average dimensions of the stone blocks are $l_b = 30$ cm and $h_b = 15$ cm, whereas the adoption of the macroblock model implies the assumption that the dimension s_b of the unit is coincident with the thickness of the cross-section of the walls ($s_b = s$). The stone blocks are arranged in a quite regular half-running bond pattern, allowing the adoption of the staggering ratio $\tan\alpha_b = l_b/(2h_b) = 1$, that is, $\alpha_b = 45°$.

Table 11.4 reports the input data needed for the rocking analysis, together with those sketched in Fig. 11.15. For the sake of simplicity, all overloads are collected in a unique uniformly distributed load applied on the vertical mid-plane of the walls. These overloads include weights transferred by the roof, r.c. ring beam, low wall, and horizontal diaphragm indicated in Fig. 11.14C. In particular, the light diaphragm of the attic is assumed to be distributed 80% on Wall 2 and 20% on Wall 1.

11.4.1 The onset of the corner failure and the corresponding geometry

The first step to assess the seismic capacity of the corner under study is the definition of the geometry of the failing wedge by means of the limit analysis procedure described in Section 11.2.4, that is, by minimizing the load factor expressed by Eq. (11.15) with respect to the two unknown angles of crack on the two intersecting walls (Eq. (11.16)). In fact, unlike the simple rocking mechanism, the geometry of the moving corner is not known a priori and needs to be evaluated by means of the kinematic approach of limit analysis.

TABLE 11.4 Main input data for the masonry corner under study.

Geometry		Loadings	
Height (H) [m]	4.35	Specific weight of walls (w) [kN/m3]	21.00
Thickness of walls (s) [m]	0.65	Overloads on Walls 1 and 2 (q_1 and q_2) [kN/m]	16.16 and 19.62
Number of rows (n_r)	29	Gravity load due to the hip rafter (W_s) [kN]	15.27
Block size (l_b x h_b) [m]	0.3 × 0.15	Static thrust due to the hip rafter (T_s) [kN]	1.87

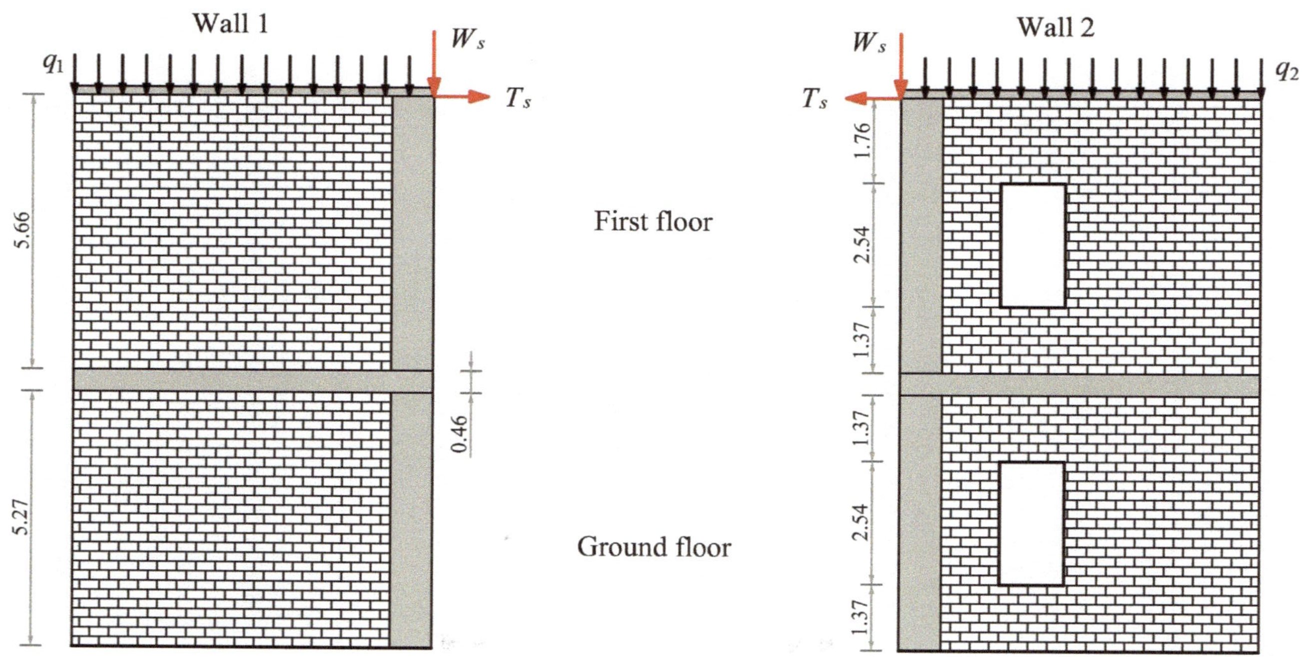

FIG. 11.15 Wall 1 and wall 2 of the failed corner at the first floor, with dimensions in meters.

The minimization procedure is developed for spreadsheet and a multi-purpose mathematical programming solver is used to solve the problem of optimum. By assuming the friction coefficient $f = 0.6$, as suitable to stone masonry walls (Casapulla & Maione, 2018; Lee et al., 2001; Lourenço & Ramos, 2004; Vasconcelos & Lourenço, 2009), the result of the minimum load factor is $\alpha_0 = 0.464$ and the resulting geometry pattern of the moving corner is a wedge characterized by the two angles of crack α_{c1} and α_{c2} equal to their maximum value given by the half-unit shape, that is, $\alpha_b = 45°$ (Fig. 11.14A). This means that zero frictional resistance is provided by minimizing Eq. (11.15) and the most likely mechanism is the pure rotation of the corner with the geometry obtained by the minimization process.

This result is in good agreement with the actual collapse pattern observed in Fig. 11.13B, where it is evident that the presence of a window in Wall 2 strongly influences the development of the cracks. In fact, these tend to localize near the vertexes of the opening, whereas a greater portion of Wall 1 is involved. By contrast, the opening in Wall 1 does not affect the configuration of the rocking wedge because quite far from the corner and the inclination of the crack line results to be at its maximum value ($\alpha_{c1} = \alpha_b$).

11.4.2 Seismic safety assessment of the rocking corner according to the simplified procedure (linear kinematic approach) from the Commentary to the Italian technical standard (CNTC19)

The simplified approach of the Italian CNTC19 (Ministero delle Infrastrutture e dei Trasporti, 2019), indicated for the analysis of complex failures, allows the seismic safety assessment of local mechanisms by only comparing the capacity and demand in terms of accelerations, with reference to the damage limit state (LS0) and the life-safety limit state (LS1). In fact, it is based on the linear kinematic approach of limit analysis and the use of the behavior factor for LS1 accounting for the displacement capacity, as is common for linear static approaches.

Using this method for the corner under study, the maximum value of the confidence factor ($F_C = 1.35$) is conservatively assumed, whereas the rate of the total mass participating in the rocking mechanism is calculated using the first part of Eq. (11.21), that is, $e^* = 0.948$. Thus, with the obtained load factor $\alpha_0 = 0.464$, the acceleration capacity is given by the first expression of Eq. (11.20), that is, $a_C = 0.363g$, considered the same for both limit states.

On the other hand, the evaluation of the acceleration demand accounts for the contributions of all the significant vibration modes of the building.

TABLE 11.5 Reference parameters of seismic hazard in Visso, soil type B, according to the NTC18 (Ministero delle Infrastrutture e dei Trasporti, 2018).

	T_R [years]	a_g [g]	F_0	T_{C^*} [s]	S
LS0	75	0.120	2.321	0.287	1.2
LS1	712	0.285	2.376	0.335	1.2

Thus, the contribution of the k-th mode to the maximum acceleration related to the floor at height z is (Ministero delle Infrastrutture e dei Trasporti, 2019):

$$a_{zk}(z) = S_e(T_k, \xi_k)\gamma_k\psi_k(z)\sqrt{1 + 0.0004\xi_k^2} \tag{11.23}$$

where S_e (T_k, ξ_k) is the elastic spectral acceleration related to the k-th mode with period T_k and damping coefficient ξ_k (expressed in %), γ_k is the participation factor of k-th mode, and $\psi_k(z)$ is the component of the modal vector at the height z from the ground. It is worth noting that S_e (T_k, ξ_k) is defined, for each limit state, on the basis of the seismic hazard parameters of the specific site, considering soil type B (Table 11.5). Then, the acceleration demand related to the floor at height z is obtained by combining the modal contributions according to the following expression (Ministero delle Infrastrutture e dei Trasporti, 2019):

$$a_{D(\text{CNTC19})} = \frac{a_z(z)}{q} = \frac{\sqrt{\sum (a_{zk}(z))^2}}{q} \tag{11.24}$$

where the parameter q is the behavior factor of the structure, which assumes the values 1 or 2 concerning the damage or the life-safety limit state, respectively.

In particular, the demand obtained for the corner is defined by considering the first three modes of vibration, to take into account the translational modes in both orthogonal directions (modes 1 and 3) and the torsional mode (mode 2), mainly affecting the 3D failure mechanism.

In Table 11.6, for each limit state, the values of the parameters required to define the acceleration demand according to the CNTC19 are listed. The periods related to the first three mode shapes are derived from a previous work on the whole building (Cattari et al., 2017).

It is worth noting that the value of the conventional viscous damping is considered as $\xi_k = 5\%$ for all the modes, and the parameters $\psi_k(z)$ and γ_k in Eq. (11.23) are assumed to have the same values for all the modes, according to the simplified formulations (Ministero delle Infrastrutture e dei Trasporti, 2019):

TABLE 11.6 Parameters defining the acceleration demand according to the CNTC19 (Ministero delle Infrastrutture e dei Trasporti, 2019).

		LS0		LS1	
k	T_k [s]	$S_e\ (T_k,\xi_k)$ [g]	$a_{zk}(z)$ [g]	$S_e\ (T_k,\xi_k)$ [g]	$a_{zk}(z)$ [g]
1	0.311	0.334	0.203	0.812	0.492
2	0.274				
3	0.266				

TABLE 11.7 Capacity/demand ratios provided by the CNTC19 (Ministero delle Infrastrutture e dei Trasporti, 2019).

Limit state	Seismic capacity (a_C)	Seismic demand (a_D)	Seismic safety index ($\zeta = a_C/a_D$)
LS0 [g]	0.363	0.351	1.034
LS1 [g]	0.363	0.426	0.851

$$\psi = \frac{z}{H_{tot}}; \qquad \gamma = \frac{3N}{2N+1} \tag{11.25}$$

where H_{tot} is the height of the building and N the number of floors (two in this case).

The results for the seismic safety of the corner in the P. Capuzi school are summarized in Table 11.7 through the index ζ, representing the capacity/demand ratio for each limit state.

From this table, it is evident that even if the predicted seismic demand is slightly lower than the capacity at the onset of the corner mechanism ($\zeta > 1$ for LS0), the safety is not assured for the life-safety limit state ($\zeta < 1$ for LS1), confirming a good agreement with the total collapse of the corner actually occurred during the 2016–2017 Central Italy earthquakes.

These results are also in good agreement with the seismic safety assessment of this corner carried out with reference to four significant 2016–17 Central Italy earthquakes, both in terms of recorded peak ground accelerations (Casapulla, Giresini, et al., 2019b) and floor accelerations (Argiento et al., 2019). In these cases, the capacity curves and the acceleration capacities were represented as their projections in different planes, to be compared with the seismic demand referred to specified directions of motion. This aspect still confirms the great potential of the macroblock model and the solution procedure presented in this chapter to define the onset and overturning conditions of any 2D and 3D rocking mechanism and to evaluate its seismic safety.

11.5 Conclusions

This work presented the seismic assessment of typical out-of-plane failure modes in traditional block masonry buildings by performing nonlinear static analyses. The adopted macroblock model, accounting for the restraining action due to the frictional resistances, is able to provide a reliable evaluation of the load factor that activates the incipient mechanism and of the capacity curve in terms of displacements, to be compared with the seismic demand.

Two typical out-of-plane modes occurred in existing buildings were investigated by means of the presented modeling approach: the rocking masonry façade of a single-nave church and the corner failure in a school building, both subjected to the seismic swarm occurred in Central Italy in 2016–2017. The former represents the single rocking of a macroblock already known a priori, whereas the latter involves a minimization process to define the most likely geometry of the complex rocking macroblock.

The seismic assessment of the two rocking structures was performed according to the Commentary to the Italian technical standard for construction (CNTC19), which included the application of CSM for the first case study. Three limit states, that are rocking initiation (LS0), moderate rocking (LS1), and near collapse (LS2) conditions were considered. LS0 is representative of serviceability limit states, whereas LS1 and LS2 are considered as life-safety and collapse limit states, respectively.

The predictive values in terms of capacity/demand ratios, obtained according to the CNTC19, confirmed a good agreement of the numerical outcomes with the real behavior of the rocking structures under the occurred earthquakes. In particular, the model was capable to provide a reliable prediction both of the onset of the façade and the corner and of the total collapse of the corner.

The accuracy of the presented assessment method in predicting the seismic response of the rocking structures under study may guarantee its potential and suitability to be applied to any kind of masonry structural/nonstructural element with reference to different geometries, constraints, and performance levels. Also, it can easily be extended to design strengthening interventions with advanced innovative systems, to prevent local out-of-plane mechanisms in masonry buildings.

Acknowledgments

The author wishes to thank her long-time collaborators, Dr. Alessandra Maione and Arch. Luca U. Argiento, for their enthusiastic support in improving the modeling approach and developing the numerical analyses for the case studies.

References

Abrams, D. P., AlShawa, O., Lourenço, P. B., & Sorrentino, L. (2017). Out-of-plane seismic response of unreinforced masonry walls: Conceptual discussion, research needs, and modeling issues. *International Journal of Architectural Heritage*, *11*(1), 22–30. https://doi.org/10.1080/15583058.2016.1238977.

Argiento, L. U., Maione, A., & Giresini, L. (2019). The corner failure in a masonry building damaged by the 2016–2017 central Italy earthquake sequence. In: *Proceedings of the 7th international conference on computational methods in structural dynamics and earthquake engineering, COMPDYN 2019*https://doi.org/10.7712/120119.6945.19268.

Baggio, C., & Trovalusci, P. (1998). Limit analysis for no-tension and frictional three-dimensional discrete systems. *Mechanics of Structures and Machines*, *26*(3), 287–304. https://doi.org/10.1080/08905459708945496.

Bagi, K. (2014). When Heyman's safe theorem of rigid block systems fails: Non-Heymanian collapse modes of masonry structures. *International Journal of Solids and Structures*, *51*, 2696–2705. https://doi.org/10.1016/j.ijsolstr.2014.03.041.

Bui, T. T., Limam, A., Sarhosis, V., & Hjiaj, M. (2017). Discrete element modelling of the in-plane and out-of-plane behaviour of dry-joint masonry wall constructions. *Engineering Structures*, *136*, 277–294. https://doi.org/10.1016/j.engstruct.2017.01.020.

Caliò, I., Marletta, M., & Pantò, B. (2012). A new discrete element model for the evaluation of the seismic behaviour of unreinforced masonry buildings. *Engineering Structures*, *40*, 327–338. https://doi.org/10.1016/j.engstruct.2012.02.039.

Casapulla, C., & Argiento, L. U. (2016). The comparative role of friction in local out-of-plane mechanisms of masonry buildings. Pushover analysis and experimental investigation. *Engineering Structures*, *126*, 158–173. https://doi.org/10.1016/j.engstruct.2016.07.036.

Casapulla, C., & Argiento, L. U. (2018). In-plane frictional resistances in dry block masonry walls and rocking-sliding failure modes revisited and experimentally validated. *Composites Part B: Engineering*, *132*, 197–213. https://doi.org/10.1016/j.compositesb.2017.09.013.

Casapulla, C., Giresini, L., Argiento, L. U., & Maione, A. (2019b). Nonlinear static and dynamic analysis of rocking masonry corners using rigid macro-block modeling. *International Journal of Structural Stability and Dynamics*, *19*(11), 1950137. https://doi.org/10.1142/S0219455419501372.

Casapulla, C., Giresini, L., & Lourenço, P. B. (2017). Rocking and kinematic approaches for rigid block analysis of masonry walls: State of the art and recent developments. *Buildings*. 7(3)https://doi.org/10.3390/buildings7030069.

Casapulla, C., & Maione, A. (2018). Modelling the dry-contact interface of rigid blocks under torsion and combined loadings: Concavity vs. convexity formulation. *International Journal of Non-Linear Mechanics*, *99*, 86–96. https://doi.org/10.1016/j.ijnonlinmec.2017.11.002.

Casapulla, C., & Maione, A. (2020). Experimental and analytical investigation on the corner failure in masonry buildings: Interaction between rocking-sliding and horizontal flexure. *International Journal of Architectural Heritage*, *14*(2), 208–220. https://doi.org/10.1080/15583058.2018.1529206.

Casapulla, C., Maione, A., & Argiento, L. U. (2019). Performance-based seismic analysis of rocking masonry façades using non-linear kinematics with frictional resistances: A case study. *International Journal of Architectural Heritage*. https://doi.org/10.1080/15583058.2019.1674944. in press.

Cattari, S., Degli Abbati, S., Ottonelli, D., Sivori, D., Spacone, E., Camata, G. , et al. (2017). *Task 4.1 Workgroup_Report di sintesi sulle attività svolte sugli edifici in muratura monitorati dall'Osservatorio Sismico delle Strutture, Linea Strutture in Muratura.* ReLUIS Report Rete Dei Laboratori Universitari Di Ingegneria Sismica (in Italian).

Drucker, D. C. (1953). Coulomb friction, plasticity and limit loads. *Technical Report no 85, Armed Services Technical Information Agency*, 1–16.

Ferrero, C., Lourenço, P. B., & Calderini, C. (2018). 2016-2017 Central Italy earthquake: Seismic assessment of "Pietro Capuzi" school in Visso (Marche). In: *Proceedings of the 9th international conference on computational methods (ICCM2018)*.

Ferris, M. C., & Tin-Loi, F. (2001). Limit analysis of frictional block assemblies as a mathematical program with complementarity constraints. *International Journal of Mechanical Sciences, 43*(1), 209–224. https://doi.org/10.1016/S0020-7403(99)00111-3.

Freeman, S. A. (1998). Development and use of capacity spectrum method. In: *Proceedings of the 6th US NCEE conference on earthquake engineering/EERI*12. Paper 269.

Gilbert, M. (2001). RING: A 2D rigid block analysis program for masonry arch bridges. In: *Proceedings of the 3rd international Arch bridges conference, Paris*.

Gilbert, M., Casapulla, C., & Ahmed, H. M. M. (2006). Limit analysis of masonry block structures with non-associative frictional joints using linear programming. *Computers and Structures, 84*(13–14), 873–887. https://doi.org/10.1016/j.compstruc.2006.02.005.

Giresini, L., Casapulla, C., Denysiuk, R., Matos, J., & Sassu, M. (2018). Fragility curves for free and restrained rocking masonry façades in one-sided motion. *Engineering Structures, 164*, 195–213. https://doi.org/10.1016/j.engstruct.2018.03.003.

Giresini, L., Fragiacomo, M., & Lourenço, P. B. (2015). Comparison between rocking analysis and kinematic analysis for the dynamic out-of-plane behavior of masonry walls. *Earthquake Engineering and Structural Dynamics, 44*(13), 2359–2376. https://doi.org/10.1002/eqe.2592.

Giresini, L., Solarino, F., Paganelli, O., Oliveira, D. V., & Froli, M. (2019). One-sided rocking analysis of corner mechanisms in masonry structures: Influence of geometry, energy dissipation, boundary conditions. *Soil Dynamics and Earthquake Engineering, 123*, 357–370.

Godio, M., & Beyer, K. (2019). Evaluation of force-based and displacement-based out-of-plane seismic assessment methods for unreinforced masonry walls through refined model simulations. *Earthquake Engineering and Structural Dynamics, 48*(4), 454–475. https://doi.org/10.1002/eqe.3144.

Griffith, M. C., Magenes, G., Melis, G., & Picchi, L. (2003). Evaluation of out-of-plane stability of unreinforced masonry walls subjected to seismic excitation. *Journal of Earthquake Engineering, 7*, 141–169. https://doi.org/10.1007/s00264-017-3709-6.

Heyman, J. (1966). The stone skeleton. *International Journal of Solids and Structures, 2*(2), 249–279. https://doi.org/10.1016/0020-7683(66)90018-7.

Lagomarsino, S. (2015). Seismic assessment of rocking masonry structures. *Bulletin of Earthquake Engineering, 13*(1), 97–128. https://doi.org/10.1007/s10518-014-9609-x.

Lagomarsino, S., & Cattari, S. (2015). PERPETUATE guidelines for seismic performance-based assessment of cultural heritage masonry structures. *Bulletin of Earthquake Engineering, 13*(1), 13–47. https://doi.org/10.1007/s10518-014-9674-1.

Lee, H. S., Park, Y. J., Cho, T. F., & You, K. H. (2001). Influence of asperity degradation on the mechanical behavior of rough rock joints under cyclic shear loading. *International Journal of Rock Mechanics and Mining Sciences, 38*(7), 967–980. https://doi.org/10.1016/S1365-1609(01)00060-0.

Livesley, R. K. (1978). Limit analysis of structures formed from rigid blocks. *International Journal for Numeridal Methods in Engineering, 12*(12), 1853–1871.

Lourenço, P. B., & Ramos, L. F. (2004). Characterization of cyclic behavior of dry masonry joints. *Journal of Structural Engineering, 130*(5), 779–786.

Ministero delle Infrastrutture e dei Trasporti (2009). *CNTC09—Circolare applicativa delle Norme tecniche per le costruzioni di cui al D.M. 14/01/2008 (NTC 2008). Gazzetta Ufficiale N. 47 del 26/02/2009 (in Italian).* Istituto Poligrafico e Zecca dello Stato.

Ministero delle Infrastrutture e dei Trasporti (2018). *NTC 2018—Norme Tecniche per le Costruzioni di cui al D.M. 17/01/2018. Gazzetta Ufficiale N. 42 del 20/02/2018 (in Italian).* Istituto Poligrafico e Zecca dello Stato.

Ministero delle Infrastrutture e dei Trasporti (2019). *CNTC19—Circolare applicativa delle Norme Tecniche delle Costruzioni di cui al D.M. 17/01/2018 (NTC 2018). Gazzetta Ufficiale N. 42 del 20/02/2018 (in Italian).* Istituto Poligrafico e Zecca dello Stato.

Olivito, R. S., & Porzio, S. (2019). A new multi-control-point pushover methodology for the seismic assessment of historic masonry buildings. *Journal of Building Engineering, 26,* 100926. https://doi.org/10.1016/j.jobe.2019.100926.

Orduña, A., & Lourenço, P. B. (2005). Three-dimensional limit analysis of rigid blocks assemblages. Part I: Torsion failure on frictional interfaces and limit analysis formulation. *International Journal of Solids and Structures, 42*(18–19), 5140–5160. https://doi.org/10.1016/j.ijsolstr.2005.02.010.

Pantò, B., Giresini, L., Sassu, M., & Caliò, I. (2017). Non-linear modeling of masonry churches through a discrete macro-element approach. *Earthquake and Structures, 12*(2), 223–236.

Portioli, F., Casapulla, C., Gilbert, M., & Cascini, L. (2014). Limit analysis of 3D masonry block structures with non-associative frictional joints using cone programming. *Computers and Structures, 143,* 108–121. https://doi.org/10.1016/j.compstruc.2014.07.010.

Sorrentino, L., D'Ayala, D., de Felice, G., Griffith, M. C. C., Lagomarsino, S., Magenes, G. , et al. (2016). Review of out-of-plane seismic assessment techniques applied to existing masonry buildings. *International Journal of Architectural Heritage, 11*(1), 2–21. https://doi.org/10.1080/15583058.2016.1237586.

Vasconcelos, G., & Lourenço, P. B. (2009). Experimental characterization of stone masonry in shear and compression. *Construction and Building Materials, 23*(11), 3337–3345.

CHAPTER

12

Mexican colonial churches: structural assessment and seismic behavior

Fernando Peña[a], *Marcos M. Chávez*[a], *and Natalia García*[b]

[a]Instituto de Ingeniería, Universidad Nacional Autónoma de México, Mexico City, Mexico [b]Universidad Autónoma del Estado de Morelos, Cuernavaca, Mexico

12.1 Introduction

The architectural heritage in Mexico is wide, rich, and varied. It has more than 50,000 archeological sites discovered until today and 110,000 buildings built between 16th and 19th centuries, as well as 35 sites inscribed on the UNESCO World Heritage List. Earthquakes have been one of the main causes of destruction of this architectural heritage. For example, the two large-scale earthquakes that occurred on September 7 (Mw 8.2) and 19 (Mw 7.1), 2017, in Mexico left 2340 damaged architectural heritage buildings throughout the center and south of the country, which corresponds to around 6% of the total of architectural heritage buildings located in the affected area.

In this context, a careful seismic assessment of the architectural heritage affected by severe earthquakes is one of the most effective strategies to understand the structural weaknesses of these constructions. Also, the interventions made in the past must be evaluated to establish which one of these have had successful behavior and must continue to be used and which did not (Peña & Manzano, 2015).

According to the behavior observed during earthquakes, the seismic response of masonry churches can be described according to recurrent

https://doi.org/10.1016/B978-0-12-821087-1.00006-5

phenomenology, which depends on the architectural and structural typologies of the church (Doglioni, Moretti, & Petrini, 1994; Lagomarsino & Podesta, 2004). Regardless if the churches vary in size and architectural sophistication, it is possible to define some basic typologies for one region. Therefore it is possible to find a relationship between the typologies of the churches and their failure mechanism (Lagomarsino, Cattari, & D Ottonelli, 2019).

Thousands of churches were built in Mexico during the Spanish colonization era (16th–18th centuries) and persist to date in rather good conditions. They vary in size and architectural sophistication but follow some basic typologies. The simplest among them are rather small parochial churches, which are found in every town and village of Mexico. One important factor that has influenced the evolution of their architectural features has been the experience of ancient builders about the damages suffered from earthquake activity (Meli & Peña, 2004).

Therefore in recent years, the study of the seismic behavior of Colonial churches has become an important issue in Mexico (Peña & Chávez, 2016; Peña & García, 2016; Preciado, Orduaña, Bartoli, & Budelmann, 2015). Hence the aim of this chapter is to present the current studies about the seismic behavior of these types of structures. The structural behavior is described, and the causes of the most common structural damages are explained.

12.2 Description of the typical colonial churches

At the beginning of the colonial era (16th century), the churches were composed of one rectangular nave with apse located east, with a simple façade and were covered by barrel vaults and some white rib vaults. These churches used to be part of conventual complexes of the Franciscan, Augustinian, or Dominican mendicant orders. Prismatic buttresses were placed along the longitudinal walls of the nave, sometimes at the ends of the west façade, and in the apse. In the first bay, an intermediate floor was built for the chorus. The principal or west façade was a flat wall with two openings, one for the door and the other for the window, which illuminated the choir (Fig. 12.1). Only few examples of churches of the 16th century remain intact. A great number of these churches were modified or rebuilt, partially or completely, along the time. This was mainly because the first builders did not take into account the high seismicity of the region, and frequently, they have been damaged by earthquakes. Another reason was that the first churches were generally small, and as the population grew, it was necessary to have bigger churches. Additionally, the appearance of new architectural styles after the 16th century (Baroque and Churrigueresque) allowed the evolution of the churches to more complex

FIG. 12.1 Typical convent church from the 16th century.

and ornate forms. This evolution and changes can be observed mainly in the plant of the church, façade, and roof (García & Meli, 2009).

Initially, façades were a flat wall without ornamentation. However, from the 17th century, as the churches began to be more elaborate, higher, and with more volume, the west façades changed to more complex forms (Fig. 12.2A). The frontispiece was higher, and the wall had more decoration. The bell towers were not very common elements of the churches during the 16th century. Most of the churches of this century lacked towers, and the bells were set in a bell gable (see Fig. 12.2). In some cases, there was

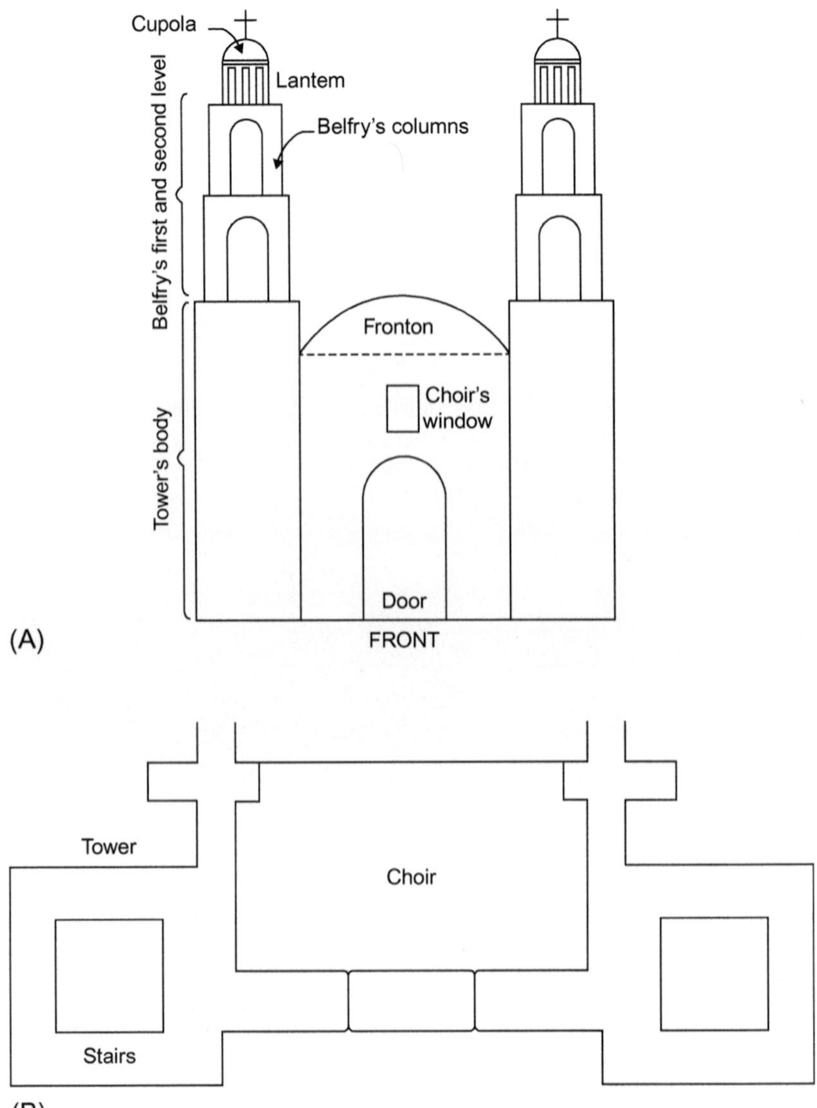

FIG. 12.2 Typical façade from the 17th century and later: (A) front view and (B) schematic representation of the plant.

a single low-rise tower, very robust and of simple architecture. This contrasts with the tall slender towers of the churches in the cities of later centuries. In many churches of the 16th century, the towers were added from the last decades of the century onward. There are some examples where the church has a bell tower and a bell gable (Meli, 2011).

The façade is composed of the main wall, which is connected with the towers and the frontispiece. It was typical to build the façade at the same time as the lateral walls, which allows a very good connection between walls. The bell tower can be divided into three parts: the body of tower or basement; the belfry that can be made with one, two, or three bodies; and a small hemispheric dome at the top of the bell tower. In plant, the body of the towers is like a staggered square pipe, which is attached to the walls of the west façade and the nave, located in the same plane or in a parallel plane of the façade, tucked outward or inward to it. Small openings can be found in the basement, whose function is to illuminate and ventilate the indoor, where the stairs to the belfry or small chapels are located (Fig. 12.2B). In general, it is common that belfries present two or three bodies; being the most top smaller than the other ones. Sometimes, the third body is more similar to a lantern-like element.

The vaults in colonial churches can be classified into four kinds (García & Meli, 2009): continuous barrel vaults, barrel vaults with cross arches, barrel vaults with lunettes, and ribbed cross vaulting. Continuous barrel vaults and ribbed cross vaults were typical of the first conventual churches of the 16th century. The last are very thick and have a dome-like shape, covering square areas. From the beginning of the 17th century, the most commonly used vaults were the barrel vaults with cross arches. Barrel vaults with lunettes were common from churches of the 18th and later centuries (Meli, 2011).

The main material of construction is a heterogeneous masonry constituted by stones agglutinated by lime-sand mortar (rubble masonry). Owing to the internal constitution of the masonry (rounded stones immersed in a matrix of mortar), this material is more similar as a cyclopean concrete than a traditional rubble masonry (Meli & Sánchez-Ramírez, 2007; Ramírez de Alba & Valdez-Medina, 2014). Therefore the material can be considered as isotropic homogeneous (Chávez & Meli, 2012). This characteristic allows building the structural elements as one leaf element. Thus no weak planes are inside the walls because of the different layers of the masonry (see Fig. 12.3). Walls, arches, and towers were made by stone masonry, while vaults and domes were mostly built with bricks (Peña & Chávez, 2016).

12.3 Typical seismic damage patterns

During September 2017, two high-intensity earthquakes hit the central and southern region of Mexico. At 23:49.17 local time on September 7, 2017, an earthquake of magnitude Mw 8.2 occurred in the Gulf of Tehuantepec, 133 km southwest of Pijijiapan, Chiapas, and 700 km from Mexico City. According to the National Seismological Service (SSN), this earthquake was catalogued as a normal fault (Servicio Sismológico Nacional,

FIG. 12.3 Details of a typical colonial masonry.

2017a). The maximum acceleration recorded near to the epicenter (PGA) was over 228 Gal. Serious damages were reported in the states of Oaxaca, Chiapas, and Tabasco, mainly to housing, ancient buildings, and some infrastructure facilities.

On September 19, a second strong earthquake with a magnitude Mw 7.1 was felt at 13:14:40 local time. The epicenter was located between Puebla and Morelos states, 12 km southeast of Axochiapan, Morelos, and 120 km from Mexico City. According to the SSN (Servicio Sismológico Nacional, 2017b), the earthquake was a normal fault intraplate with a maximum acceleration recorded over 220 Gal, measured near to the epicenter area.

Owing to both strong earthquakes, 2340 heritage buildings suffered some kind of damage because of the occurrence of the September earthquakes. Only the States of Oaxaca, Puebla, and Morelos concentrated 62.7% of the total damages. Of these, Oaxaca and Puebla represented just over half of the damages (51.6%). For this reason, a campaign of reconnaissance of damages was performed to assess the damage in historical temples in these states. A total of 58 temples were visited (Fig. 12.4): 22 in Oaxaca, around the Mixteca Alta and Tehuantepec regions; 11 inside the Mixteca region, Puebla; 15 in Morelos, along the route called *Ruta de los Conventos* (Convents Rout). Some of them are included in the World Heritage List of the UNESCO as "Earliest 16th Century monasteries on the slopes of Popocatepetl volcano" and 10 in Mexico City. The damage observed is described according to the type of structural elements of the church. For this purpose, the temples were divided into facades and bell towers; domes and vaults; and side walls and apse.

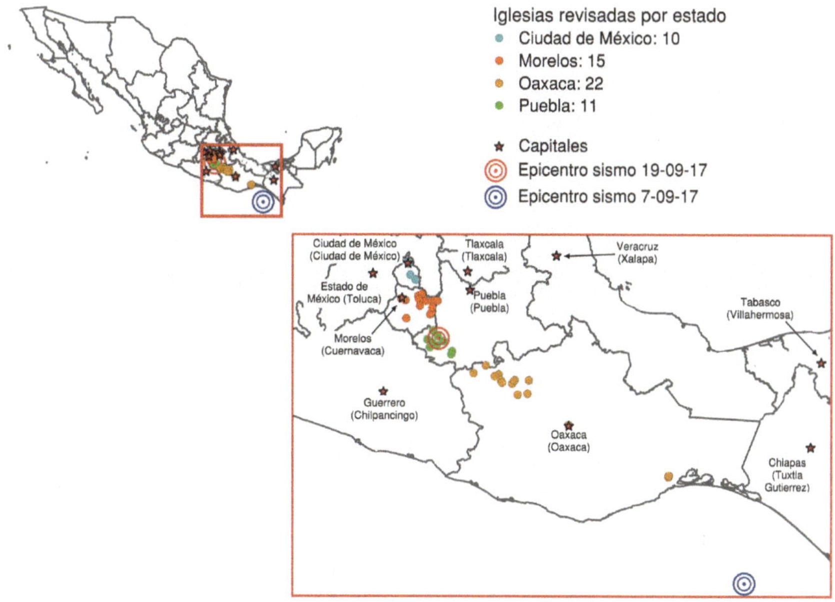

FIG. 12.4 Location of the temples visited after the earthquake of September 19, 2017.

The damages observed in the façades were produced mainly because of higher levels of shear stresses into the plane of this element. These damages were displayed with the formation of diagonal cracks and/or with the separation of the main body of the façade or the separation of the side walls. However, some damages were associated with an out-of-plane behavior, mainly due to the overturning of the frontispiece and belfries. The bell towers were one of the most damaged elements. In most cases, a partial or total collapse of the belfries was observed.

Some typical examples are showed in Fig. 12.5, where it is observed that the façade of Tlaquiltenango temple, in Morelos, presented a separation between the façade and sidewalls, as well as a diagonal crack in the middle of this element (Fig. 12.5A). In Tlayacapan, some diagonal cracks were located over the façade, near to the door and windows. The bell gable showed a partial collapse due to the out-of-plane behavior of this element (Fig. 12.5B). A similar behavior was observed in the Sagrario temple, in Huajuapan de León, Oaxaca (Fig. 12.5C). The façade of Chila de la Sal chapel showed some cracks between the side walls, in the middle, and the towers (Fig. 12.5D).

The towers were the structural elements with more damages presented in the temples, located mainly in Puebla and Morelos. In San Andrés de la Cal chapel, Morelos, the arches of the lower body of the bell tower collapsed, and horizontal cracks appeared in the upper and base of the

FIG. 12.5 Damages patterns in facades: (A) Tlaquiltenango, Morelos; (B) Tlayacapan, Morelos; (C) Huajuapan de León, Oaxaca; and (D) Chila de la Sal, Puebla.

pilasters (Fig. 12.6A). A similar behavior was observed in the pilasters of the bell towers of Santa Ana Jolalpan (Fig. 12.6B) and Pilcaya, Puebla (Fig. 12.6C).

The temples of Hueyapan, Jiutepec, and Ocuituco located into the Convents Route, in Morelos, have only one bell tower. In all these cases, this element presented a different level of damage, from only cracks until a partial or total collapse (Fig. 12.7). Hueyapan tower showed cracks in the upper and the base of the pilasters of the first body of the belfry. This is because the upper body, being more rigid and smaller, worked mainly as a rigid body with a considerable mass, where the seismic shear was concentrated (Fig. 12.7A). The temples of Jiutepec and Ocuituco presented collapses of the bell towers, while the main walls of the facades had slight damage, which was concentrated in the opening of the window of the choir (Fig. 12.7B and C).

FIG. 12.6 Damage patterns in bell towers: (A) San Andrés de la Cal, Morelos; (B) Santa Ana Jolalpan, Puebla; and (C) Pilcaya, Puebla.

FIG. 12.7 Damage of belfries in churches of Morelos: (A) Hueyapan; (B) Jiutepec; and (C) Ocuituco.

Although most of the temples in Mexico have a vaulted roof, as stone or brick masonry, some façades were damaged because of the presence of the out-of-plane behavior. For example, in Pilcaya, Puebla (Fig. 12.8), the façade was separated from the vault because of a deficient anchoring of the roof to the façade wall. Likewise, the frontispiece collapsed due to movement out of plane. In Tayacapan, Morelos (Fig. 12.9), vertical cracking was present between the slender west façade and the side walls of the nave due to an out-of-plane behavior despite diagonal buttresses at its ends, which presented the crushing of material at its base.

The roofing system is mainly made by domes and vaults. The most common are the barrel vaults without or with lunettes. Many cases of damages in this system were observed, from longitudinal and transversal cracks until a partial or total collapse. Fig. 12.10 shows the typical cracking of continuous barrel vaults, where longitudinal cracks are present along

FIG. 12.8 Temple of Pilcaya, Puebla: (A) Overturning of the frontispiece and (B) separation of the façade from the nave.

the vault. Usually one or more cracks appear at the top of the vault, in the intrados and extrados, ridding the fillings that commonly cover the kidneys of the vault in the extrados (Fig. 12.10). In some cases, the cracks were large enough to cause the collapse of the vault and the arches that cross.

Typical damages were observed in domes: radial cracks at the base of the dome and meridian cracks on the body of the dome (Fig. 12.11). In some cases, the damage was located only in the drum. In others, the dome collapsed and the drum remained standing, apparently undamaged (Fig. 12.12). In some apses of the temples visited, a sail vault was located, only one of them collapsed totally where old cracks that were previously intervened were observed.

(A) (B)

FIG. 12.9 Façade of the Tlayacapan temple, Morelos: (A) General view and (B) vertical cracking between the façade and the wall of the nave.

(A) (B)

FIG. 12.10 Damage in vaults: (A) Tlayacapan, Morelos and (B) Tlaquiltenango, Morelos.

FIG. 12.11 Radial and meridian cracks in the dome of the Jiutepec temple, Morelos.

(A) (B)

FIG. 12.12 Damage in the main dome in temples of Morelos: (A) Totolapan and (B) Zacualpan.

The temples are massive structures. The weight and geometry of their structural elements contribute to the stability. The walls usually are robust and can hardly collapse. However, damage may occur mainly by shear forces that the masonry cannot resist as shown in Fig. 12.13.

In some cases, the side walls have horizontal cracks that cross the thickness of the wall and generally are located below the windows (Fig. 12.14). This type of cracking is due to bending moments outside the plane of the wall, which occur mainly in churches that have other lower bodies attached to the side walls, such as convents, chapels, or parish offices. These attached bodies function as struts on the side walls.

Buttresses with added sections at different stages of construction collapsed by separating mainly in areas with old cracks that had been repaired in previous works. In flying buttresses, shear cracks occurred even when their arches had been bricked up, in addition to the separation of the walls (Fig. 12.15).

12.4 Mechanical characterization of masonry

The September earthquakes caused a partial or total collapse of elements in many colonial temples. Because of this, a significant amount of debris was generated, which opened the possibility of collecting samples of material for testing in the laboratory to obtain their basic mechanical properties. A campaign to collect samples of masonry from different temples in the state of Morelos was performed. Twelve temples were visited, and a total of 35 samples of different sizes were obtained. To minimize the alteration of the samples, all of them were cut using a diamond chain concrete saw (Fig. 12.16). Because the collected samples had an irregular shape, it was necessary to cut them trying to obtain specimens with prismatic shapes with an aspect ratio 2:1. Finally, to ensure a uniform distribution of stresses on the contact surface of the specimens, they were capping with a layer of plaster-cement mortar of minimum thickness.

A total of 33 uniaxial compression tests were performed. To measure the longitudinal and transversal deformations of the specimens during the application of the load, displacement transducers were placed at the points of interest. Initially, a combined system of LED sensors and displacement transducers were used, with the main objective to compare the results obtained between both kinds of instruments. The differences between both measurements of the instruments were minimal; hence it was decided only to use LEDs because its placement in the specimens is easier than the displacement transducers. The instrumentation setup of the specimens is shown in Fig. 12.17A and B. The test setup of one of

(A)

(B)

FIG. 12.13 Damage on side walls: (A) El Calvario temple, Huajuapan, Oaxaca and (B) Ayoxutla, Puebla.

FIG. 12.14 Horizontal wall cracks in the temple of Tlayacapan, Morelos.

the specimens is shown in Fig. 12.17C. Testing began with the application of three preload cycles of around 1 and 5 kN to achieve a rearrangement of the specimen and instrumentation. Subsequently, the load continued to be applied at a rate of 30 kN/min until specimen failure was reached.

The density for each masonry specimen was determined. The density values of the specimens were between 1.3 and 1.7 kg/m^3.

The compressive strength was obtained for each of the specimens tested and was calculated by performing the ratio of the maximum weighted load between the cross-sectional area of the corresponding specimen. The resistance obtained varied from one temple to another, including significant variations in specimens obtained from the same temple. This may be because in the same temple, there were areas intervened by high-strength mortar injections or additions of other material outside the original construction, or the long construction periods, common in old large buildings; also because of the inherent heterogeneity of the materials. The calculated resistances vary from 1.0 to 2.7 MPa. Some high strengths of up to 14.3 MPa were also obtained, as is the case of Hueyapan. This was because the size of the stones was very large, which covered virtually the entire cross-sectional area of the specimen. In the cases of Jantetelco and Tlayacapan, the high resistance was due to the presence of more resistant grout injections than the original due to previous interventions. The average compressive strength for each temple is shown in Fig. 12.18.

(A)

(B)

FIG. 12.15 Typical damage to buttresses in Morelos: (A) Tlayecac and (B) Tlayacapan.

FIG. 12.16 Extraction of masonry samples collected: (A) selection of the sample; (B) cutting; and (C) final sample.

The results of the modulus of elasticity were variable. The values were between 0.2 and 16 GPa. The average values have a high coefficient of variation, so these values should be used with reserve. In the case of Hueyapan samples, they showed relatively close values to each other, 0.6 GPa on average. On the other hand, the values of the Poisson coefficient were in the range of 0.2–0.3.

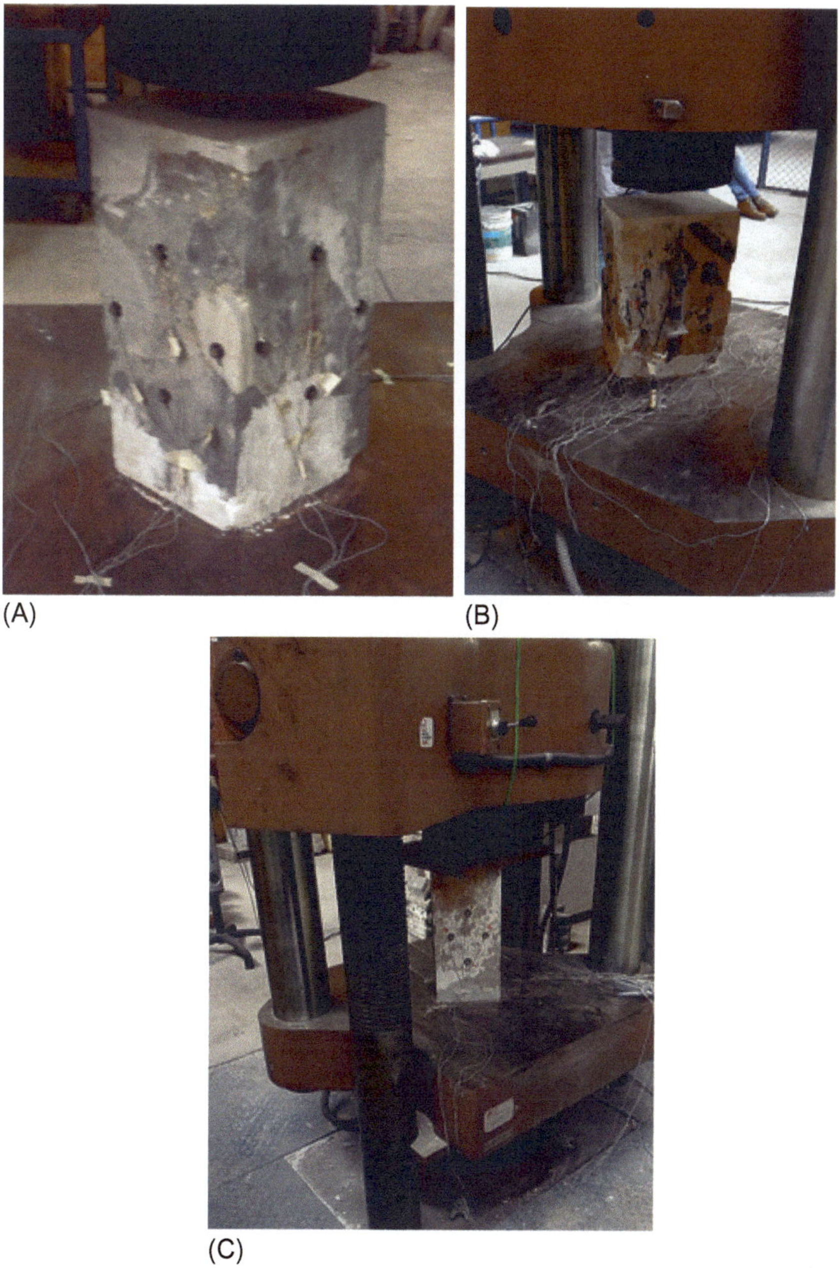

FIG. 12.17 Uniaxial compressive test: (A) instrumentation based with LEDs; (B) instrumentation based with LEDs and transducers; and (C) example of a compressive test.

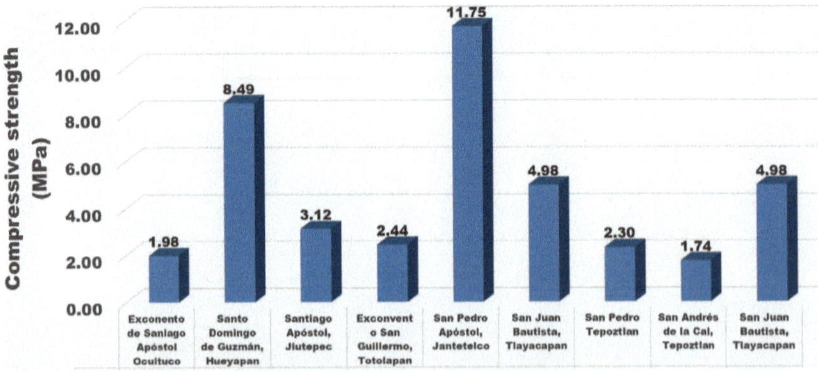

FIG. 12.18 Average compressive strength for each temple.

12.5 Numerical evaluation of three typical temples

To study the seismic behavior of these types of structures, three typical colonial temples of Morelos were selected: San Guillermo, Totolapan; Santiago Apostle, Jiutepec; and Santiago Apostle, Ocuituco. All the three temples have a single nave and one bell tower. They have typical elements such as buttresses, facades, battlements, and vaults. They were selected mainly because they have not been subjected to previous invasive interventions and remain practically unchanged (García & Meli, 2009).

The former convent of San Guillermo, Totolapan, was built by Augustinian friars. It has a small central courtyard or cloister, surrounded by two-story arcades with semicircular arches, where there are still remnants of mural painting. The temple is of small proportions compared to others in Morelos; the nave is covered by a barrel vault with lunettes, and the apse is covered by a sail vault; it also has botarel buttresses on the longitudinal wall opposite the convent (Fig. 12.19A). The construction of the convent of Santiago Apostol, Jiutepec, began after 1570; from the chronicles of the time, we know that in 1586 there was a convent without church, which was built at the end of the 16th century. It has a simple façade and battlements on the frontispiece (Fig. 12.19B). It is one of the simplest temples in its structural elements. The Augustinian convent of Santiago Apostol in Ocuituco was the first convent of that order outside of Mexico City. The temple has one bell tower and rectangular buttresses attached to the lateral walls. Fig. 12.20 shows the plan view of each temple.

Finite element models of each temple were performed by means of the commercial code Ansys. Fig. 12.21 shows the numerical models. The mechanical properties of the material were obtained from the experimental tests described in the previous section. Table 12.1 shows the mechanical properties used for each model.

FIG. 12.19 Selected temples: (A) Totolapan; (B) Jiutepec; and (C) Ocuituco.

Eigenvalue analyses were performed. Table 12.2 shows the main modal periods for the first mode in each global direction. It can be seen that the most flexible part of each church are the belfries, as well as the lateral direction. For each model, about 250 modes are needed to reach a participating modal mass of 90%, thus giving an indication of the complexity of their response to ground motions. The most rigid temple is Totolapan, which is because of the type of botarel-style buttresses (the more robust among the three buildings).

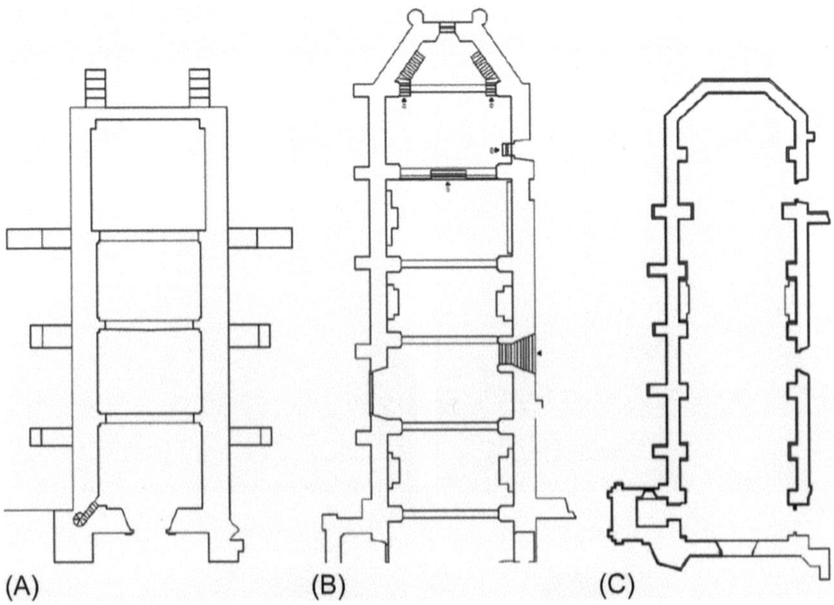

FIG. 12.20 Plan view of the selected temples: (A) Totolapan; (B) Jiutepec; and (C) Ocuituco.

When comparing the behavior of the temples analyzed, the similarity between the temples of Ocuituco and Jiutepec is evident, which, despite being of different materials, have coincidences in their geometry. In addition, they have similar dynamic properties. Only the dome in Jiutepec is a little less stiff, which increases the period in the longitudinal direction compared to Ocuituco, which lacks a dome. Fig. 12.22 shows the behavior of the main modes of the three models. As can be seen, the first mode corresponds in the three temples to the transverse movement of the tower. In the longitudinal mode, the bell towers move in this direction. Finally, in the vertical mode, the belfries of all the temples present torsion.

For the dynamic analysis, a real earthquake record was used. The accelerogram considered was registered on September 19, 2017, in the location of Yautepec, Morelos, near the selected temples. The peak ground acceleration values were 1.17 m/s^2 (0.11 g) and 0.97 m/s^2 (0.09 g), for the horizontal and vertical directions, respectively (Fig. 12.24). The frequency content of this record was near to the frequencies of the selected temples and gave rise to large spectral amplitudes.

A very interesting result is that the most critical direction for all models is the transversal direction, while the action of the earthquake in the longitudinal direction is not critical. Fig. 12.23 shows the principal stress map for each temple. Stresses are concentrated in the belfries, which tend to collapse. On the other hand, in the vault, it was observed

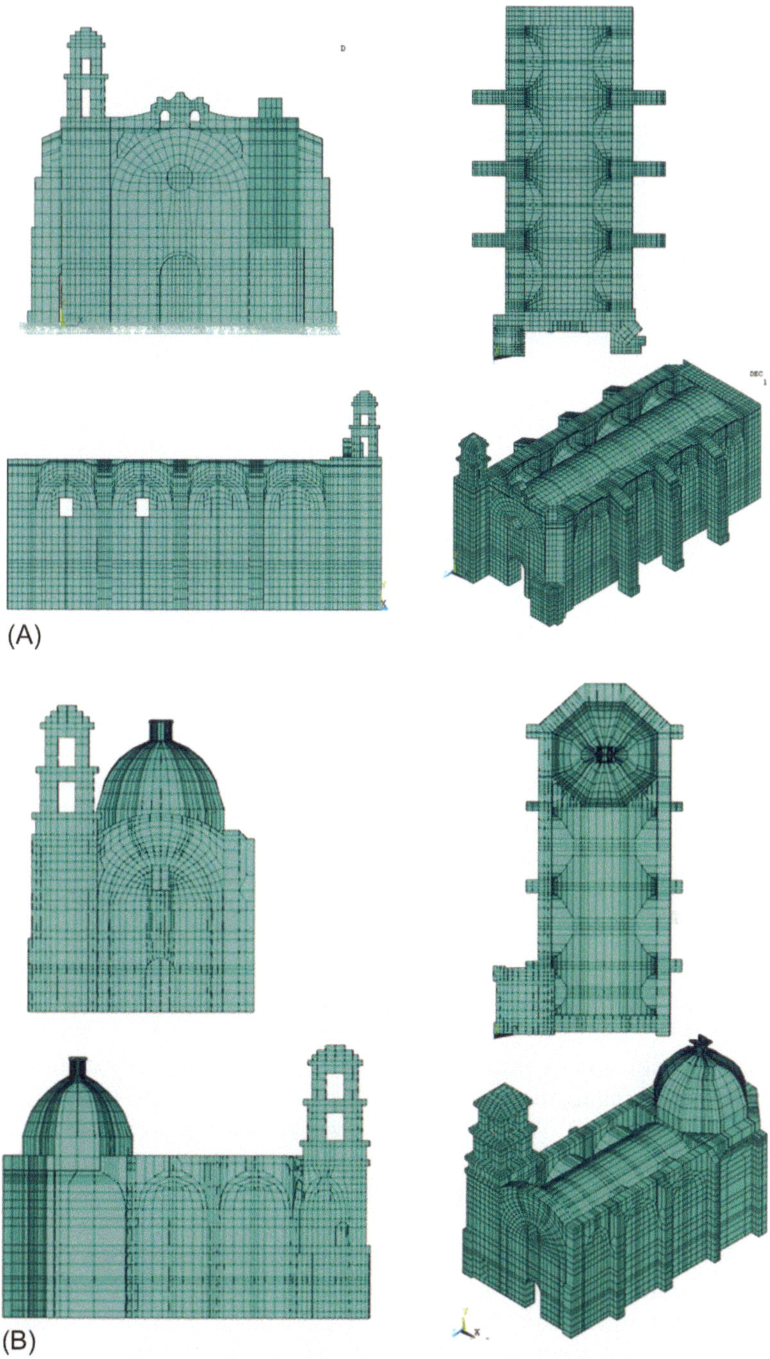

FIG. 12.21 Numericalmodels of the selected temples: (A) Totolapan; (B) Jiutepec; and *(Continued)*

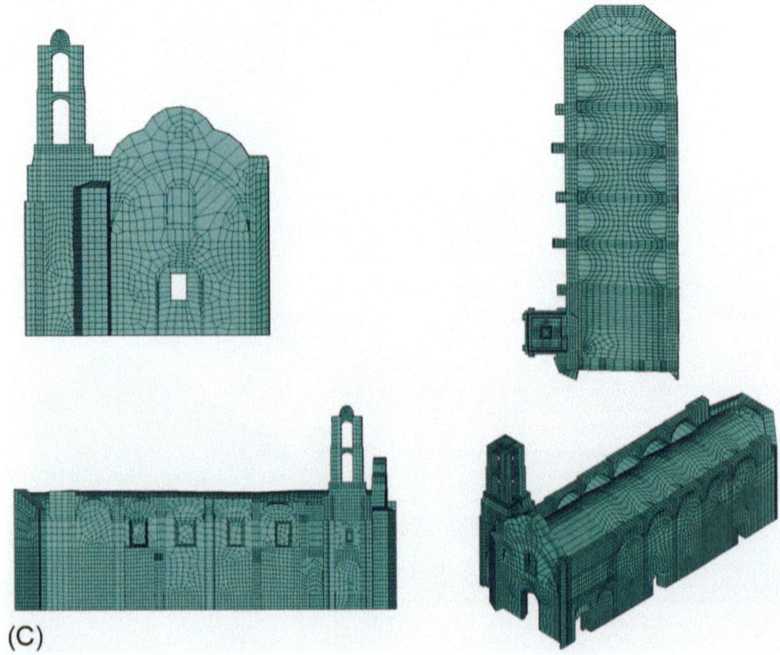

FIG. 12.21, CONT'D (C) Ocuituco.

TABLE 12.1 Mechanical characteristics of the stone masonry for each temple.

Temple	Density (kg/m^3)	Modulus of elasticity (GPa)	Poisson's coefficient	Compressive strength (MPa)
Totolapan	1460	0.34	0.31	1.10
Jiutepec	1340	0.48	0.29	3.00
Ocuituco	1760	0.46	0.29	1.90

that the distribution of stresses was similar in all cases; only differences were observed regarding the magnitude of the stresses and their areas of influence (Fig. 12.24). The presence of great tensile stresses in the arches is appreciated, which could cause the detachment of some segments. This behavior was observed in the damage of the temples of Jiutepec and Ocuituco.

When comparing the behavior of the three temples, it was observed that the type of vault and shape of the buttresses are of great importance for the walls, because in the models where they have larger dimensions of these elements, the displacements in the vault and frontispiece are considerably reduced, also reducing the magnitude of the damage.

TABLE 12.2 Main modes of vibration for transversal, longitudinal, and vertical directions.

	Transversal direction			Longitudinal direction			Vertical direction		
Model	**Period (s)**	**Mode**	**Mass participation**	**Period (s)**	**Mode**	**Mass participation**	**Period (s)**	**Mode**	**Mass participation**
Totolapan	0.20	1	0.604	0.11	6	0.490	0.05	24	0.182
Jiutepec	0.57	1	0.455	0.26	6	0.341	0.11	28	0.104
Ocuituco	0.55	1	0.589	0.25	9	0.312	0.13	26	0.129

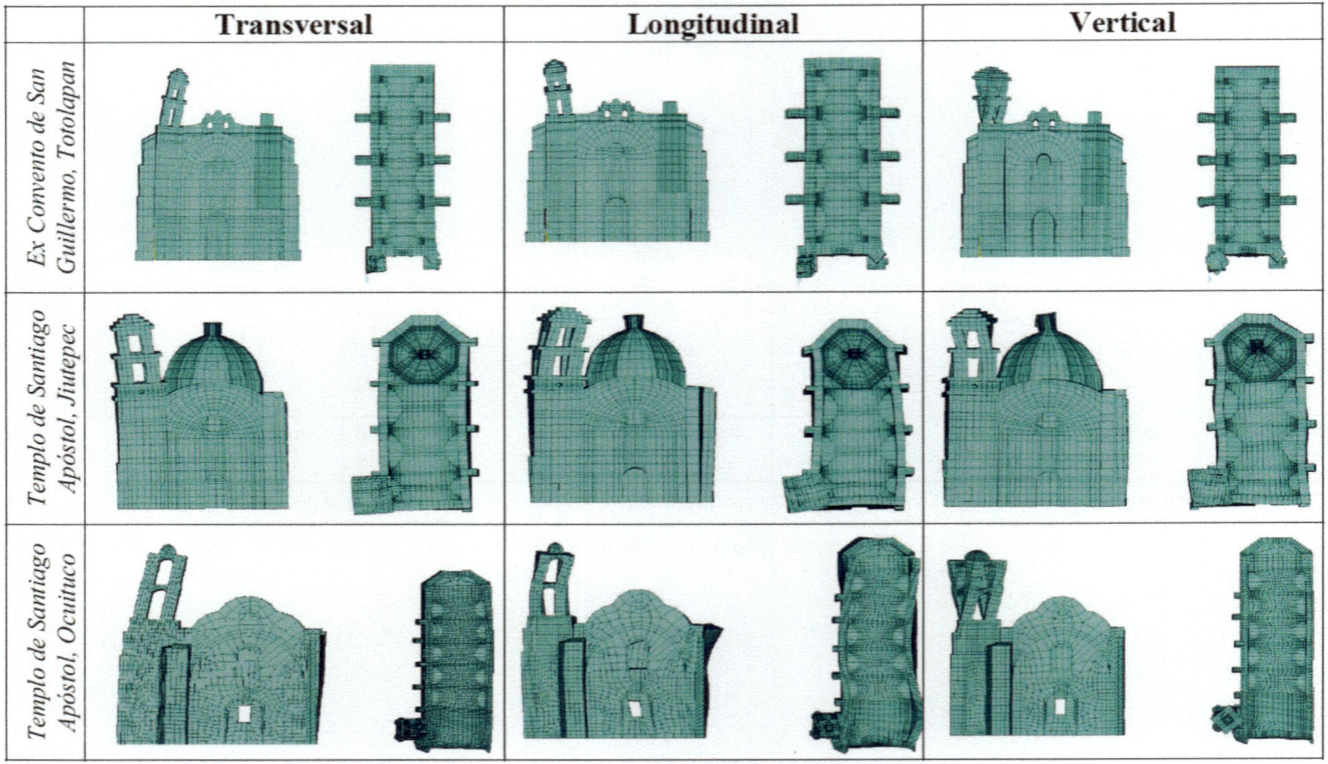

FIG. 12.22 Main vibration modes for selected temples.

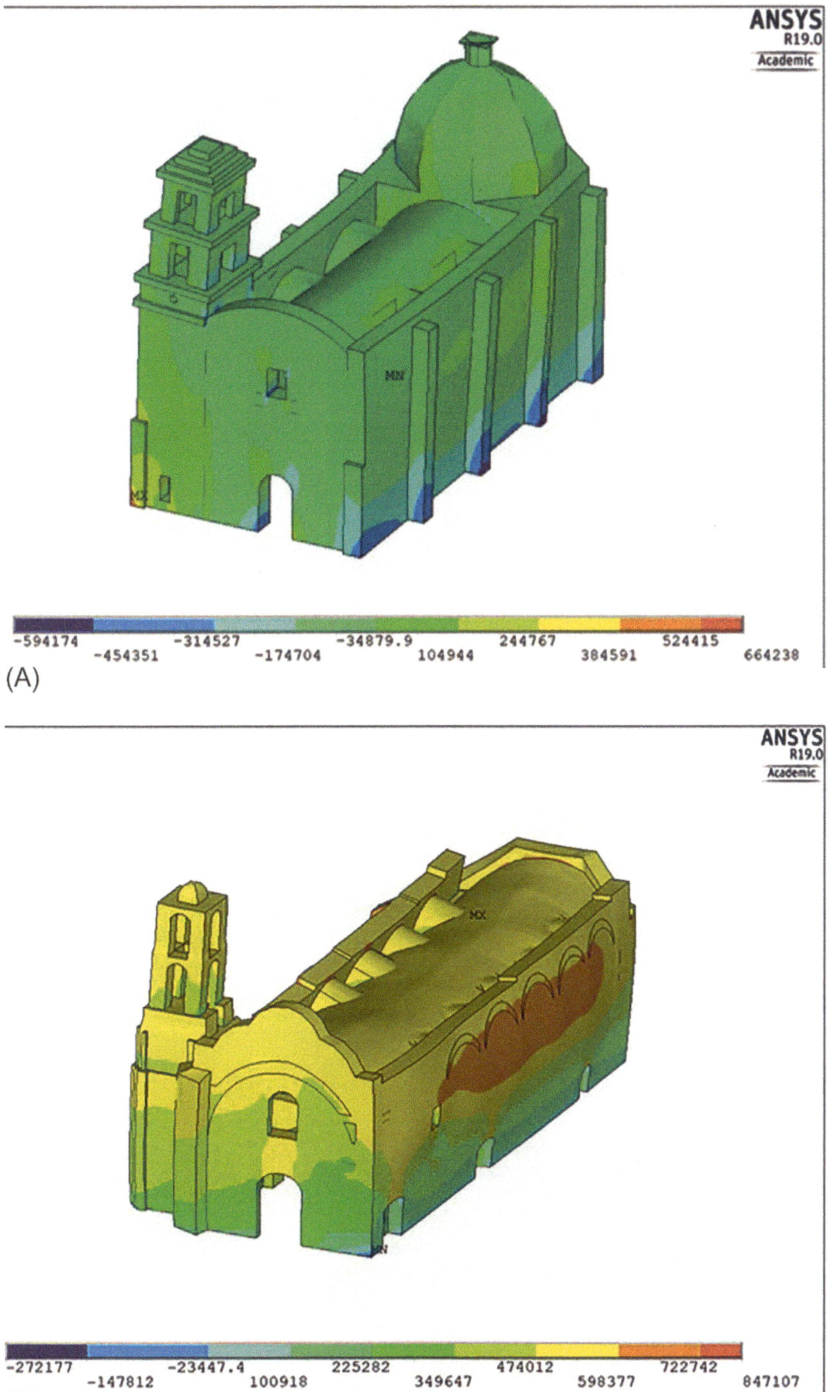

FIG. 12.23 Maps of principal stresses: (A) Jiutepec; (B) Ocuituco;

(Continued)

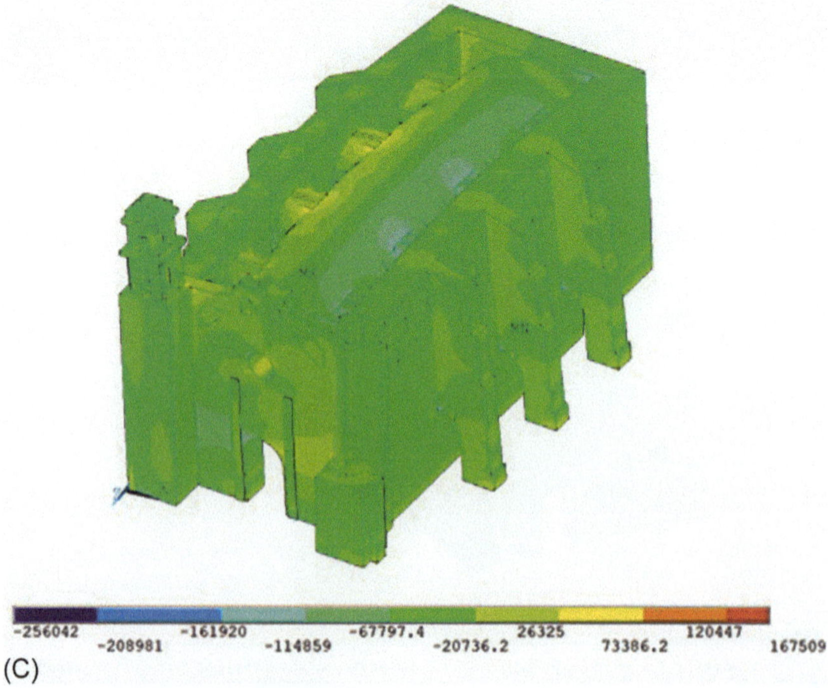

FIG. 12.23, CONT'D and (C) Totolapan.

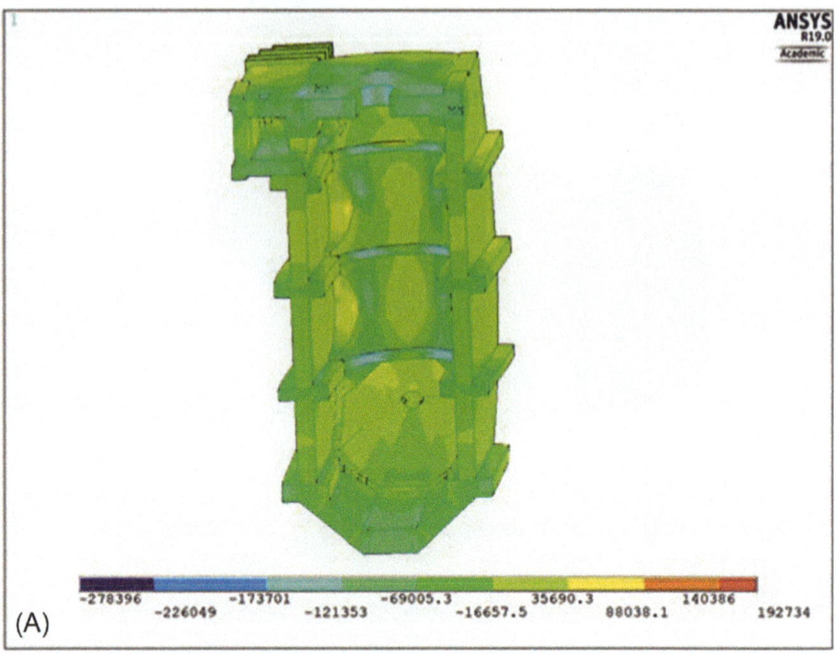

FIG. 12.24 Maps of stresses in vaults: (A) Jiutepec;

(Continued)

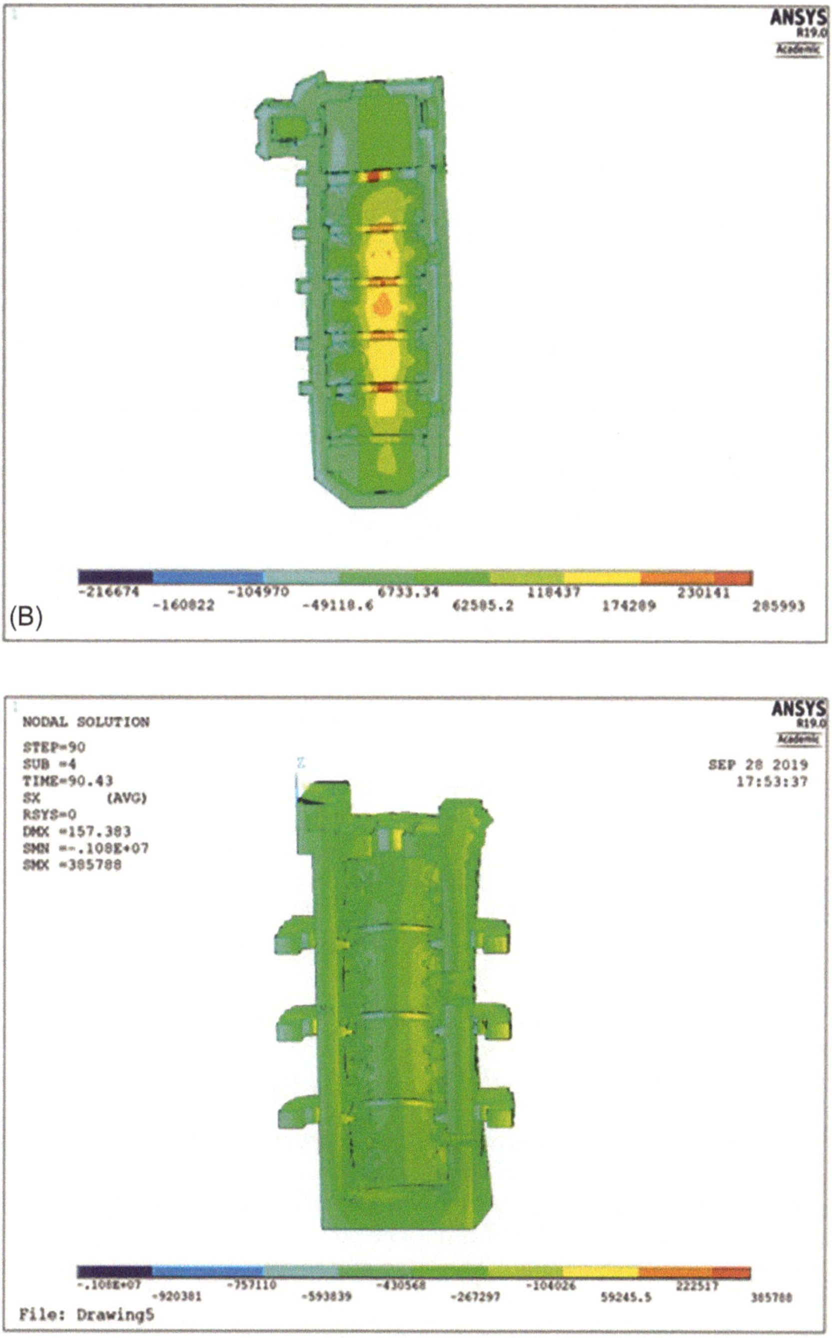

(C)

FIG. 12.24, CONT'D (B) Ocuituco; and (C) Totolapan.

12.6 Final remarks

More information about the seismic behavior of masonry churches can be obtained by observing their response and damage due to a seismic event. The damages reported in this chapter correspond to the observations in the Colonial temples of Oaxaca, Puebla, and Morelos states, as well as Mexico City, due to the earthquake of September 2017. The damage patterns and collapse mechanism described are similar than those observed in previous earthquakes. The bell towers, vaults, and domes still remain as the most vulnerable elements, while the out-of-plane damage of façades is mostly less important than the in-plane damage. The body of the tower generally is damaged by the seismic forces, which, in some cases, tend to separate from the main wall of the façade. Generally, the out-of-plane damage is only with regard to the detachment of the façade from the nave but without reaching collapse.

Lack of maintenance, poor quality of the masonry, structural modifications, and inadequate previous interventions are the main causes of the damages observed in these kinds of structures. Continuous and correct maintenance of the structure is indispensable to reduce the possible damages due to earthquakes. A detailed structural assessment of the temples is always necessary to perform an adequate intervention program according to the criteria of restoration.

The occurrence of the earthquake of September 19, 2017, gave an opportunity to collect samples from the damaged temples to determine with laboratory tests some of their mechanical properties: compressive strength, modulus of elasticity, and Poisson's ratio. This type of masonry can be modeled as a homogeneous isotropic material.

In some temples, it was detected that the masonry was formed by pebbles linked by lime-sand mortars and, in other cases, by earth mortars. This masonry, of origin, is of poor quality and disintegrates easily. The deterioration of the masonry is due to various factors, among which are: degradation, lack of maintenance, and presence of harmful fauna and flora.

The results of the numerical analyses have reproduced the observed damages and have confirmed that for seismic actions, the most critical direction is the transverse one, while the out-of-plane failure of the façades in not critical. The roof system cannot be considered as rigid diaphragm, and the thrust of the system roof over the façade and the apse can be neglected.

References

Chávez, M. M., & Meli, R. (2012). Shaking table testing and numerical simulation of the seismic response of a typical Mexican colonial temple. *Earthquake Engineering and Structural Dynamics, 41*, 233–253. https://doi.org/10.1002/eqe.1127.

Doglioni, F., Moretti, A., & Petrini, V. (1994). *Le chiese e il terremoto*. Trieste, Italy: Lint Press. (in Italian).

García, N., & Meli, R. (2009). On structural bases for building the Mexican convent churches from the sixteenth century. *International Journal of Architectural Heritage, 3*, 24–54. https://doi.org/10.1080/15583050701842344.

Lagomarsino, S., Cattari, S., & D Ottonelli, S. (2019). Giovinazzi, earthquake damage assessment of masonry churches: Proposal for rapid and detailed forms and derivation of empirical vulnerability curves. *Bulletin of Earthquake Engineering, 17*, 3327–3364. https://doi.org/10.1007/s10518-018-00542-8.

Lagomarsino, S., & Podesta, S. (2004). Damage and vulnerability assessment of churches after the 2002 Molise, Italy, earthquake. *Earthquake Spectra, 20*(s1), s271–s283. https://doi.org/10.1193/1.1767161.

Meli, R. (2011). *Los conventos mexicanos del siglo XVI*. Mexico: Porrúa. (in Spanish).

Meli, R., & Peña, F. (2004). On elastic models for evaluation of the seismic vulnerability of masonry churches. In P. Lourenço, P. Roca, & C. Modena (Eds.), *Proceeding of structural analysis of historical constructions* (pp. 1121–1131). Italy: Padova.

Meli, R., & Sánchez-Ramírez, R. (2007). Criteria and experiences on structural rehabilitation of stone masonry buldings in Mexico City. *International Journal of Architectural Heritage, 1*, 3–28. https://doi.org/10.1080/15583050601123118.

Peña, F., & Chávez, M. M. (2016). Seismic behaviour of Mexican colonial churches. *International Journal of Architectural Heritage, 10*, 332–345. https://doi.org/10.1080/15583058.2015.1113341.

Peña, F., & García, N. (2016). Numerical evaluation of the seismic behaviour of facades of Mexican colonial churches. *Engineering Failure Analysis, 62*, 164–177. https://doi.org/10.1016/j.engfailanal.2016.01.011.

Peña, F., & Manzano, J. (2015). Dynamical characterization of typical Mexican colonial churches. In Psychariset al. (Ed.)*Seismic assessment, behaviour and retrofit of heritage buildings and monuments*. Switzerland: Springer International Publishinghttps://doi.org/10.1007/978-3-319-16130-3_12. Chapter 12.

Preciado, A., Orduaña, A., Bartoli, G., & Budelmann, H. (2015). Façade seismic failure simulation of an old Cathedral in Colima, Mexico by 3D limit analysis and nonlinear finite element method. *Engineering Failure Analysis, 49*, 20–30. https://doi.org/10.1016/j.engfailanal.2014.12.003.

Ramírez de Alba, H., & Valdez-Medina, E. A. (2014). Cementitious material and concrete of ancient cultures in America and Yaxchilan bridge. In: F. Peña, & M. M. Chávez (Eds.), *Proceeding of structural analysis of historical constructions, Mexico*.

Servicio Sismológico Nacional (2017a). *Sismo de Tehuantepec (2017-09-07 23:49 Mw 8.2), Reporte Especial. Available from(2017a). http://www.ssn.unam.mx/sismicidad/reportes-especiales/2017/SSNMX_rep_esp_20170907_Tehuantepec_M82.pdf.*

Servicio Sismológico Nacional (2017b). *Sismo del día 19 de septiembre de 2017 (Mw 7.1), Reporte Especial. Available from(2017b). http://www.ssn.unam.mx/sismicidad/reportes-especiales/2017/SSNMX_rep_esp_20170919_Puebla-Morelos_M71.pdf.*

CHAPTER

13

Features and seismic response of large masonry structures: A case study of Singh Durbar main building, Nepal

Rishi Ram Parajuli
University of Bristol, Bristol, United Kingdom

13.1 Introduction

Masonry structures are one of the most traditional structural forms that are extensively used across the globe. These are affordable as well as sustainable even though some risks in the seismic region are higher for such typologies. The 2011 census showed that nearly 45% of the houses are made up of stone or brick masonry with mud mortar and another 15% of masonry constructions were built with cement mortar (Central Bureao of Statistics, 2012); however, recent construction practices, availability of materials, and specifically reconstruction after the 2015 Gorkha earthquake changing the scenario to lower numbers of masonry buildings. Nepal, a culturally rich country, has its history not only as an independent country but also hosts several monuments and heritage values, including masonry buildings and arts. There are several masonry structures built specially during the Malla Dynasty (1200–1768) as temples, *dharamshalas* (rest places), etc. (Parajuli, 1986). Later on, Ranas had also added up large neoclassical structures over Nepal mainly influenced by European architecture. This chapter mainly focuses on features and observational seismic responses of large masonry structures in Nepal. Large masonry structures, rather than setting a distinct boundary to categorize its size, structures with larger floor area and at least three storied are considered as large structures, mostly categorized as historic architecture. Especially, observation-based seismic

Masonry Construction in Active Seismic Regions
https://doi.org/10.1016/B978-0-12-821087-1.00013-2

response of Singh Durbar, one of the major government administrative buildings, will be discussed along with other structures.

The historic architecture reflected in historic monuments, mansions, castles, and palaces represent the engineering knowledge, techniques, and skills together with cultural patterns of the time and the authentic flavor of the patrons of the created work. There are several dots of historic structures carrying over the cultural heritage in Kathmandu valley and the outskirts. World heritage sites, historical settlements, Bagmati riverbank hosts several historic structures where large masonry structures are dominant. Temples, palaces (*durbar*), and *dharamshalas* built centuries ago are of large masonry structures (Tamrakar & Parajuli, 2019).

13.2 Archaeological values of neoclassical structures

The Rana palaces that dotted the Kathmandu city landscape after the late 19th century till the end of the First World War stood in sharp contrast to the pagoda or stupa style medieval temple and residential architecture of the valley. The new Rana palaces embraced the new vision of grandeur on scale and concept, creating huge living structures using entirely imported materials from Europe to herald a new era of comfort and luxurious living. These palaces in walled compounds used European designs and values both in interior and exterior aspects with functional courtyards, green lawns, and foliage plants around. These opulent mansions were modeled on noble chateau houses of France, Belgium, England, and Italy and remarkably there were also attempts at putting in some Nepali design features in them.

The first palace to rise in Kathmandu with the grand touch of European style to set the tone for subsequent Rana palaces was the Tribhuvan Gallery in the Hanuman Dhoka Durbar complex. This new mansion was raised for use by the Prince Trailokya, the son-in-law of Prime Minister Jung Bahadur Rana. Needless to say, this mansion was commissioned after Jung Bahadur's return home from his European tour.

Of course, the culmination of the series of the famed Rana palaces came with the Singh Durbar (Fig. 13.1) built by Prime Minister Chandra Shamsher in 1903. Two Nepali architects Kumar and Kishor Narsingh Rana were the principal architects to design and oversee the construction. The Singh Durbar was built as residence-cum-administrative command center of the all-powerful Rana Prime Minister. It was set on 50 ha of manicured lawn and comprised seven connected courtyards. It had over 1000 rooms and could be counted among the largest palaces in South Asia. It was seriously damaged by fire in 1973 and most parts of the palace were lost with only the front courtyard remaining for use. This too was

FIG. 13.1 Singh Durbar: One of the largest masonry structures in Asia (Bristish Library, 1948).

devastated by the 2015 Gorkha earthquake which now is under construction for conservation.

The magnificent facade of well-proportioned archways, between which rise double Corinthian colonnades, gives a tremendous sense of space and splendor. This magnificent building is a pride in the context of the Rana period architecture of Nepal. The Durbar Hall at the Singha Durbar was where the Prime Minister held his formal meetings and banquets for foreign dignitaries and also used for family ceremonies. This Durbar was supposed to be fashioned after the Chateau of Versailles, France. Well-known historian Greta Rana opines "Singha Durbar always reminds me of the huge Chatsworth House, Woburn Abbey and I think it owes much to Maharaja Chandra's visit to some of the noble houses of England" (Sever, 1992). It is not the ultimate copy of anything and has a distinctive royalty of its own. The grand hall has Italian marble floors, colored mirrors with embossed decorative patterns from Belgium, crystal chandeliers from Murano, and stained-glass door from England. The Crystal clock was one of the only two produced, and the lead crystal light fountain, which stands above 20 ft. tall, was specially designed in France. Now, the crystal clock is taken to the national museum for safekeeping, which will be redeployed in its original place after the palace conservation is completed in a proper manner.

13.3 Overview of Singh Durbar

This chapter mainly focuses on the Singh Durbar to provide insights of the building. The outer to outer dimension is 110 m and 21 m in longitudinal and transverse directions, respectively. However, two major load-bearing walls are 6.85 m apart and 97.5 m long. Outer walls in the front and sides hold major architecture of the decorative pillars. Figs. 13.2 and 13.3 show four floor plans and elevations of the building, respectively.

The building is symmetric in the short axis and asymmetric in the long axis in general view but is asymmetric in both directions. Some short walls are missing in different floors; hence, when looking at detail, this structure is asymmetric. Figs. 13.4 and 13.5 show the sectional views of the building in longitudinal and transverse directions, respectively. Detailed dimensions (in mm) of building height are shown in Fig. 13.5.

One of the major features of this building is a big hall (State hall) with double floor height in center having 33 m unsupported length of longer

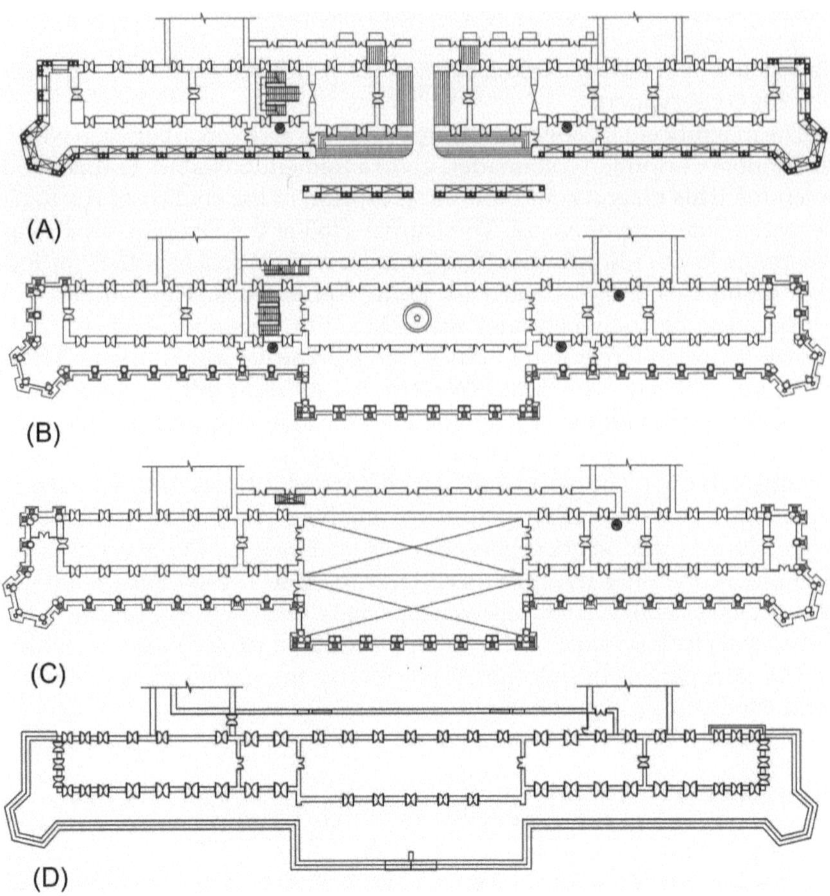

FIG. 13.2 Building plan: (A) ground floor, (B) first floor, (C) second floor, and (D) top floor of Singh Durbar.

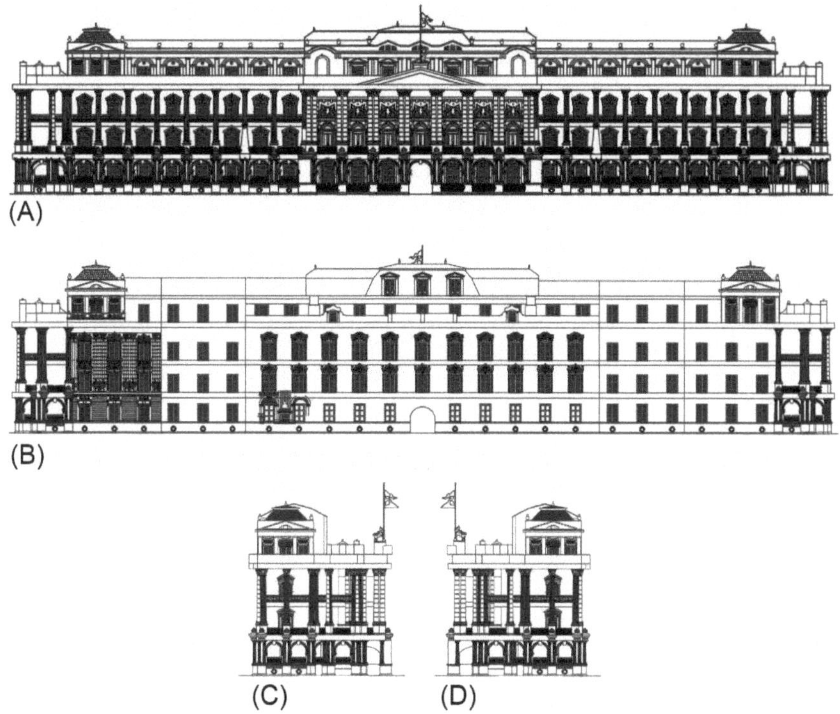

FIG. 13.3 Elevations of Singh Durbar: (A) front, (B) back, (C) left, and (D) right.

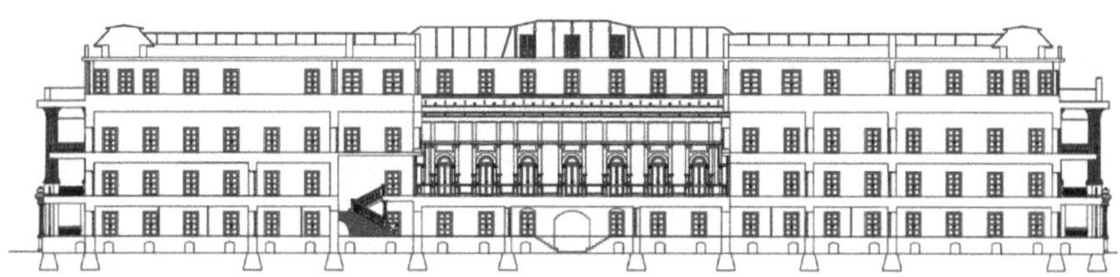

FIG. 13.4 Sectional view along longitudinal axis of the building.

walls. Such a major hall with double story height is found in almost all neoclassical buildings. There are integrated decorative ceiling and wall claddings of timber panels and boards inside the hall, which provides inner confinement.

The other major feature in the building is vaults in the basement (*Hawas*), which is mainly constructed to provide the air circulation route below the plinth level of the structure. These are mainly the tunnels with the arch shape at the top with dimensions of 0.914 m in breadth and 1.040 m in height (crest height). These tunnels are running through longitudinal as well as in transverse direction as shown in Fig. 13.6. Fig. 13.7

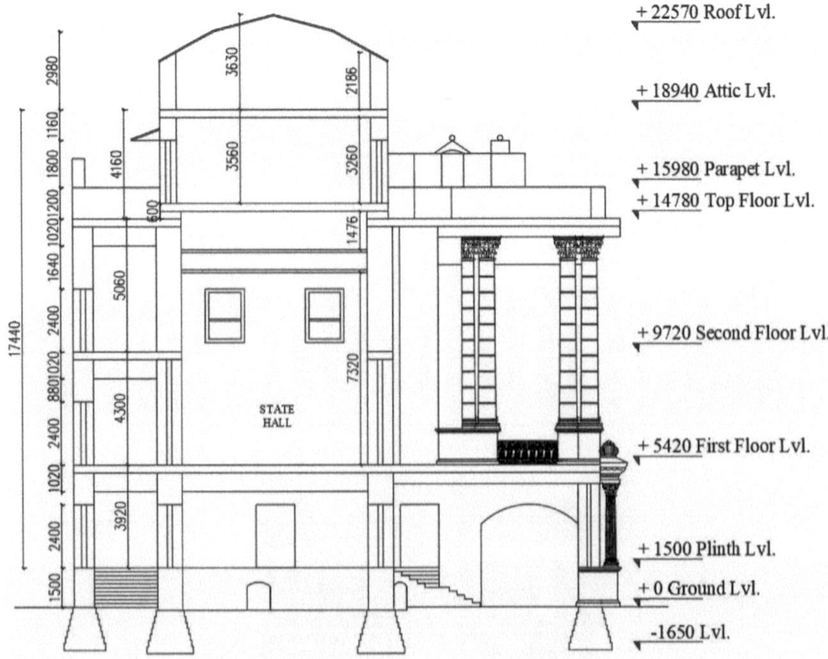

FIG. 13.5 Cross-sectional view along transverse direction through center of building.

shows the photos of vaults inside the building and its end at the outer wall. The air circulation system in the building is not only under the plinth but is also provided between the upper floor and ceiling of each floor. There are spaces provided between beams across the walls and for in and out of natural air, where perforated sheets are also attached in the ceiling to provide access to fresh air and release of hot air.

13.4 Construction materials and methodology

Structures built centuries ago are mostly made up of brick masonry with mud mortar (few cases with lime mortar); in many cases, we have heard there was the use of natural admixtures like urad bean (*Vigna mungo*) pastes in mortar. However, the investigation of walls in Singh Durbar, supposed use of such admixtures, cannot find any evidence of the use of such materials. Bricks of higher strength, mud mortar with the proper ratio of different grain sizes, timber, and steel beam sections are used in neoclassical structures, whereas only timber sections are used as beam and column in medieval structures. Fig. 13.8 shows the brick units used in Singh Durbar and the closeup view of the wall with bricks and mortar.

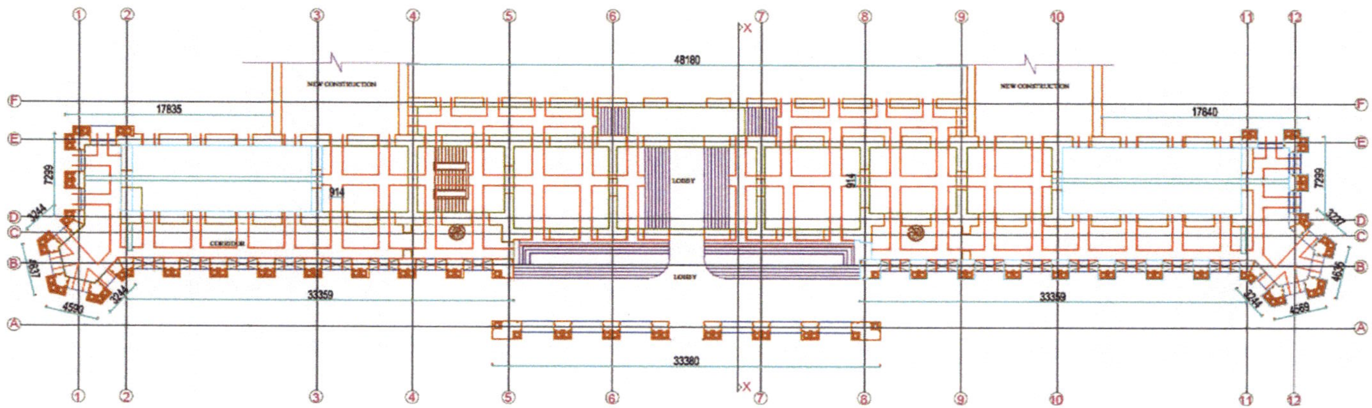

FIG. 13.6 Basement vault plan *(red lines)*.

FIG. 13.7 Basement vault (*Hawas*) (L) and opening provided for vaults in outer wall (R).

FIG. 13.8 Brick units on left and wall section with brick and mud mortar on right.

Test results of the materials from the different structures showed that high strength bricks were used in Singh Durbar, much lower in Shreemahal but higher than the average commercial bricks. Average compressive strength of brick samples from Singh Durbar and Shreemahal is 19.64 and 6.4 MPa, respectively, where mortar strength is found to be 1.32 MPa from cube test (Parajuli, Furukawa, & Gautam, 2020).

Materials used in nonstructural components are timber planks, mud for flooring; lime *surkhi*, mud with rice husk and cow-dung for plaster; cast iron and timber for staircases; cast iron for railings; timber planks and boards for the ceiling. Corrugated galvanized iron sheets are used to cover the building with timber trusses in neoclassical buildings where *Jhingati (roof tile)* are commonly found in medieval structures. Fig. 13.9 shows the typical timber roof trusses used in monumental buildings.

FIG. 13.9 Roof timber trusses in Singh Durbar.

13.5 Structural system

Singha durbar is the load-bearing structure of brick masonry with mud mortar. It has load-bearing wall as the major structural component to transfer the vertical load where beams are transferring loads from slab to the wall. Beams of timber are used in short spans where steel sections of various sizes are used in longer spans and the top floor. This structure is symmetrical in longer side where the main horizontal wall is of 97.5 m. The central part of the front façade is 5.43 m wider than the sides for a length of 33.38 m. On the ground floor, there is a lobby to cross the building, which breaks the continuity of the main walls.

There is another wall parallel to the main wall, 3.28 m apart, in the central, backside of the building to provide passage of length 48.18 m. Structural irregularities in horizontal as well as vertical plane are also present in this structure. In the central part, the extended balcony and 1.35 m off the grid of front main wall shows some irregularities in structural system. The Central Hall, the state hall is more than double story high, where the bottom of the lower false ceiling level is about 7.5 m high. Another about 3 m vertical space is there to reach the top floor level. This contributed to vertical irregularities.

13.5.1 Foundation

The foundation of the structural wall is a stepped foundation of brick masonry with mud mortar as shown in Fig. 13.10. The depth of the foundation is 1.61 m from the level of the basement, where height of basement to the ground floor level is 1.524 m.

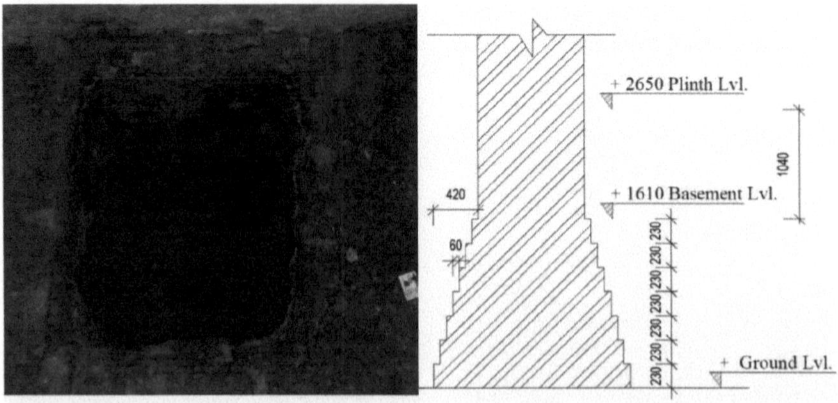

FIG. 13.10 Foundation observation pit (L) and sectional sketch of foundation of main wall (R).

13.5.2 Wall

This building is rectangular with an overall dimension of 110 × 21 m. L/B ratio is more than five. Major two long walls with larger thickness are placed to transfer the load of the building where relatively thinner outer walls are also in place, to provide architectural aesthetics. However, these walls are also subjected to loading more than the usual buildings. Distance between two major longitudinal walls is about 6.8 m where a number of cross walls are present to minimize the unsupported length, which varies in each floor.

The thickness of the major and minor walls is different, major walls have the wall thickness of 1.1 m in between ground level and plinth where the thickness reduced and varied up to 0.67 m at the top floor. Other walls are of varying thickness in different floors and grids. Typical cross-sections of major walls are shown in Fig. 13.11 with varying thickness throughout the height.

The main walls (longitudinal) of the building are 1.0 m wide in the ground floor, 0.9 m on the first floor and second floor, cross walls are of 0.87 m thick where the walls on the side of the lobby in the ground floor are 0.66 m thick. On the backside, wall for passage is 0.81 m thick. On the top floor, main walls are 0.67 m thick where cross walls are 0.5 m thick. Parapet walls are 1.220 m high and 0.5 m thick. Height of the walls in the ground floor is about 3.92 m, first floor is about 4.3 m except the Central (State) Hall, where the height is about 9.35 m. Mezzanine floor height is about 4.3 m and the top floor height is about 3.55 m.

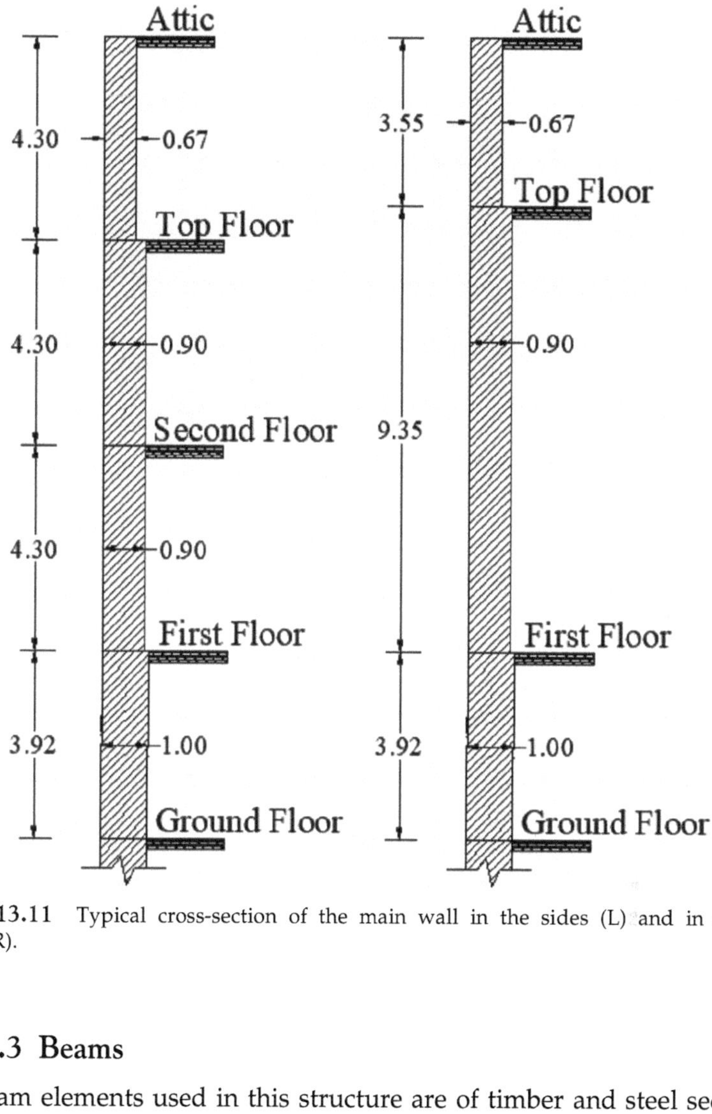

FIG. 13.11 Typical cross-section of the main wall in the sides (L) and in central part (R).

13.5.3 Beams

Beam elements used in this structure are of timber and steel sections where jack arch beams are also supporting the slabs above passages on back side. Iron (I-sections) of depth 380, 280, 150, and 100 mm is used. The long span in the center of the structure is supported by I-sections. Timber sections of 200×200 mm are used to support the floor on a shorter span. There are three layers of beams above the state hall, two layers of timber beams and a top layer of steel sections to support the floor as shown in Fig. 13.12.

FIG. 13.12 Beam and ceiling arrangements above State Hall.

13.5.4 Wall-beam connection

The major path of load in load-bearing structure is beam to the wall, which ultimately drags major point of interest for structural engineers. The practice of providing wall plates above the wall to distribute the load uniformly is a common practice even though only a base plate or stone is used in stone masonry residential structures. In this building, wall plates are used only in some locations. Timber elements of cross-sectional size of 125 × 100 mm size, at the top of the wall, are holding timber beams. Two wall plates are running on each side of the solid wall top. Even though the wall plates are used on both sides, the sitting of the beam is in only one side, where the beam enters into the wall. Typical wall, wall plate, and timber beam layouts are shown in Fig. 13.13 and Fig. 13.14A and B. Fig. 13.14C and D show the wall and iron beam connection where instead of continuous timber wall plates, large stones are used. In some locations, I-Beams are found just rest on the brick as shown in Fig. 13.14D.

13.5.5 Floor system

This building has typical flexible flooring system with different floor finishes. In ground floor, where the open space is available under the plinth level, there is single layer of wooden planks flooring, but in other parts, there are either wooden planks or bricks rest on the joist, on top of that mud of various thicknesses are under the floor finish. Total thickness of the floor varies from 240 mm to 300 mm above the timber joist. Typical flooring system is shown in Fig. 13.15.

FIG. 13.13 Wall and timber beam connection layout.

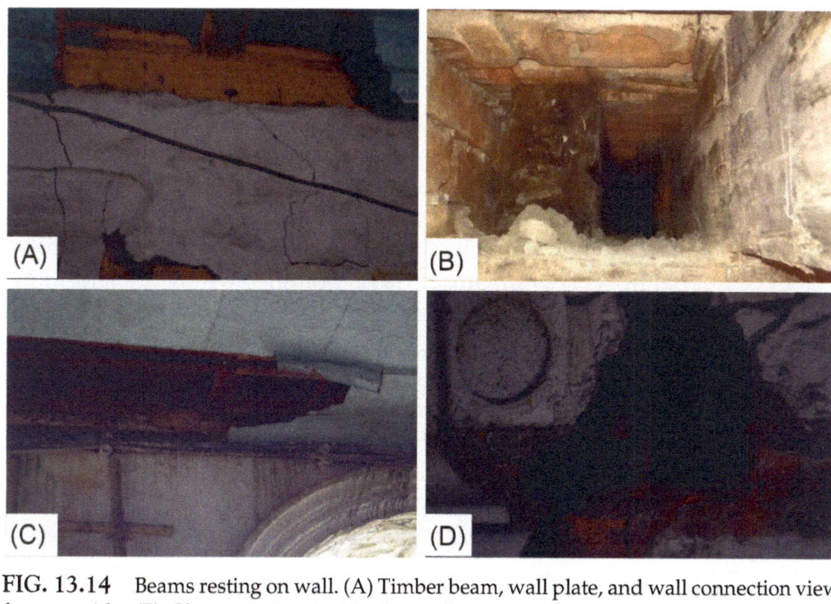

FIG. 13.14 Beams resting on wall. (A) Timber beam, wall plate, and wall connection view from outside. (B) Close-up view inside the wall. (C) Connection of steel beam and wall. (D) close-up view of connection.

13.5.6 Roof

The roof of this structure is supported by the timber frame truss on the top. Roof truss has beams resting on the attic floor to tie up long span beams, those running on top of the major walls in few locations. In the center, where the attic level is also used as a service floor, the height of the

FIG. 13.15 Floor system in the building having wooden plank above joist (L) and brick above joist (R).

truss is larger than the other sides. Fig. 13.9 shows the roof truss system on the central top of the building.

13.6 Seismic response of large masonry structures

On April 25, 2015, Central Nepal was rocked by the Gorkha earthquake 2015 of magnitude 7.8, the largest earthquake in the region after the 1934 Great Nepal-Bihar Earthquake. More than 8979 people, including 70 foreigners, lost their lives where hundreds of thousands of structures got collapsed or damaged severely due to this earthquake sequence. Nepal hosts several structural typologies where ancient monumental to low strength masonry and modern framed structures serve the country. Ground motion recorded at Kathmandu shows a different characteristic of lower frequency dominance (Parajuli & Kiyono, 2015). Damage patterns, scenarios, geological, topographical effects along with structural deficiencies are reported by several studies (Gautam, Rodrigues, Bhetwal, Neupane, & Sanada, 2016; Parajuli & Kiyono, 2015; Sharma, Subedi, Parajuli, & Pokharel, 2017).

Historical monument structures faced varying damage grades depending on different circumstances, like maintenance, use, and site condition. Damages in historical monumental structures are reported in few studies (Bhagat et al., 2018; Gautam, 2017; Parajuli, 2018). Vulnerability analysis, comparison of vibration characteristics, damage detection through ambient vibration had also been done in some of the historical buildings in Patan (Furukawa, Hanafusa, Kiyono, & Parajuli, 2019; Furukawa, Kiyono, Parajuli, Parajuli, & Toki, 2017; Kiyono, Furukawa, Parajuli, Maskey, & Parajuli, 2017). Damage assessment and retrofitting proposal of large structures, Bagh durbar and Shree Mahal, are reported by Adhikari et al. (Adhikari, Jha, Gautam, & Fabbrocino, 2019) and Parajuli (Parajuli, 2018). Failure mechanisms observed in Singh Durbar are discussed in the following section.

13.6.1 Failure mechanism

Masonry structures are brittle in nature and are the most vulnerable to earthquakes. The structural integrity, perfect load path system, floor rigidity, and box action are important to sustain earthquake shaking; consequently, to prevent the collapse. The connection between the masonry walls should have a good interlocking system. The floor diaphragm is another major component of the structure to control seismic behavior. There are mainly two major failure mechanisms in the structure as global and local failure. Furthermore, more damage types are discussed.

13.6.1.1 Global failure mechanism

In this system, we look at the holistic approach where the failures in the parts of the building are associated or not. The major weak portion of the building in the global aspect is bending of slender columns; damages in the major walls; those exist on first to top floor level. There are few cracks observed on the base of the column, minor damages in major walls but not in the state of failure.

Shear cracks on the walls in the ground floor are under the global mechanism, which are mostly concentrated on the corners of the building. There are few shear cracks on the short walls on the ground floor and the severity is reducing in the upper floors. When looking at the locations and size of cracks on the wall, the hammering from the next building may have contributed to those cracks.

Out of plane bending effect on the building is not observed substantially. The vertical plumb line of the building is in good condition, only on the top of the wall of the state hall have some damages associated with the out of plane bending. Vertical and horizontal separation joints are seen but not in a large scale.

13.6.1.2 Local failure mechanism

There are several local failures observed in the building, which are not only due to the earthquake but the deterioration of members (structural or nonstructural) also have contributed in the damage. The collapse of the floor in several locations is due to the deteriorated timber beams and joists. In some cases, the absence of the wall plate and weak beam column connections is responsible for the local damage. There is severe damage on the long walls in upper floors where the adjacent building was hit during the earthquake shaking.

13.6.2 Observed damage patterns

This structure has almost all types of failures that can appear in the masonry structure. Probable reasons for those associated with the current damages are also discussed in this section.

13.6.2.1 *Shear failure*

There are several diagonal shear cracks observed throughout the wall, mainly in short walls. In some cases, not only the head and bed joint failure but also bricks are found cracked in the failure plane. Some locations witnessed large cracks (>5 mm wide) on the wall at the brick mortar joint due to the weak mortar without cracking of brick, resulting in stepped failure pattern. Typical failure is shown in Fig. 13.16.

One of the major reasons for such cracks is the lack of shear strength of the masonry. The thickness of the short walls is also less than that of longer walls. The shear strength of these walls should be enhanced during retrofitting. Few other failures are shown in Fig. 13.17.

13.6.2.2 *Vertical and horizontal joint failure*

Another failure pattern observed in the structure is vertical and horizontal joint separations. There are few vertical joint separations through the joints between the major wall and partition wall. Along with the vertical separation horizontal floor, separation joints are also observed. The flexible floor diaphragm having some separations in the wall floor joint is not that severe, but in the verandah, major cracks on floor wall joints are alarming. Typical separations are shown in Fig. 13.18.

FIG. 13.16 Shear failure in short wall at first floor.

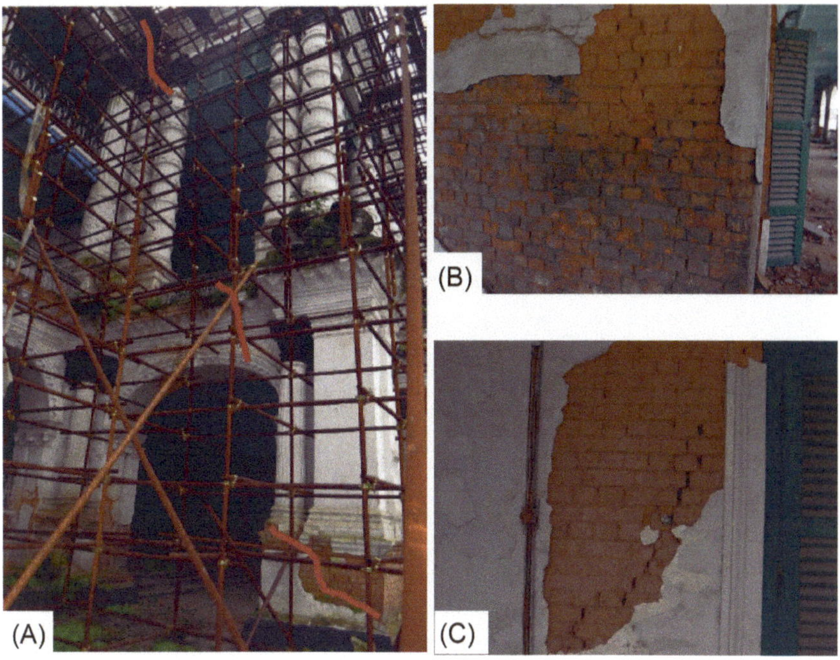

FIG. 13.17 (A) Shear failure seen in the extended wall in front of the building (B) and (C) Shear cracks observed on ground floor.

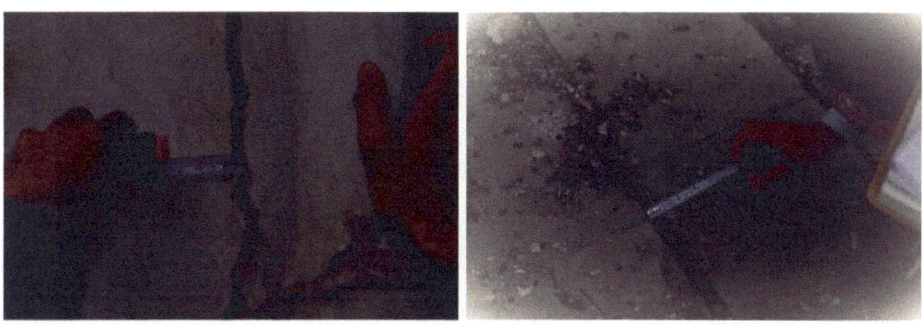

FIG. 13.18 Vertical (L) and horizontal (R) joint separation observed in the building.

13.6.2.3 Bending failure

Bending failures of structural elements are also found in this building. Mainly in the long columns from the first floor to the top floor, we can find the cracks on the bottom of the column due to bending of it; in another way, we can consider the failure at bottom of this slender column as a rocking failure. On the top of the tall wall, outside of the state hall, out of plane bending failure was observed. Major walls at the top floor are also

FIG. 13.19 Damages due to bending in bottom of column (L) and damage on top of the tallest wal (out of state hall) (R).

subjected to bending failure. Bending failures of column and wall top are shown in Fig. 13.19.

13.6.2.4 Toe crushing

Bricks at the outer skin of the major walls in the ground floor are found cracked, mainly due to the rocking of the building. Such a phenomenon of damage can be categorized as toe crushing failure. Such failure locations are observed on outside of the walls only. It has not continued throughout the wall cross-section, hence cannot be considered as compression failure. Similar failures were observed in the outmost wall with decorative pillars in the ground floor.

13.6.2.5 Arch failure

Masonry arch component parts of the structures also witnessed cracks on its crest. Where ever the shape of the arch is not perfect and the opening is large in the ground floor lobby, brick crushing on sides are also observed. Mainly, cracks seen on the arch are associated with shear and the separation along with compression. Typical cracks observed in masonry arch are shown in Fig. 13.20.

13.6.2.6 Damages due to hammering

Singh Durbar complex is of four buildings having square courtyard at center, built at the same time. Fire in 1973 damaged other three sides of the complex except the Western part. Reconstructed structure in three sides of the complex is of load-bearing brick masonry with cement sand mortar, and reinforced concrete beam and slabs. Hence, the stiffness of adjoining structure is higher.

There are extensive damages observed in the walls that are adjacent to the other buildings in two locations on the East part as shown in rectangles in Fig. 13.21. Damages on the major walls in the contact area are severe in the top floor; however, it is minor in the ground floor. Maximum residual

FIG. 13.20 Typical damages on masonry arch.

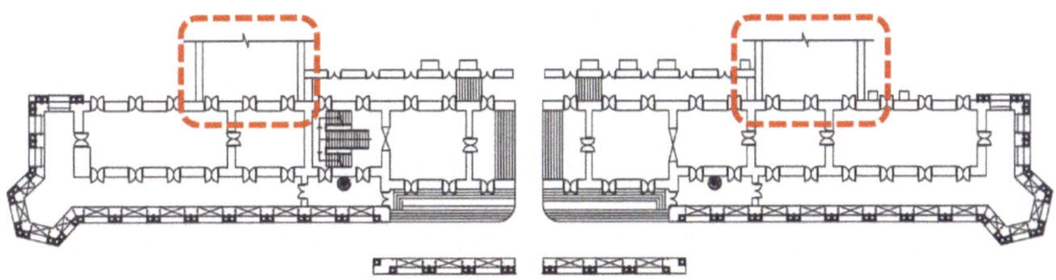

FIG. 13.21 Locations of building connected to other stiff structures.

displacement measured on the punched wall is more than 15 cm on the top floor. Damage patterns and locations indicate the effect of other building that hammered on the surface during the shaking.

Horizontal cracks, like bending failure, occur due to the hit on the top of the wall. In cross walls near the junctions, there are large diagonal cracks, which have also the contribution from hammering. Other cross walls have not suffered from such large cracks.

13.6.3 Other damages

There are other several failures like damages above the openings, where shear cracks are observed in most of the locations. Few vertical cracks are also seen above the openings where the punching effect of the wall plate is found on the wall. Nonstructural damages inside the buildings are also one part of damage observation. Utilities and services are mostly out of order.

13.7 Discussions

Large masonry structures suffer several damages where the archaeological values and restoration are crucial. The detailed survey of the Singh Durbar shows the weaker parts of the structure. Openings, arches, the outer skin of load-bearing walls in the ground floor should be the focus of seismic strengthening. Control of drift could minimize most of the damages. Tall walls, walls on the top floor without proper connection to the roof, and top beam are other areas of damages. There is an extended part in the building at the center, which has severe damage in the top floor extended part, which is an obvious location due to the irregular shape.

In large masonry structures, there are redundant components that contribute to maintain the load path system. The load path of this building is not heavily affected due to the earthquake except in a few locations where the beam was failed, which does not seem actually due to the earthquake only, but due to deterioration. Hammered locations by the adjacent building on the top floor are affected, which is on the load path system. Cross walls having the larger diagonal cracks are not in the load path system; hence, those damages are not serious for the overall structural safety.

Most of the buildings that have archaeological values are damaged more due to lack of proper maintenance. Similar to the cases in Bagmati river corridor as reported by Tamrakar and Parajuli (2019), Singh Durbar has also witnessed several damages due to lack of proper maintenance. Rotted timber beams, joists are found in the locations where floors are collapsed.

13.8 Final remarks

Large masonry structures, having larger thickness, are weaker in bending and shear but if the confinement of such walls could provide properly, the larger thickness of wall would be a great asset to control the drift, which ultimately limits the total damage. Seismic gaps to prevent the pounding/hammering should be provided. Slender walls and columns should be stiffed against rocking. Wallplates should provide to distribute the loads through beam. Seismic strengthening of the Singh Durbar is ongoing using the steel frame confinements.

There is a number of such large masonry structures in Nepal; insight on the Singh Durbar, a large masonry structure, provides a piece of good knowledge on the structural system, materials, seismic response, and damages. It can be further used to evaluate and strengthen other alike structures all around the globe.

Acknowledgments

The author is grateful to archaeologist Mr. Bishnu Karki, Ex-Director General of Department of Archaeology, Nepal, for providing information on historical and archaeological values of the building. The author would also like to extend his thanks to architects Anami Bohara and Anita Tamrakar, and engineers Kusum Sharma and Surendra Katuwal.

References

Adhikari, R., Jha, P., Gautam, D., & Fabbrocino, G. (2019). Seismic strengthening of the Bagh durbar heritage building in Kathmandu following the Gocrkha earthquake sequence. *Buildings, 9*(5), 128.

Bhagat, S., Samith Buddika, H. A. D., Kumar Adhikari, R., Shrestha, A., Bajracharya, S., Joshi, R. , et al. (2018). Damage to cultural heritage structures and buildings due to the 2015 Nepal Gorkha earthquake. *Journal of Earthquake Engineering, 22*(10), 1861–1880.

Bristish Library (1948). *Monuments and Rana Palaces [1921–1948]. EAP166/2/1/12. Retrieved: (1948). https://eap.bl.uk/archive-file/EAP166-2-1-12.*

Central Bureao of Statistics (2012). *National population and housing census 2011 (National report) Government of Nepal.* (Vol. 01), .

Furukawa, A., Hanafusa, R., Kiyono, J., & Parajuli, R. R. (2019). Investigation of natural frequency reduction mechanism of a historic masonry building in Patan following the 2015 Gorkha earthquake in Nepal. *Journal of Japan Association for Earthquake Engineering, 19*(8), 19–37.

Furukawa, A., Kiyono, J., Parajuli, R. R., Parajuli, H. R., & Toki, K. (2017). Evaluation of damage to a historic masonry building in Nepal through comparison of dynamic characteristics before and after the 2015 Gorkha earthquake. *Frontiers in Built Environment. 3*(23) https://doi.org/10.3389/fbuil.2017.00062.

Gautam, D. (2017). Seismic performance of world heritage sites in Kathmandu Valley during Gorkha seismic sequence of April–May 2015. *Journal of Performance of Constructed Facilities, 31*(5)6017003.

Gautam, D., Rodrigues, H., Bhetwal, K. K., Neupane, P., & Sanada, Y. (2016). Common structural and construction deficiencies of Nepalese buildings. *Innovative Infrastructure Solutions, 1*(1), 1.

Kiyono, J., Furukawa, A., Parajuli, H. R., Maskey, P. N., & Parajuli, R. R. (2017). Vulnerability assessment of historic structures in Kathmandu, Nepal. In: *Proceedings of the 16th world conference on earthquake engineering, Santiago, Chile*, pp. 9–13.

Parajuli, Y. K. (1986). *Experiences in Preservation and Restoration in Medieval Town (1974–1985).* Deutsche Gesellschaft fuer Technisches Zuzammenarbeit (GTZ).

Parajuli, R. R. (2018). Damages due to Gorkha earthquake 2015 and deficiencies in Shreemahal. *Journal on Rural Infrastructure Development, 9*(9), 127–132.

Parajuli, R. R., Furukawa, A., & Gautam, D. (2020). Experimental characterization of monumental brick masonry in Nepal. *Structures, 28*, 1314–1321.

Parajuli, R. R., & Kiyono, J. (2015). Ground motion characteristics of the 2015 Gorkha earthquake, survey of damage to stone masonry structures and structural field tests. *Frontiers in Built Environment, 1*(23).

Sever, A. (1992). *Nepal under the Ranas.* Asia Publishing House.

Sharma, K., Subedi, M., Parajuli, R. R., & Pokharel, B. (2017). Effects of surface geology and topography on the damage severity during the 2015 Nepal Gorkha earthquake. *Lowloand Technology International, 18*(4), 269–282.

Tamrakar, A., & Parajuli, R. R. (2019). Conservation of cultural heritage: Issues along the Thapathali-Teku stretch of the Bagmati River in Kathmandu, Nepal. *Heritage, 2*(3), 2228–2242.

CHAPTER

14

An engineering view on the traditional timber frames with infills in Romania

Andreea Duțu

Technical University of Civil Engineering Bucharest, Bucharest, Romania

14.1 Introduction

In 2009, this topic was little studied around the world, and due to the complexity of the structures' behavior (masonry and timber interaction), many researchers avoided it. Another reason is because most of the people thought of the system as being old and not used anymore, at least in Romania. But after several research stages around the world (Portugal, Turkey, Japan, China, France) and coming back to Romania, I found the buildings with such structures not only behave well during earthquakes, but they also represent an important cultural identity for all the countries they can be found in.

In this chapter, some of the timber frames around the world are described briefly to understand the differences in their characteristics. Many interesting details are discovered looking at the past construction methods, which are useful also for new structures. The seismic behavior of such buildings is presented in relation to earthquakes in the recent past, and it is surprising how such old structures can withstand moderate to strong earthquakes without collapse.

Because the earthquake proofs are not available in all the countries that have such buildings, where they are available, several experimental campaigns are briefly described and their results summarized to have an idea on how traditional buildings may behave when a future imminent earthquake will occur.

https://doi.org/10.1016/B978-0-12-821087-1.00005-3

The conclusions show how a type of structure, which at least in Romania was not considered interesting until now, is both beautiful and seismic resistant. Recent tendency is to take care of the vernacular architecture and promote it for the future. So, timber-framed masonry buildings are planned to be considered in Eurocodes. This is a tremendous step for the topic. Ten years ago, some would laugh and say they are too few (at least in Romania).

These buildings are easy to construct and the use of local materials makes them a cheap and earthquake-resistant typology. Even though the seismic predictions are still complicated when compared to other types of structures (reinforced concrete, steel, etc.), accurate models exist and can be used in the future to check whether such houses (with different details) can resist to earthquakes. The conclusions can also be used when considering building new houses with traditional architecture.

14.2 Timber frames with infills in the world

In many countries in the world, the timber-framed masonry (TFM) structures are a part of local cultural identity (Romania, Portugal, Italy, Greece, Turkey, China, etc.) (Dutu, Gomes Ferreira, Guerreiro, Branco, & Gonçalves, 2012; Gülkan & Langenbach, 2004; Langenbach, 2006; Mascarenhas, 2005; Makarios & Demosthenous, 2006; Ruggieri & Tampone, 2015; Vasconcelos et al., 2013). Seismic prone country or not, the structural system is made of timber and infilled with masonry. This masonry can be made of stone, clay brick, or adobe. The timber connections can be crosshalved (Gonçalves, Ferreira, Guerreiro, & Branco, 2012), mortise-tenon with nails (Ali et al., 2012), metal plates (Vieux-Champagne et al., 2014), etc. Within the timber frame, the braces can be either one or two, as will be described later for every country considered in this chapter. The visibility of the main timber structure (covered or uncovered by finishing) is also depending on the country, but it can also differ within the same country. The later statement is valid for all the above-mentioned details.

Most of the buildings with TFM structure are not engineered, although the local seismic culture was a good advisor in what details are necessary to withstand earthquakes without collapse. Many are considered heritage buildings, but some of them are only residential houses. A big percentage represents old existing buildings, but some are built new as a wish of the owner to cherish the local traditional architecture.

TFM houses specific to several countries are briefly presented further, and where available, the experimental results are mentioned so they can be later compared with the Romanian ones.

14.2.1 Portugal—The "pombalinos" and the "gaioleiros" buildings

This type of building is characterized by its specific internal walls (called frontais) having braces (Fig. 14.1, right) and a small type of brick masonry infill. Besides these, which have the main structural role, there are also exterior walls made of only masonry and also some partition walls made from only timber (called tabiques). They can be found in many cities in Portugal (Lisbon—Fig. 14.1, left, Porto, etc.) and they are heritage buildings.

There are several studies, both experimental (Dutu, Gomes Ferreira, & Gonçalves, 2012; Ferreira, Teixeira, Duţu, Branco, & Gonçalves, 2014; Goncalves, Guerreiro, Candeias, Ferreira, & Campos Costa, 2018; Poletti & Vasconcelos, 2012) and numerical (Ferreira et al., 2014; Kouris & Kappos, 2014; Kouris, Meireles, Bento, & Kappos, 2014; Lukic, Poletti, Rodrigues, & Vasconcelos, 2018). The experimental studies confirmed the estimated good seismic behavior, showing a great capacity to deform, thus ductility, and a good collaboration of both component materials: timber and masonry. The numerical models are also becoming more accurate, and efforts are being taken to make them as simple as possible, but due to the complicated structure, the models are still scientific and hard to reproduce by practical engineers.

FIG. 14.1 A street in downtown Lisbon with Pombalino buildings (left) and a part of a Pombaline "frontais" wall being rehabilitated (right).

FIG. 14.2 Experimental tests on Pombalino buildings: (A) one cell level; (B) four-cell wall; and (C) two-storeys building level (Goncalves et al., 2018).

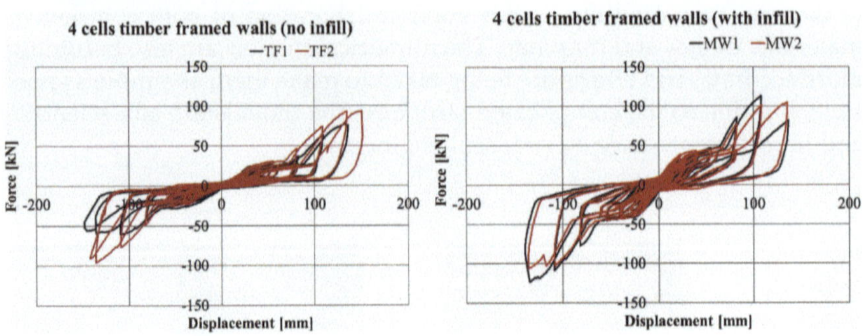

FIG. 14.3 Histeresis of the four-wall tests conducted in IST with (right) and without (left) masonry infill.

The experiments started from a small cell within the "frontais" wall (Ferreira et al., 2014), continued to a four-cell wall (Gonçalves et al., 2012), and then to the whole building tested on the shaking table (Goncalves et al., 2018). Representative photos from the experiments are shown in Fig. 14.2, whereas in Fig. 14.3, the wall tests' histeresis conducted in Lisbon are represented.

14.2.2 Turkey—The himis buildings

These buildings were first built in the 15th century and they were used until the end of the 18th century when the *bağdadi* buildings took over, with plaster replacing the masonry infill (Dişkaya, 2007). *Himis* (Fig. 14.4) can still be found in many places in Turkey, one of the most famous area is Safranbolu.

FIG. 14.4 A timber-framed masonry building in Bursa city.

The postdisaster investigations of *himis* after the Kocaeli earthquake in 1999 showed that the structure could dissipate seismic energy and exhibited good behavior through the straining and sliding of the masonry and timber elements (Doğangün, Tuluk, Livaoğlu, & Acar, 2006; Gülkan & Langenbach, 2004). It was courios to see that the traditional buildings sometimes behaved better than the reinforced concrete ones nearby. And actually, this earthquake was the trigger of the research on such type of house. Many scientists started to wonder why these traditional buildings behaved better than some poorly executed reinforced concrete buildings.

One experimental campaign was conducted so far on himis buildings to validate the typology in the laboratory and understand more about its seismic behavior. The details are reported in (Aktaş, 2017; Aktaş, Akyüz, Türer, Erdil, & Güçhan, 2014), and Fig. 14.5 shows the summary of the experimental campaign conducted on *himis* timber frames with and without several types of infills (brick infill, adobe infill, *bağdadi* cladding, *şamdolma* cladding).

14.2.3 Greece—The timber-framed buildings

Timber-framed masonry buildings in Greece usually have as first load-bearing system the masonry, whereas the second is the timber frame, as shown schematically in Fig. 14.6, left (Makarios & Demosthenous, 2006). A house example is presented in Fig. 14.6, right from Lefkada Island, the island known to have the highest seismicity in Greece. After the 2003 earthquake, the survey done by Vintzileou and Touliatos (2005) showed that even if poorly maintained, the houses with such structure behaved well during several earthquakes.

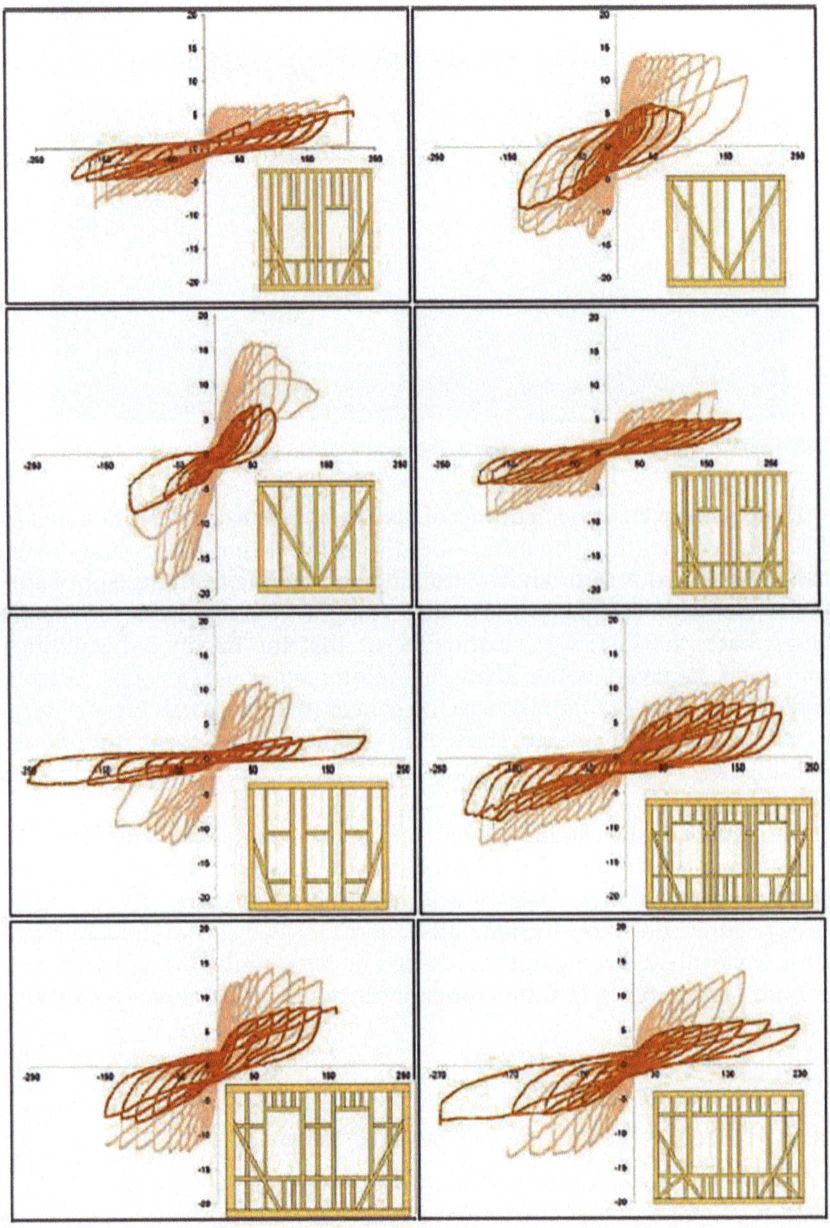

FIG. 14.5 Experimental results (hysteresis) for all the tested walls in METU (*dark red line* is the specimen without infill/cladding, whereas *light red line* is the one with infill/cladding) (Aktaş et al., 2014).

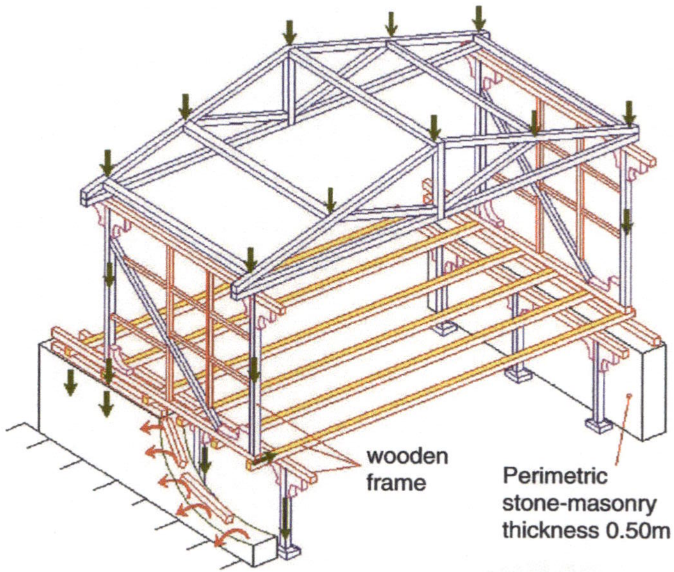

FIG. 14.6 Sketch (left) and timber-framed building in Lefkada Island (right) (Makarios & Demosthenous, 2006).

14.2.4 Italy—The casa baraccata buildings

The *casa baraccata* system was born in the 14th century in central Italy, and after the 1783 earthquake in Calabria, it was used for reconstruction in the affected areas, being included in the antiseismic construction standards issued by the government of that time (Tobriner, 1997; Vivenzio, 1873). An example is shown in Fig. 14.7.

Two types of Italian timber frames were tested, one by Ceccotti, Faccio, Nart, Sandhaas, and Simeone (2006) and the other by Ruggieri, Tampone, and Zinno (2015) to study deeper the seismic analysis of such system. The tested wall and results for the first experimental campaign are shown in Fig. 14.8. For the later, Fig. 14.9 shows both hysteresis and wall specimen tested by University of Calabria in collaboration with IVALSA.

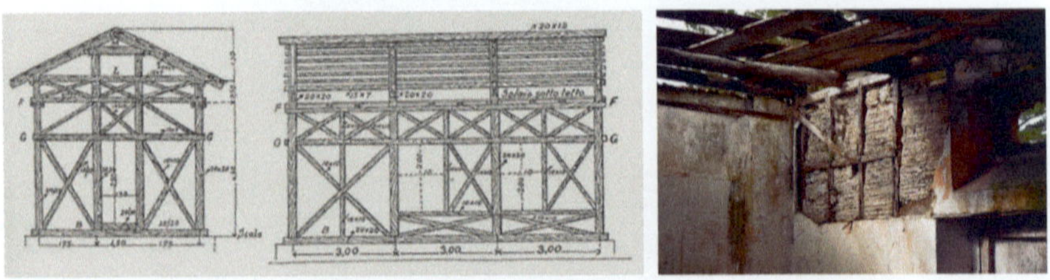

FIG. 14.7 *Casa baraccata* system with timber-framed masonry, a sketch by Vivenzio (1873) and a real example from Mileto (Calabria region).

FIG. 14.8 Tested specimen with diagonal metal angle bars, pulled by two pairs of hydraulic jacks (Ceccotti et al., 2006).

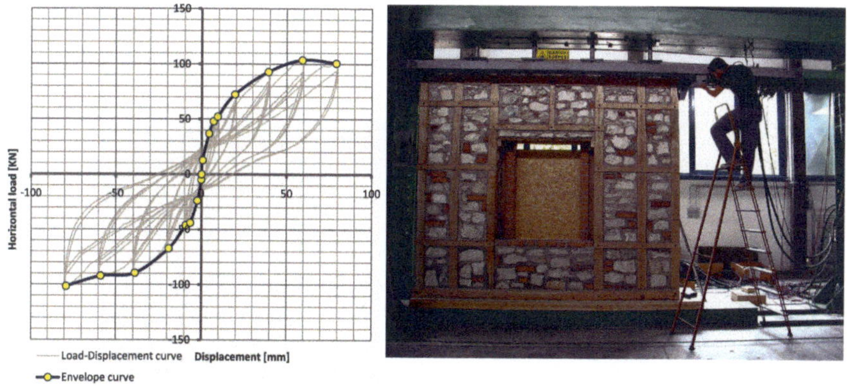

FIG. 14.9 The hysteresis and photo of the wall test conducted by Ruggieri et al. (2015).

14.2.5 France—The colombage buildings

This building style has been used in France at least since the Middle Age and until the 19th century (de Strasbourg, 2004). The Alsace region is famous for such houses, examples are shown in Fig. 14.10. Although the area experiences small earthquakes, the details of such houses indicate the purpose of the layout of the structure is not for earthquake resistance. In Fig. 14.11, the diagonals are missing on the right part of the house, and the same details in the other direction. The aesthetic value of these houses

FIG. 14.10 The *Colombage* building type in Strasbourg: may be brick masonry infilled (left), a retrofit work with the timber structure visible (center) and an example without diagonals from Provence area in France (right).

FIG. 14.11 Lack of symmetric details in some houses in Strasbourg.

is obvious and well appreciated in France; this can be seen from the good maintenance of such traditional architecture and from the fact that in certain areas only this typology is allowed to be built.

14.2.6 Germany—The fachwerk buildings

Germany is only affected by small earthquakes and only in some parts of the country. But it has an interesting architectural style, called *fachwerk*, which was not explicitly designed to carry lateral forces, although sometimes it shows a complete system for transmitting horizontal loads to the ground. In Fig. 14.12, the braces can be found in some parts, but not necessarily symmetric. This means that they are chosen like that for aesthetic or for constructive reasons.

Fachwerk buildings can be found both in southern Germany, where there is moderate seismic activity, and in the north, where seismic activity is minimal. They are widespread and there are about two million houses built with wooden frames filled with masonry, adobe, or plaster. They were built until 1970. There are documents that testify to the occurrence of this type of construction in 1320, but most of them date from the end of the 18th century (Dutu, Gomes Ferreira, Guerreiro, et al., 2012).

14.2.7 China—The chuandou buildings

This type of buildings is spread in China, and some of them are infilled with masonry, whereas others are not, so just having the timber frame with the specific connections and configuration of the posts and beams.

FIG. 14.12 *Fachwerk* building.

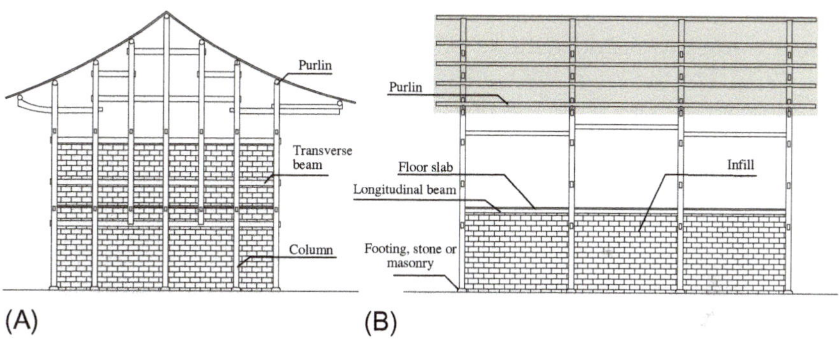

FIG. 14.13 Structural configuration of Chuandou timber houses with infills: (A) transverse elevation and (B) longitudinal elevation (Qu, Dutu, Zhong, & Sun, 2015).

A two-story house is shown in Fig. 14.13 (Qu et al., 2015). Round timber columns are erected on top of independent footings made of stone, or in a few cases, brick masonry. Usually, the columns are simply put on the top surface of the footing without anchorage (Qu et al., 2015).

After the Lushan earthquake in 2013, these houses were investigated by Qu et al. (2015) and the results showed that such traditional houses exhibit damages, but rarely collapse. In general, they behave better than URM houses, but less than reinforced masonry ones. The field study also showed that even if masonry infills collapse, the lateral load can be carried by the timber frame alone.

This result inspired further studies of such houses, and experimental tests were recently conducted by Huang, Wu, Li, Sun, and Chen (2018)

FIG. 14.14 Wall test specimens tested in IEM China: Short length—1.63 m (left) and long length—3.1 m (right) (Qu et al., 2020). *(© Prof. Zhe Qu.)*

on Chuandou with/without wattle and daub infill, by Xue and Xu (2018) on timber frames with/without wallboard infill, and by prof. Zhe Qu on timber frames with and without brick masonry infill (Qu, Fu, Kishiki, & Cui, 2020). For the later, the wall specimens are presented in Fig. 14.14.

14.2.8 Myanmar—The traditional timber-framed masonry buildings

Timber frames are widely observed in Bagan and Chauk, one example is shown in Fig. 14.15. After the earthquake in August 2016, the observed damages resumed only for out-of-plane falling of the infills, especially in the transverse direction of the houses. However, no full collapse was observed for such a house (Htwe Zaw, Ornthammarath, & Poovarodom, 2019).

14.2.9 Nepal—The centibera buildings

The *Centibera* buildings are constructed with the stone masonry walls up to the plinth level or the first floor, on top of which the walls comprised of wooden frames together with woven bamboo mesh and cement mortar plaster or plaster composed of a mixture of mud, cow dung, and rice husk are built (Fig. 14.16). After the earthquake in September 2011, it was found that in *Centibera* buildings, there were no cracks (Fig. 14.17); however, in the brick masonry building close to the *Centibera* building, many cracks were visible in the masonry walls. Based on the onsite investigation of buildings in the Kathmandu valley, the structural system of architecture in Nepal is mainly timber-masonry composite structure. The timber

FIG. 14.15 TFM houses in Myanmar, undamaged after the August 24, 2016 earthquake. *(© Dr. Matsutaro Seki.)*

FIG. 14.16 *Centibera* buildings in Nepal (Wijeyewickrema, Shakya, Pant, & Maharjan, 2011).

column and beam elements are often installed inside the masonry walls, not visible from the exterior (Wijeyewickrema et al., 2011).

14.2.10 Haiti—The gingerbread houses

The Haitian Gingerbread Houses may have three primary construction systems: braced timber frame, *Colombage* (braced timber frame with masonry infill), and masonry bearing wall. Usually, the construction types are combined, resulting in hybrid typologies (Langenbach et al., 2010).

FIG. 14.17 *Centibera* buildings after the 2011 earthquake in Nepal (Wijeyewickrema et al., 2011).

Braced timber frame construction is composed of vertical wood members mortised into wood sills and top plates of each story, and mechanically connected with wooden pegs known as trenails. Diagonal timbers are placed at corners and other locations to brace the frame assembly (Langenbach et al., 2010).

The timber frame with masonry infill is called *kayi pey* and is shown in Fig. 14.18. One can see that the construction details are poor and with no engineering influence.

FIG. 14.18 The *kayi pey* structure (Vieux-Champagne et al., 2014).

The *Colombage* building technique is similar to the French one, as the name suggests it, and is a timber frame infilled with masonry. Timber framing includes top and bottom plates, vertical studs, and diagonal braces. Again, the frame construction generally uses mortise-and-tenon joinery (Langenbach et al., 2010).

After the Port-au-Prince earthquake in 2010, the Gingerbread houses behaved well and even if the infills sometimes collapsed, this did not cause the general collapse of the building. This means that the timber frame is the main structure, whereas infills only increase the stiffness. An interesting detail is also the presence of timber boards on the inside of the walls; this was also observed in the *Colombage* houses in Strasbourg (Fig. 14.10, center).

An extensive experimental program was conducted for the *kayi pey* type of structure (Fig. 14.19), starting from cell level and walls in a static cyclic regime and then a full house on the shaking table (Caimi et al., 2013; Vieux-Champagne et al., 2017, 2014).

The walls presented a high ductility and the infill's contribution was obvious. The shaking table test on the building provided valuable results for the engineering community, showing the relatively good seismic resistance of such buildings and allowing the local government to promote the use of TFM also for new buildings.

14.3 Romanian traditional timber frames with braces and infills

14.3.1 Field investigation

An initial field investigation was necessary to identify the specific construction details of the traditional houses in Romania. At the beginning, it was not known which type of house is predominant. Many people expressed their different opinions. Specialists thought that timber frames with masonry infills are not frequent in Romania, and even if they were, the current design code does not allow them to be built anymore, because they do not correspond to the current seismic requirements for new houses. There was no engineering published document stating anything about the traditional houses' behavior during earthquakes.

So, an area was selected, in several counties near the main seismic source in Romania, Vrancea, to see which type of houses can be found. Details on the investigation are presented in Dutu, Niste, Spatarelu, Dima, and Kishiki (2018). The summary of it is further presented.

A total of 129 traditional houses were investigated and the following typologies were found: houses with timber skeleton and brick masonry infill structure ("*Paiantă*") (Fig. 14.20); houses with timber skeleton and strips applied at 45 degrees and clay plaster (Fig. 14.21); houses with timber skeleton and wattle and daub (Fig. 14.22); houses with timber skeleton

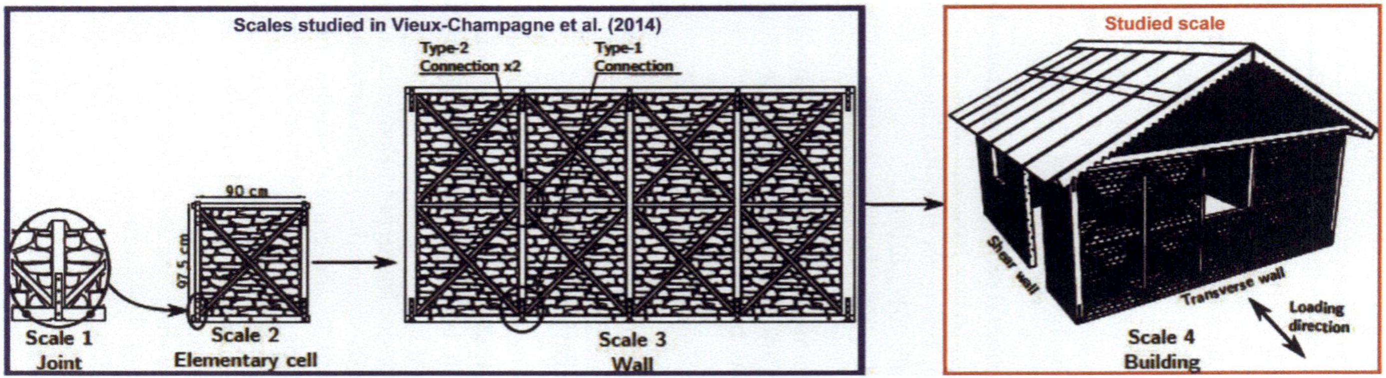

FIG. 14.19 Experimental tests on the Haitian *kayi pey* (Vieux-Champagne et al., 2017).

FIG. 14.20 House with timber frames and masonry infills (Type 1) "*Paiantă*": general view (left), and lateral view (right).

FIG. 14.21 House with timber skeleton and strips applied at 45 degrees and clay plaster (not present here) (Type 2).

and horizontal strips, infilled with earth and straw ("*In grădele*") (Fig. 14.23). Even though they are not traditional ones, an interesting adaptation of traditional houses to modern materials are houses with timber skeleton and AAC (autoclaved aerated concrete) masonry infill (Duțu, Dima, & Bulimar, 2017) or timber planks (Fig. 14.24). Few of them were found, though.

Most of the old traditional buildings have stone foundations (generally damp river rock). These types of foundations are made by stones (laid up

FIG. 14.22 Detail of a house with timber skeleton and wattle and daub infill (Type 3).

FIG. 14.23 House with timber skeleton and horizontal strips, infilled with earth and straw ("*In grădele*") (Type 4).

FIG. 14.24 House with timber skeleton and AAC/timber planks infill (Type 5).

one over the other), generally without any mortar, and usually with similar stone sizes, but not necessarily (Fig. 14.25). Sometimes, the stones are round shaped, and in other cases, they are rectangular. Usually, the foundations are continuous under the walls, as shown in Fig. 14.26. Over the stone foundation, the lower members of the wooden frames were placed, called "soles" and are bearing elements of the superstructure of the house. These were made of hardwood (oak, acacia, walnut, hornbeam, etc.). The soles were connected by cross-halving joints (Fig. 14.25). More details on

FIG. 14.25 Stone foundation of a *Paiantă* building and cross-halved connection of the soles (bottom beam).

FIG. 14.26 Stone foundation continuous under the walls and soles.

the constructive systems are described in Dutu, Niste, Spatarelu, Dima, and Kishiki (2018) and Dutu et al. (2018).

Generally, the walls are made of a main timber frame, which can vary in the existence of braces or not, and different kind of infills. Among the four types of houses that were found in the investigation, the only one not having braces is Type 3. The others usually have braces that do not connect to the upper joint, but directly to the column at the upper part, and are also not necessarily connected to the columns at the bottom part (Fig. 14.20, right).

The main timber frame is made of posts, beams, and braces. The connections between the posts and beams can be cross-halved or mortise-tenon type. Nails can also be present in both types of connections. Occasionally, steel clamps are added to increase the resistance and stiffness of the joint.

The main frame is always infilled with something, depending on the type of house. If it is Type 1 (*Paiantă*), the infill is made of mud brick masonry. The quality of such bricks is variable, depending on their position in the stack that is burned in the oven and also on the quality of the mud (sandy or clayish) in the corresponding area.

If it is Type 2, the infill is usually just mud mixed with straw. For Type 3, the wattle and daub infill has most of the times branches of acacia tree, on which mud mixed with straw is applied. This type of house does not have braces, but between the main posts there are smaller posts, helping to knit the acacia branches around them. The branches are connected with small nails (3 × 60 mm) to the main posts. As for Type 4, the main frame and infill are usually similar to Type 2, with the difference in the position of the exterior strips, in this case being laid horizontally. The connection

FIG. 14.27 The floor above the ground level.

between the strips and the main frame is also usually made with small plain nails (3 × 60 mm).

The floors are made of with wood planks (the ones above the foundations), and the ones above the ground floor are usually made of timber beams with wood planks (Fig. 14.27), or wattle and daub, or sometimes reed. Above the main structure of the floor, there is a layer of earth (mud) on top of the planks. The planks are connected to the timber beams by nails, which gives some stiffness to the floor in-plane, but still not enough to be considered a rigid diaphragm (Dutu, Niste, Spatarelu, Dima, & Kishiki, 2018; Dutu, Niste, Spatarelu, Kober, et al., 2018).

The roof type is usually hip or gable roof, with several king post trusses, but rarely with struts. An example shown in Fig. 14.28. The roof covering of a traditional house, in traditional solution, was made by wood shingle, but now most of them are replaced because they were damaged (biological decay) (Dutu, Niste, Spatarelu, Dima, et al., 2018; Dutu, Niste, Spatarelu, Kober, et al., 2018). The replacement is usually made of steel sheet (Figs. 14.20 and 14.21).

14.3.2 Experimental study

Based on the field investigations, wall specimens, which are representative for some of the house types that were found, were built for static cyclic in-plane loading tests. Due to the variation of the details of the houses even when they were part of the same typology, it was decided to take one representative house and follow its details. The height of the houses usually was around 2.7 m (from foundation until underneath

FIG. 14.28 Example of roof structure.

the roof). However, the most convenient height for the reaction frame, on which the tests were conducted, was 2.4 m. The maximum in-plane dimension of specimens is 300 × 240 cm, and the scale was slightly reduced (1:0.88) so they can fit into the reaction frame. The types of the tested specimens are shown in Figs. 14.29–14.32.

Timber skeleton of the main frame is composed of post, beams, and bracing timber elements (Pinaceae) with 12 × 12 cm cross-section. Both S1 and S2 specimens have masonry infill (mud bricks with rough dimensions of 24 × 11.5 × 6.3 cm and earth mortar), and the difference between the two is the position of the diagonal timber brace (connected to the upper timber joint or not). S3 and S4 have wattle and daub and mud mixed with straw infills, respectively. Except S3, all the specimens have braces, with the same cross-section as the posts and the beams.

The bottom connections are mortise-tenon type with nails and the upper ones are cross-halving type with nails. The bricks are traditional, made from the local clay in Viperesti area. In this respect, most of the used materials were collected from the Viperesti area (earth and bricks). The construction was done by first erecting the timber frame and fixing it to the position, and afterward, applying the infills.

Tests were conducted in a static cyclic regime on a reaction frame provided with 2 horizontal hydraulic jacks with the maximum capacity of 1000 kN (228 mm stroke) and 1 vertical hydraulic jack with the maximum capacity of 2000 kN (100 mm stroke). Test setup is shown in Fig. 14.33.

Axial load was applied at the top of the wall, with the value of 26 kN, representing the roof's weight and snow load. Lateral load was applied manually. To be able to compare with previous experimental tests, the CUREE Caltech standard protocol for wood frames was used. Details about the test setup can be found in Dutu, Niste, Spatarelu, Dima, and

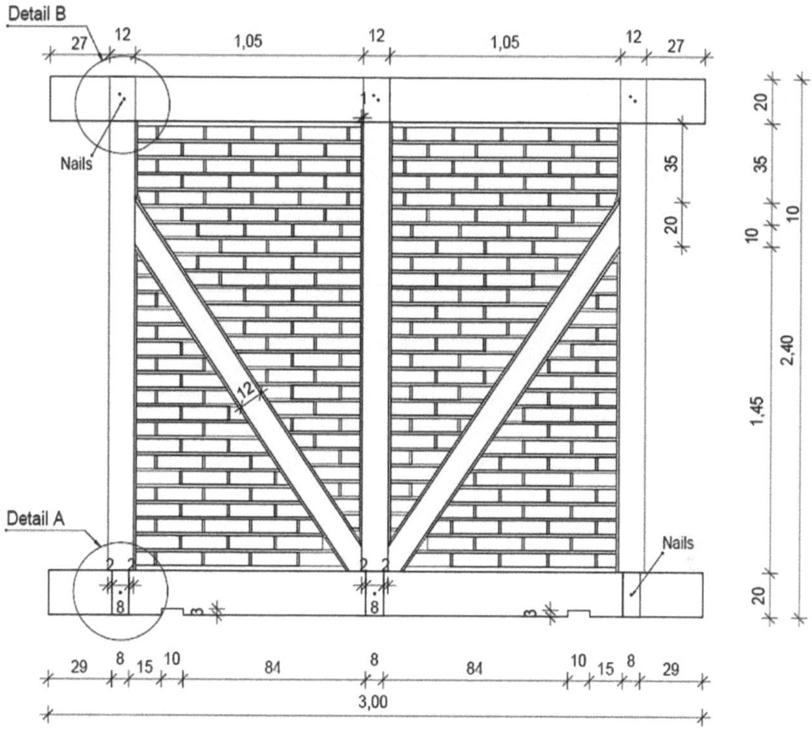

FIG. 14.29 Specimen S1 (from Type 1)—timber and brick masonry infill structure with the diagonal bracing lower than the upper timber joint (cm).

Kishiki (2018) and Dutu, Niste, Spatarelu, Kober, et al. (2018). The material properties determined by tests are summarized in Table 14.1.

The flexibility of the tested traditional walls is proven by the few damages observed after each test. Although similar damages appeared in all the specimens, they are not usually leading to collapse. This was also seen in real earthquakes, when such traditional houses rarely collapsed, unlike poorly constructed reinforced concrete houses. To get an idea on what kind of damages occurred in the tests, Table 14.2 centralizes them and shows which specimen exhibited each type of damage. Then, corresponding to each damage type, examples are shown in Figs. 14.34–14.41.

The hysteretic curves are shown in Figs. 14.42–14.45, and comparison of envelope curves on the positive loading is presented in Fig. 14.46.

It is interesting to notice the comparison between S1 and S2, which shows a slight lower stiffness and yielding strength for the specimen with lower diagonal (not connected to the upper connection). However, they had similar ultimate force. As for the overall comparison among all specimens, the role of diagonal braces in the traditional walls that were tested was clearly visible. When no brace is present (S3), for these walls with mud

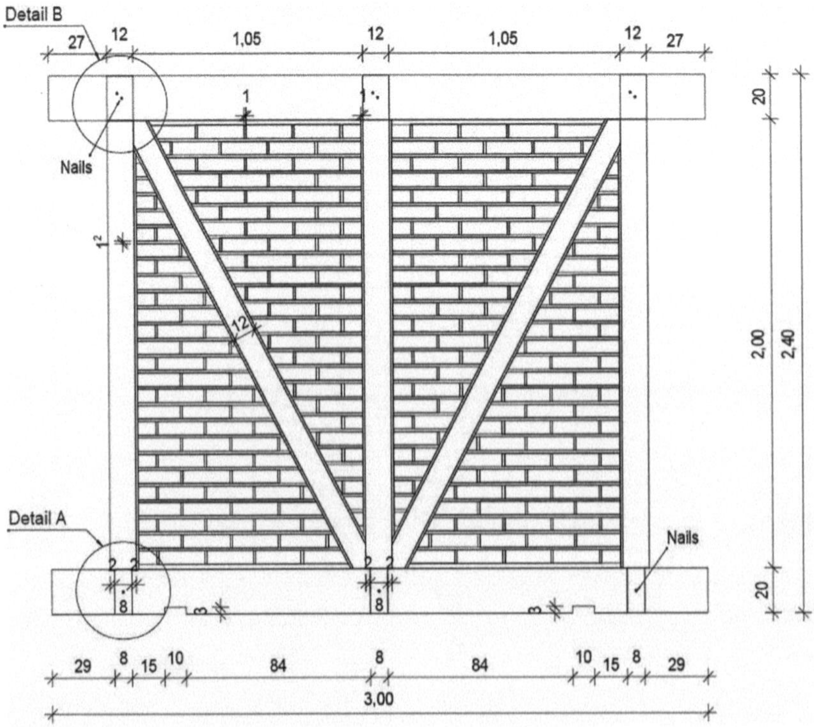

FIG. 14.30 Specimen S2 (from Type 1)—timber and brick masonry infill structure with the diagonal bracing connected to the upper timber joint (normal) (cm).

infill, the capacity of the wall when subjected to lateral load is half as the ones with braces. This is due to the configuration of the main timber frame. As will be discussed in the next section, when the layout includes an intermediary beam (along the height of the wall), and infill is stronger, braces are not so important anymore.

As a tool for future predictions of seismic behavior, when construction details may vary even in the same village and we cannot conduct experiments for every new detail found, based on the existing experiments, a numerical model was developed in OpenSees software and detailed in Parisse (2019).

14.4 Timber-framed masonry with no braces

Timber walls with no braces, but with and without masonry infill, were tested in Japan and later in Romania. The first series of tests (in Japan) was done before starting the research on *Paiantă* houses in Romania, and it was inspired from the already conducted experimental campaign in Portugal

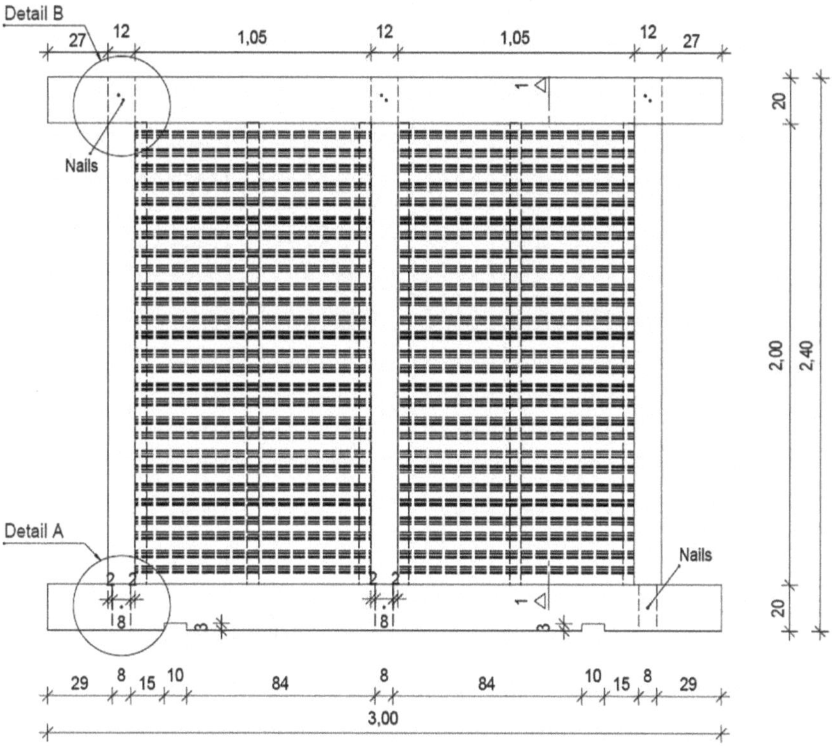

FIG. 14.31 Specimen S3—timber structure with wattle and daub infill, plastered with mud mixed with straw infill.

on *Pombalino* structures (Gonçalves et al., 2012). After proving that the influence of the masonry infill, for *Pombalino*, is without question benefic, the role of the braces within the overall behavior was still intriguing. This motivated the study in Japan.

After the experiments conducted in Japan, where the application of the axial load raised many concerns related to the real behavior of the wall, another test was done in Romania, on a specimen similar in dimensions to the ones tested in Japan, but made with Romanian materials and with a different vertical load application setup. Both tests are briefly presented further. The materials' properties comparison and axial loading influence are presented in Dutu, Sakata, Yamazaki, Niste, and Spatarelu (2018).

14.4.1 Japanese test

The layout of the timber frame and masonry infill of the tested walls is shown in Fig. 14.47, having four masonry panels with the same characteristics (Dutu, Sakata, Yamazaki, & Shindo, 2016). The frame had

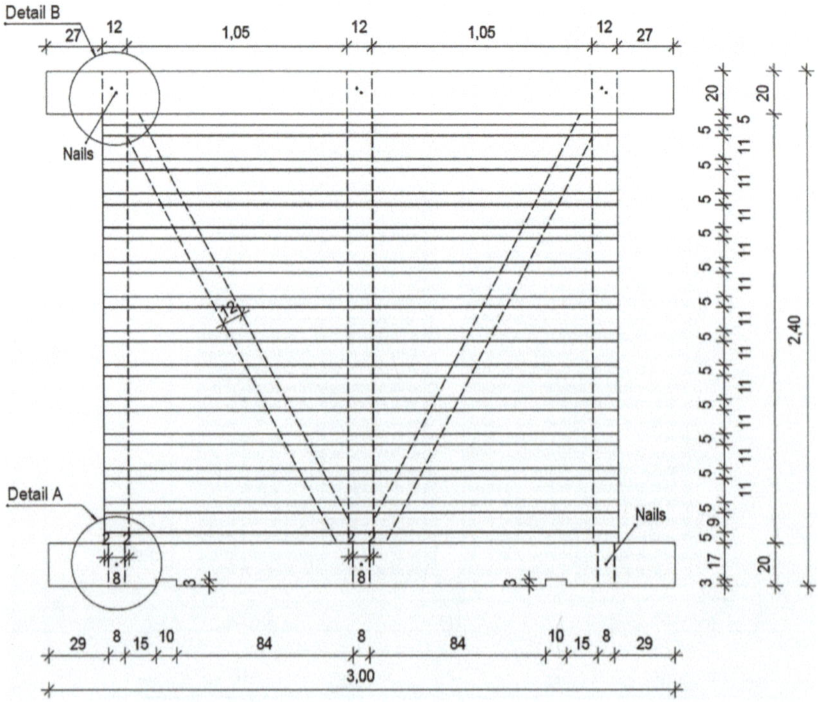

FIG. 14.32 Specimen S4—timber structure and horizontal strips with earth and straw infill.

FIG. 14.33 Setup of wall specimen in the reaction frame.

TABLE 14.1 Material properties determined by tests.

Mud compressive strength (MPa)	2.6
Masonry compressive strength (MPa)	1.86
Masonry Young's modulus (GPa)	0.6
Timber average specific weight (kg/m^3)	385
Timber average moisture content (%)	15
Timber Young's modulus (GPa)	8.9

TABLE 14.2 Failure modes seen in wall tests.

Damage description/specimen	S1	S2	S3	S4
Local compression perpendicular to grain in the upper cross-halved connections (Fig. 14.34)	o	o	o	o
Pull-out of the braces (Fig. 14.35)	o	o	×	×
Uplift in the bottom connections (Fig. 14.40)	Δ	o	×	o
Out of plane behavior of the infill (Fig. 14.37)	o	o	×	×
Mud mortar was initially cracked due to the shrinkage from the drying (Fig. 14.41)	o	o	o	×
Cracks in the infills at the bottom (Fig. 14.38)	o	o	×	o
Infill separated from the timber frames after the first cycle of the loading protocol (Fig. 14.36)	×	o	Δ	o
Upper marginal timber connection detachment (Fig. 14.39)	×	o	o	o

o, means present; ×, means not present; Δ, means very small occurrence, but present.

FIG. 14.34 Local compression perpendicular to grain in the upper cross-halved connections.

FIG. 14.35 Bottom middle connection with significant pull-out of one diagonal brace.

FIG. 14.36 Separation of the infill from the timber frame after the initial cycle.

FIG. 14.37 Out-of-plane of the infill.

FIG. 14.38 Cracks in the infills at the bottom.

cross-halved connections, reinforced with screw nails having a 6-mm diameter and 90-mm length, chosen based on the available materials in Japan and on the fact that the nails shear capacity was previously determined by Sakata, Suyama, and Matsuda (2009). The results of the test with and without infill are thoroughly presented in Dutu et al. (2016) and will only be shown briefly hereby.

FIG. 14.39 Upper marginal timber connection detachment.

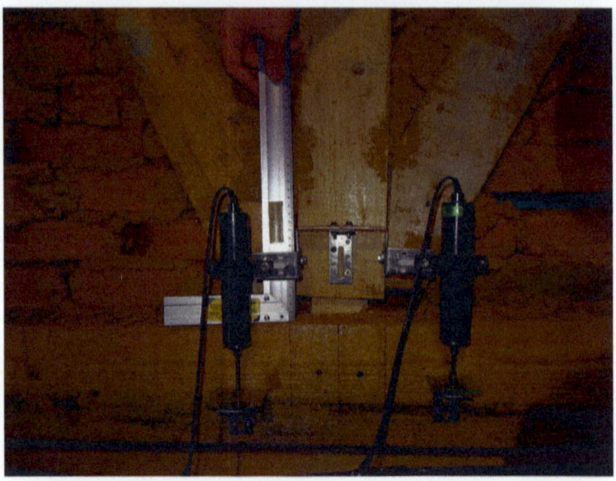

FIG. 14.40 Uplift in the bottom connections.

A vertical force of 60 kN was initially introduced through steel tie rods and uniformly distributed on the top of the specimen using steel plates connected to the upper beam with screw nails. This value was calculated as the equivalent force acting on a first-floor wall in a four-storey tall building (having as reference a *Pombalino* building). The loading protocol is CUREE Caltech.

The hysteresis comparison is shown in Fig. 14.48. As some would say, the comparison has no meaning. The timber frame without infills is almost a mechanism in theory (because the connections are usually considered

FIG. 14.41 Initial cracks in the mud mortar (S3).

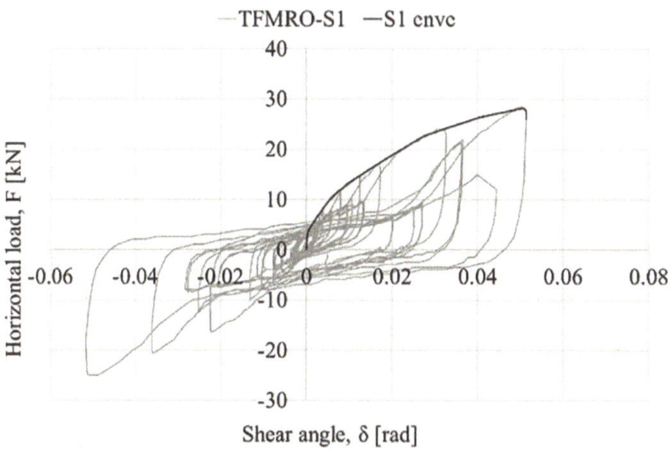

FIG. 14.42 Hysteresis curve of S1.

hinges), although in practice the specimen was standing by itself and also showed a maximum capacity of 18 kN, which is a low value when compared with the specimen with infill (maximum capacity of 118 kN), but still is an important value for a light timber structure. Thus, the cross-halved connections cannot be considered in numerical analysis as hinges.

For this test, a shear spring model was developed based on analytical-empirical calculations and is detailed in Dutu, Yamazaki, and Sakata (2018). The model can be made with simple software such as Excel and requires some material test results.

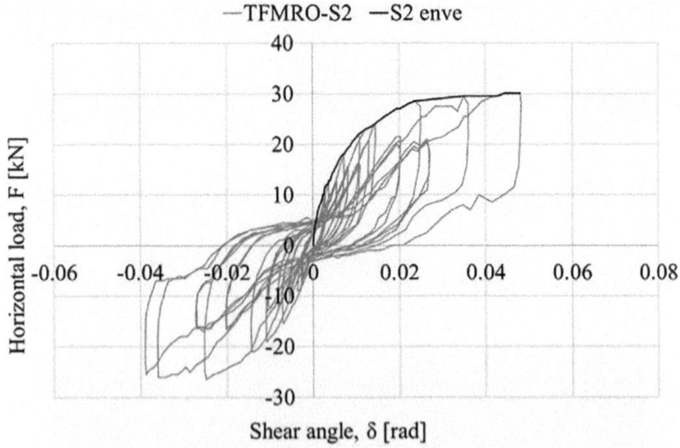

FIG. 14.43 Hysteresis curve of S2.

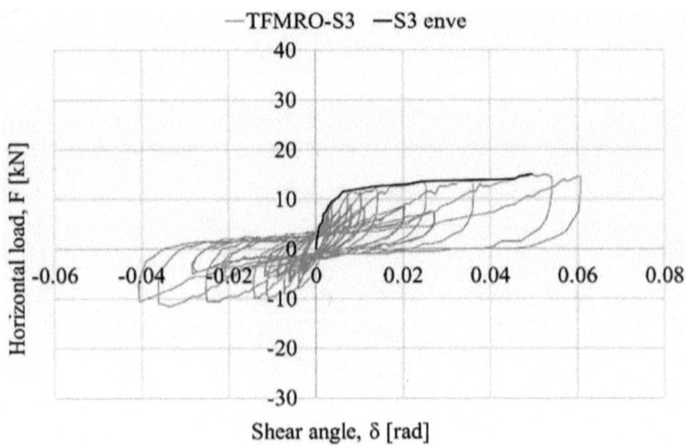

FIG. 14.44 Hysteresis curve of S3.

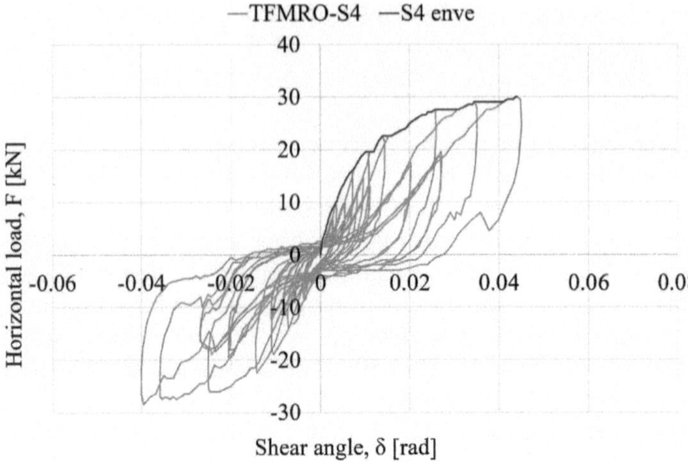

FIG. 14.45 Hysteresis curve of S4.

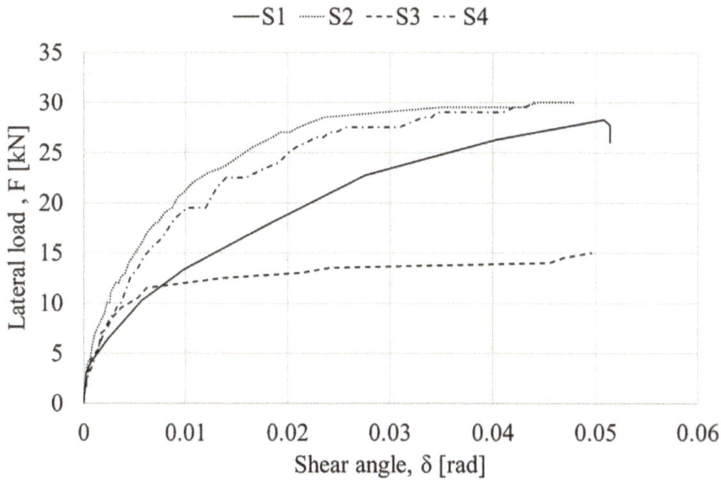

FIG. 14.46 Comparison of all envelope curves

FIG. 14.47 The wall specimen tested in Japan.

14.4.2 Romanian test

The test performed in Romania was on a specimen that was slightly smaller than the one tested in Japan. The materials were different, specific to the ones available in the Romanian market. The nails in the connections were screwed, specific for timber structures. The timber was quality B with a humidity around 10%, softwood Pinaceae type (Dutu, Sakata, Yamazaki, Niste, & Spatarelu, 2018).

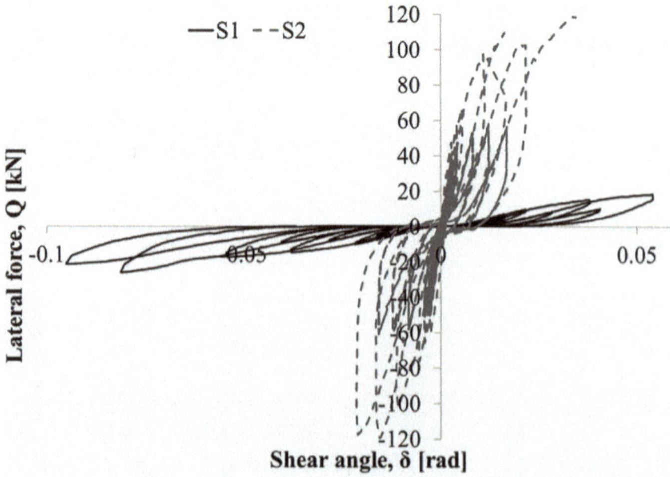

FIG. 14.48 Hysteretic curves of the wall specimens with (S2) and without (S1) infill.

The test was conducted on the same reaction frame as the traditional walls presented in the previous section. The rotation of the upper beam was allowed by applying the axial load on the columns of the specimen.

The axial load applied was 55 kN, and the value was exactly as the weight of the loading beam. Even though the target value was 60 kN, the setup was easier for the 55 kN value, because this was exactly the steel loading beam's weight. The test specimen is shown in Fig. 14.49 and hysteresis is presented in Fig. 14.50.

FIG. 14.49 S5 test in Romania (TFMRO project).

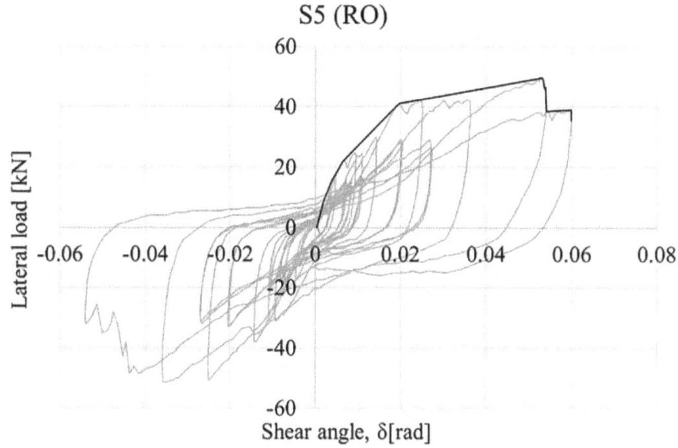

FIG. 14.50 Hysteresis of S5 test in Romania (TFMRO project).

The comparison of envelope curves of Japan vs Romanian tests is shown in Fig. 14.51. The stiffness difference is observed. The strength of the Japanese wall was 2.6 times higher than the one obtained in the Romanian test. Although the tests were not identical, due to the difference of materials and reaction frame, the comparison is still valid and it was proven that the application of the axial force influences a lot the maximum capacity of the wall. One interesting finding is that axial load applied in such a manner is actually increasing the capacity significantly. And this finding can be further investigated as a strengthening method, when necessary (because usually these buildings have enough resilience and do not

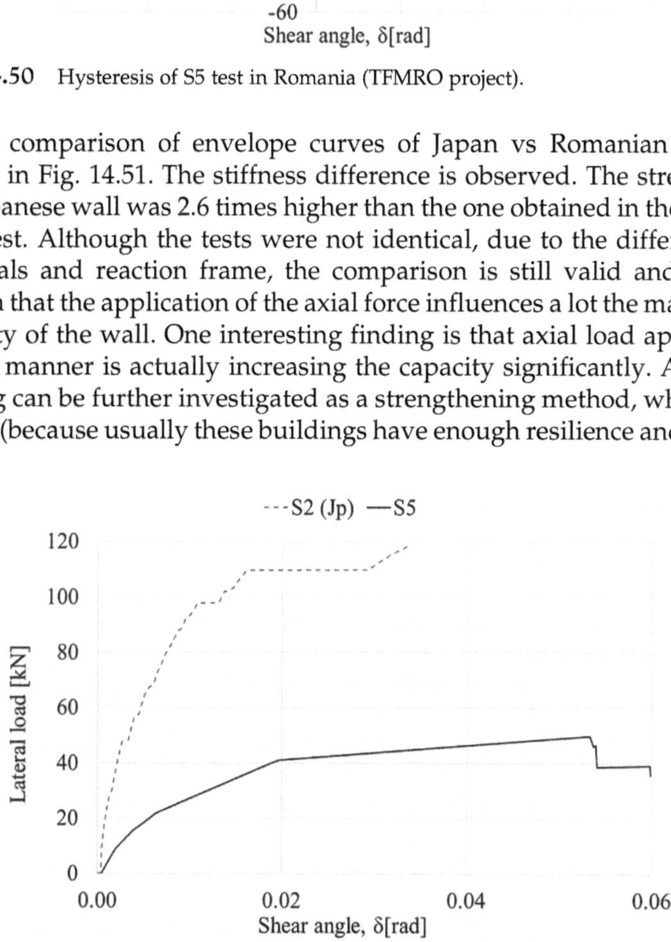

FIG. 14.51 Comparison of envelope curves for Romanian specimen and Japanese specimen.

need increase of the maximum capacity). Of course, the application of the possible strengthening solution should be simplified and take into account the disruption of activities inside a house to be as little as possible. Nevertheless, when looking at the Romanian test, 50 kN maximum capacity is not small. When checking the results of traditional walls (S1–S4, Fig. 14.46), the contribution of the strong masonry infills is obvious. The infills of S5 are made of burned clay bricks, instead of mud brick as in the case of S1 and S2. The latter type of infills can be considered to have no influence in the behavior of the wall, because they are vulnerable to out-of-plane failure even in static in-plane tests. Not the same situation can be observed for the burned clay brick masonry infills. The panels are rigid, and even though the cracks appear on the diagonal strut or horizontally in the joints, the panel still exhibits local compression perpendicular to the grain on the main timber frame. This later property is the key of this structural system. Together with the sliding in the joints of the masonry infill, an interesting mechanism is formed, allowing significant deformation of walls (almost 10 cm at the top) with no failure. This would not be possible for unreinforced masonry structures. The confining effect of the timber frame is of tremendous help in seismic behavior of masonry structures. This also can be a great strengthening solution for unreinforced masonry houses.

14.5 Comparison with other infilled timber frames

As seen in Section 14.2, some of the other countries' timber frames were also investigated experimentally. The setup is never the same even if dimensions might be similar, thus the comparison sometimes is hard to be trustworthy. But to get a general idea on the difference in the tested systems, a comparison is shown in Table 14.3.

The *Paiantă* seems comparable with the *Kayi pey*, although slightly less strong than the latter. It is stronger than the *Himis* system, which resulted with small capacity, no matter the infill. The ductility of the Romanian system is large, as proven by the large corresponding displacement to the maximum force. The bracings contribute to the maximum capacity, in case of the Romanian system, but although the layout is similar to the Turkish one; for the later case, the capacity is not significant.

A strong system seems the *Casa baracatta*, which also presents a specific layout, with many small panels of masonry, although braces are lacking. By looking at it and considering the tests presented in Section 14.4, even if braces are not present, when the infills have an appropriate height to length ratio (about 1 m^2) and are made of brick or stone, they give the system enough stiffness to withstand earthquake loads without general collapse. This would be the primary loading system, which would fail out-of-plane in case of strong earthquakes and poor adherence between

TABLE 14.3 Comparison among several TFM in different countries.

Country/ref.	Type and dimensions L × H (m)	Specimen *L × H (m)	Infill	Vertical force	Max. force (kN)	Corresp. displ. (mm)
Portugal/(Gonçalves et al., 2012)	Timber frame with St. Andrews cross 2.4 × 2	MW1 and MW2	Brick masonry	30 kN/m	45	55
Turkey/(Aktaş et al., 2014)	Timber frame with braces	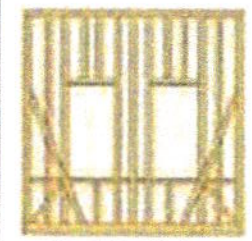 *3.10 (L) × 3.25 (H)	Brick masonry	0	~9	~170
		 *4.85 × 3.40			~13	~90
		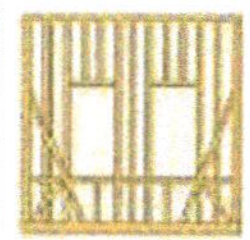 *3.10 × 3.25	Adobe		~8	~210
		 *3.30 × 3.60			14	~30
		 *5.20 × 3.40	Samdolma		~12	~190

Continued

TABLE 14.3 Comparison among several TFM in different countries—cont'd

Country/ref.	Type and dimensions L × H (m)	Specimen *L × H (m)	Infill	Vertical force	Max. force (kN)	Corresp. displ. (mm)
		 *3.30 × 3.60			17	~50
		 *3.70 × 3.30	Bagdadi		11	~60
		 *4.0 × 3.0			~13	~160
Italy/(Ceccotti et al., 2006)	Timber frame with no braces, and with horizontal branches 5.07 × 1.9	M2	Branches + plaster	Present, but not specified	20.46	25.81
		M3			35.7	28.4
		M6			35.52	38.57
	Timber frame with braces (no branches) 5.07 × 1.9	M4	Stones + plaster		69.56	78.86
		M5	Bricks + plaster		98.4	65.6
Italy/(Ruggieri & Zinno, 2014)	Timber frame with no braces 3.39 × 2.95	1	Stone	18.6 kN/m	103.64	59.18

China/ (Qu et al., 2020)	Timber beams and posts 3.1/1.63 × 2.35	Long50	Brick masonry	33 kN	17	47
		Short50			15	23
		Long110			30	23
		Short110			13	94
Haiti/(Vieux-Champagne et al., 2014)	Timber frame with St. Andrew's cross 3.6 × 2.0	1	Stone	0	42.5	126
		2	Stone		39.5	100
Kashmir/(Ali et al., 2012)	Timber frame with no braces	DW1	Stone (Strong)	6	15	~45
	3.1 × 2.5	DW2	Stone (soft)		12	~55
Romania	Timber beams and posts 3.00 × 2.40	S1	Mud brick masonry	19 kN/m	29	117
		S2	Mud brick masonry		30	124
		S3	Wattle and daub		15	118
		S4	Mud and straw		30	117

the infill and timber frame. But then, the secondary loading system comes into action, the main timber frame, which usually does not collapse as seen in postdisaster reconnaissance reports. For the adherence, several methods were observed to be used, such as nails inputted in the frame and embedded in the mortar of the masonry panel, carvings of the frame so the mortar enters the thus created small cavities and barbed wire connected at both ends to the frame by nails and embedded in the masonry joints.

The infill panels that have other shapes than approximately squared may show a high initial stiffness, but the rocking phenomenon and the small deformation capacity of the masonry itself make it a low-performance system. This was seen in the Chinese tests (Qu et al., 2020) and confirmed after the Lushan earthquake (Qu et al., 2015). With 1 m^2 panels, even if the rocking phenomenon is present, it will not influence the in-plane failure of the infill. Moreover, the infill may crack in the joints dissipating energy until its potential out-of-plane failure.

For new houses or reconstruction projects, when choosing between mud (brick, mixture with straw, etc.) and burned clay brick infills, the selection should depend on the easier maintenance. The mud infill needs yearly attention, because it is normally cracking as it dries. The burned clay brick one does not need such attention.

Between the St. Andrew's cross bracing (*Pombalino*) and the regular one (such as the Romanian *Paiantă*), it seems the first one gives more stiffness to the system, transforming the wall into a truss. However, in this case, the infills may be harder to build. An interesting observation is evident when looking at the comparison between the *Pombalino* (Fig. 14.2B) and the wall with no braces tested in Romania (Fig. 14.49). The dimensions were similar for the timber frame, the bricks were different though. However, Fig. 14.52 shows that the behavior of the two specimens is similar, at least until the displacement of 65 mm, when the vertical loading setup starts to influence the results of the *Pombalino* wall. This is useful, especially when no skilled workmanship is available and owners want to build a timber frame with masonry infill. The braces may not be necessary if one uses burned clay brick of $240 \times 115 \times 63$ mm dimensions.

14.6 Conclusions

When looking at traditional old buildings, one may say they are not responding well to the requirements of nowadays. But when checking them after major earthquakes, not few cases were surprisingly doing much better than other typologies such as poorly executed concrete frames or unreinforced masonry walls. They may exhibit damages (infills are prone to out-of-plane failure), but the whole house usually does not

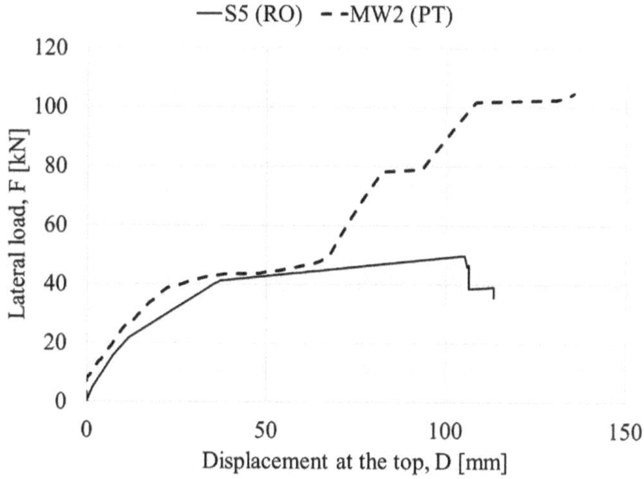

FIG. 14.52 Comparison of envelope curves of one specimen tested in Portugal (Fig. 14.3) and the specimen tested in Romania (Fig. 14.50).

collapse. This was the case even if some houses were poorly maintained. This means they are resilient. Although they may have complicated mechanical behavior and it is hard to predict their seismic performance by calculations, the structural system (timber frame and some kind of infill) is an interesting topic for study and research.

This chapter briefly mentioned three experimental campaigns in three different countries, together with conclusions after their comparison. It is interesting to see that even though the timber frames with infills are reported since BC, research proves that even nowadays the system can still be improved (braces can be replaced by square masonry panels) and still be used to withstand successfully major earthquakes.

In Romania, the old traditional buildings are usually considered not modern and often strengthened with reinforced concrete jacketing, which completely ruins the collaboration of the timber frame with the infill and turns the house into an unpredictable building in case of an earthquake. Efforts are being made, generally by architects, but not necessarily (Dutu, Niste, Spatarelu, Dima, et al., 2018; Dutu, Niste, Spatarelu, Kober, et al., 2018), to revive the traditional architecture and adapt it to the current needs of space and comfort. But the success of such actions depends on many factors such as timber factories should provide strength classes for the products they sell and also reliable dry timber, appropriate workmanship, and owners' sympathy toward tradition and cultural identity.

Further efforts will concentrate on methods to rehabilitate such houses, while keeping their advantages (e.g., timber frame's flexibility working together with the infill's stiffness, in this case). And also a simple evaluation method is necessary, because the scientific ones developed so far are still complicated to be applied by regular engineers.

Acknowledgment

A part of this work was supported by a grant of the Romanian National Authority for Scientific Research and Innovation, CNCS—UEFISCDI, project number "PN-II-RU-TE-2014-4-2169."

The Japan Society for Promotions of Science (JSPS) is also acknowledged for the financial support through the postdoctoral research grant no. PE 13092 for experimental tests in Tokyo Institute of Technology, Japan.

The author acknowledges the collaborators in the above-mentioned projects, which led to the opinions expressed in this chapter: Prof. Hiroyasu Sakata, Dr. Yoshihiro Yamazaki, Prof. Joao Gomes Ferreira, Assist. Prof. Mihai Niste, and Lect. Iulian Spatarelu. They helped a lot with the research previously conducted by the author.

References

Aktaş, Y. D. (2017). Seismic resistance of traditional timber-frame hımış structures in Turkey: A brief overview. *International Wood Products Journal, 8*, 21–28. https://doi.org/10.1080/20426445.2016.1273683.

Aktaş, Y. D., Akyüz, U., Türer, A., Erdil, B., & Güçhan, N. Ş. (2014). Seismic resistance evaluation of traditional ottoman Timber Frame Himiş houses: Frame loadings and material tests. *Earthquake Spectra, 30*(4), 1711–1732. https://doi.org/10.1193/011412EQS011M.

Ali, Q., Schacher, T., Ashraf, M., Alam, B., Naeem, A., Ahmad, N., et al. (2012). In-plane behavior of the dhajji-dewari structural system (wooden braced frame with masonry infill). *Earthquake Spectra, 28*(3), 835–858. https://doi.org/10.1193/1.4000051.

Caimi, A., Guillaud, H., Moles, O., Vieux-Champagne, F., Garnier, P., Grange, S., et al. (2013). Traditional and scientific knowledge for a sustainable vulnerability reduction of rural housing in Haiti. In *Structures and architecture: Concepts, applications and challenges—Proceedings of the 2nd international conference on structures and architecture, ICSA 2013* (pp. 1807–1815). https://doi.org/10.1201/b15267-247.

Ceccotti, A., Faccio, P., Nart, M., Sandhaas, C., & Simeone, P. (2006). Seismic behaviour of historic timber-frame buildings in the Italian dolomites. In *ICOMOS international wood committee 15th international symposium, Istanbul and Rize, Turkey, 1–16*. Retrieved from c:%5CUsers%5CUsuario%5CDesktop%5CPhD%5CPapers%5CCasa Baraccata%5CCeccotti A., Faccio P., Nart M., Sandhaas C., Simeone P. - Seismic Behaviour of Historic Timber-Frame Buildings in the Italian Dolomites.pdf.

Dişkaya, H. (2007). Damage assessment of 19th century traditional timber framed structures in Istanbul. In *From material to structure—Mechanical behaviour; Venice and Vicenza 11th–16th November 2007*.

Doğangün, A., Tuluk, Ö. I., Livaoğlu, R., & Acar, R. (2006). Traditional wooden buildings and their damages during earthquakes in Turkey. *Engineering Failure Analysis, 13*(6), 981–996. https://doi.org/10.1016/j.engfailanal.2005.04.011.

Duţu, A., Dima, D. I., & Bulimar, E. G. (2017). Materials and techniques for traditional romanian residential houses. In *3rd International conference on protection of historical constructions (prohitech) Lisbon, Portugal, 12–15 July* (pp. 12–15).

Dutu, A., Gomes Ferreira, J., Guerreiro, L., Branco, F., & Gonçalves, A. M. (2012). Timbered masonry for earthquake resistance in Europe. *Materiales de Construcción, 62*(308), 615–628. https://doi.org/10.3989/mc.2011.01811.

Dutu, A., Niste, M., Spatarelu, I., Dima, D. I., & Kishiki, S. (2018). Seismic evaluation of Romanian traditional buildings with timber frame and mud masonry infills by in-plane static cyclic tests. *Engineering Structures*. https://doi.org/10.1016/j.engstruct.2018.02.062.

Dutu, A., Sakata, H., Yamazaki, Y., & Shindo, T. (2016). In-plane behavior of timber frames with masonry infills under static cyclic loading. *Journal of Structural Engineering (United States), 142*(2). https://doi.org/10.1061/(ASCE)ST.1943-541X.0001405.

Dutu, A., Yamazaki, Y., & Sakata, H. (2018). Shear spring model proposed for seismic evaluation of a timber framed masonry infilled wall. *Engineering Structures*. https://doi.org/10.1016/j.engstruct.2018.02.061.

Dutu, A., Gomes Ferreira, J., & Gonçalves, A. M. (2012). The behaviour of timber framed masonry panels in quasi-static cyclic testing. In *9th international conference on urban earthquake engineering/ 4th Asia conference on earthquake engineering* (pp. 8–13).

Dutu, A., Niste, M., Spatarelu, I., Kober, D., Barbu-Mocanescu, D., Nicula, V., et al. (2018). *TRAROM project. Retrieved January 4, 2020, from trarom.uctb.ro.*

Dutu, A., Sakata, H., Yamazaki, Y., Niste, M., & Spatarelu, I. (2018). Influence of axial force application in the behavior of timber framed masonry walls under in-plane static cyclic loading. In *WCTE 2018—World conference on timber engineering* (pp. 5–6).

Ferreira, J. G., Teixeira, M. J., Duţu, A., Branco, F. A., & Gonçalves, A. M. (2014). Experimental evaluation and numerical modelling of timber-framed walls. *Experimental Techniques, 38* (4). https://doi.org/10.1111/j.1747-1567.2012.00820.x.

Gonçalves, A. M., Ferreira, J. G., Guerreiro, L., & Branco, F. (2012). Experimental characterization of pombalinos "frontal" walls cyclic behaviour. In *15th international conference on experimental mechanics, (September)* (pp. 1–13).

Goncalves, A. M. N., Guerreiro, L. M. C., Candeias, P., Ferreira, J. G., & Campos Costa, A. (2018). Characterization of reinforced timber masonry walls in "pombalino" buildings with dynamic tests. *Engineering Structures, 166*(April 2017), 93–106. https://doi.org/10.1016/j.engstruct.2018.03.036.

Gülkan, P., & Langenbach, R. (2004). *The earthquake resistance of. (2297).*

Htwe Zaw, S., Ornthammarath, T., & Poovarodom, N. (2019). Seismic reconnaissance and observed damage after the Mw 6.8, 24 August 2016 Chauk (Central Myanmar) earthquake. *Journal of Earthquake Engineering, 23*(2), 284–304. https://doi.org/10.1080/13632469.2017.1323050.

Huang, H., Wu, Y., Li, Z., Sun, Z., & Chen, Z. (2018). Seismic behavior of Chuan-Dou type timber frames. *Engineering Structures, 167*, 725–739. https://doi.org/10.1016/j.engstruct.2017.10.072.

Kouris, L. A. S., Meireles, H., Bento, R., & Kappos, A. J. (2014). Simple and complex modelling of timber-framed masonry walls in Pombalino buildings. *Bulletin of Earthquake Engineering, 12*(4), 1777–1803. https://doi.org/10.1007/s10518-014-9586-0.

Kouris, L. A. S., & Kappos, A. J. (2014). A practice-oriented model for pushover analysis of a class of timber-framed masonry buildings. *Engineering Structures, 75*, 489–506. https://doi.org/10.1016/j.engstruct.2014.06.012.

Langenbach, R. (2006). From opus craticium to the Chicago frame Earthquake resistant traditional construction. In *Structural analysis of historical constructions*. New Delhi.

Langenbach, R., Kelley, S., Sparks, P., Rowell, K., Hammer, M., Jean Julien, O., et al. (2010). *Preserving Haiti's gingerbread houses: 2010 earthquake mission report.*

Lukic, R., Poletti, E., Rodrigues, H., & Vasconcelos, G. (2018). Numerical modelling of the cyclic behavior of timber-framed structures. *Engineering Structures, 165*, 210–221. https://doi.org/10.1016/j.engstruct.2018.03.039.

Makarios, T., & Demosthenous, M. (2006). Seismic response of traditional buildings of Lefkas Island, Greece. *Engineering Structures, 28*(2), 264–278. https://doi.org/10.1016/j.engstruct.2005.08.002.

Mascarenhas, J. (2005). *Sistemas de Construção. V – O edifício de Rendimento da Baixa Pombalina de Lisboa*. Livros Horizonte.

Parisse, F. (2019). *Numerical modelling of the seismic performance of romanian traditional timber-framed buildings*. University of Minho. Retrieved from. http://hdl.handle.net/1822/62517.

Poletti, E., & Vasconcelos, G. (2012). Seismic behaviour of traditional half-timbered walls: cyclic tests and strengthening solutions. *Wiadomosci Lekarskie—Journal of Heritage Conservation, 32*, 137–142.

Qu, Z., Dutu, A., Zhong, J., & Sun, J. (2015). Seismic damage to masonry-infilled timber houses in the 2013 M7.0 Lushan, China, earthquake. *Earthquake Spectra, 31*(3). https://doi.org/10.1193/012914EQS023T.

Qu, Z., Fu, X., Kishiki, S., & Cui, Y. (2020). Behavior of masonry infilled Chuandou timber frames subjected to in-plane cyclic loading. *Engineering Structures, 211*, 110449. https://doi.org/10.1016/j.engstruct.2020.110449.

Ruggieri, N., Tampone, G., & Zinno, R. (2015). In-plane versus out-of-plane "behavior" of an Italian timber framed system—The Borbone constructive system: Historical analysis and experimental evaluation. *International Journal of Architectural Heritage, 9*(6), 696–711. https://doi.org/10.1080/15583058.2015.1041189.

Ruggieri, N., & Tampone, G. (2015). Historical earthquake-resistant timber frames in the mediterranean area. In *Historical earthquake-resistant timber frames in the mediterranean area*. https://doi.org/10.1007/978-3-319-16187-7.

Ruggieri, N., & Zinno, R. (2014). Seismic assessment of " baraccato " system : Constructive analysis and experimental investigations. In *Second European conference on earthquake engineering and seismology, Istanbul Aug 25–29*.

Sakata, H., Suyama, T., & Matsuda, K. (2009). Study on hysteresis model of joints using screws and holddown bolts for conventional post-and-beam wooden house (in Japanese). *Journal of Structural and Construction Engineering, 74*(645), 2061–2067.

de Strasbourg, S. é. d. m. d. l. V. (2004). *Habiter une maison alsacienne, Mur à colombage décoré, Issenhausen, Bas-Rhin, Musée Alsacien, Strasbourg*.

Tobriner, S. (1997). *La casa baraccata: un sistema antisismico nella Calabria del XVIII secolo, 115 Costruire in laterizio*.

Vasconcelos, G., Poletti, E., Salavessa, E., Jesus, A. M. P., Lourenço, P. B., & Pilaon, P. (2013). In-plane shear behaviour of traditional timber walls. *Engineering Structures, 56*, 1028–1048. https://doi.org/10.1016/j.engstruct.2013.05.017.

Vieux-Champagne, F., Sieffert, Y., Grange, S., Belinga Nko'ol, C., Bertrand, E., Duccini, J. C., et al. (2017). Experimental analysis of a shake table test of timber-framed structures with stone and earth infill. *Earthquake Spectra*, 33(3), 1075–1100.

Vieux-Champagne, F., Sieffert, Y., Grange, S., Polastri, A., Ceccotti, A., & Daudeville, L. (2014). Experimental analysis of seismic resistance of timber-framed structures with stones and earth infill. *Engineering Structures, 69*, 102–115. https://doi.org/10.1016/j.engstruct.2014.02.020.

Vintzileou, E., & Touliatos, P. (2005). Seismic behaviour of the historical structural system of the island of Lefkada, Greece. *WIT Transactions on the Built Environment, 83*, 291–300. https://doi.org/10.1016/j.conbuildmat.2005.04.002.

Vivenzio, G. (1873). *Historia de' tremuoti ..., Stamperia Regale, Napoli, (dal volume di Clementina Barucci, La Casa Antisismica)*.

Wijeyewickrema, A. C., Shakya, K., Pant, D. R., & Maharjan, M. (2011). *CUEE Report 2011-1 earthquake reconnaissance survey in nepal of the magnitude 6.9 sikkim earthquake of september 18, 2011*.

Xue, J., & Xu, D. (2018). Shake table tests on the traditional column-and-tie timber structures. *Engineering Structures, 175*, 847–860. https://doi.org/10.1016/j.engstruct.2018.08.090.

CHAPTER

15

Dynamic characteristics of Inca's stone masonry

Carlos Cuadra

Akita Prefectural University, Yurihonjo, Japan

15.1 Seismicity and historical sites in Peru

The west coast of South America is affected by frequent earthquakes primarily due to tectonic interactions between the Nazca Plate and the South American Plate. The Peruvian Institute of Geophysics (IGP) has published the catalog of historical earthquakes in this region, consisting of information from the year 1471. Old records of this earthquake catalog have information of probable magnitude and estimated location inferred from historical reports on damages in affected zones. However, earlier records prior to instrumental measurement are not representative of the actual distribution of hazard because at earlier times, only earthquakes occurring around populated areas seem to have been reported. If only records with complete information of location, magnitude, depth, etc. are selected, the distribution of these instrumental recorded earthquakes appears, as can be observed in Fig. 15.1. It can be noted that the epicenters are concentrated along the coastline. Moreover, it can be observed that the southern part of Peru exhibits more activity than the northern and central parts. To study the nature of the earthquakes of the Northern zone, the Central zone, and the Southern zone, sectionals that show the earthquakes distributed in elevation were constructed. Locations of these sections are indicated in the Fig. 15.1 as section A-A for the Northern part, section B-B for the Central part, and section C-C for the Southern part of Peru. In all cases, earthquakes located within a 200-km band centered on the corresponding section line were considered.

Fig. 15.2 shows the distribution of earthquakes in elevation for each section described in the previous paragraph. Although the distribution pattern presents the same trend for each zone with shallow near-shore

Masonry Construction in Active Seismic Regions
https://doi.org/10.1016/B978-0-12-821087-1.00002-8

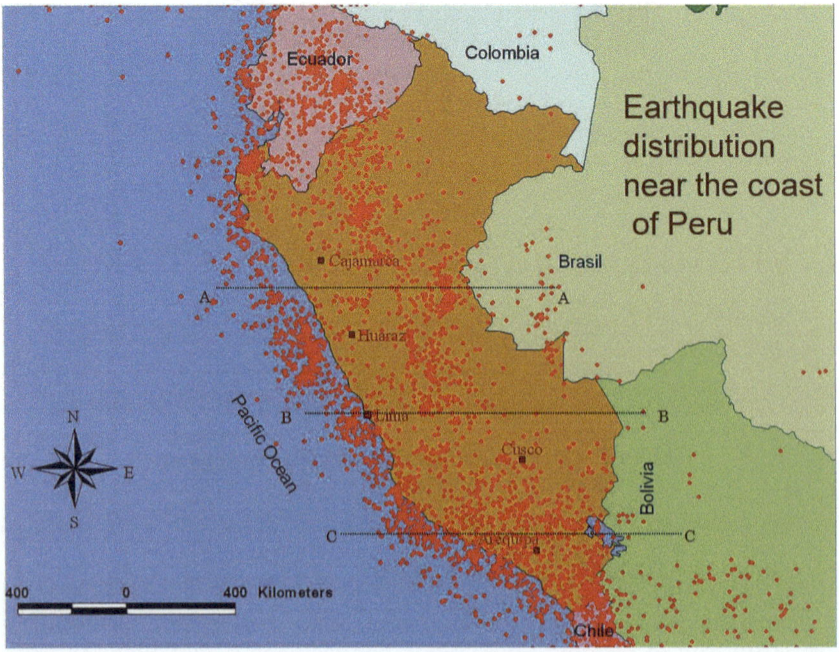

FIG. 15.1 Earthquake distribution near the coast of Peru.

earthquakes and deeper inland earthquakes, the depth distribution and the number of earthquakes for each zone are quite different.

Section A-A shows the distribution of shallow earthquakes near the coast. Here, a concentration of epicenters at a focal depth of 100–200 km is observed. In addition, a concentration of deep epicenters is observed in the Amazon jungle area (eastern part), with depths of the order of 500 km.

In the section B-B, the distribution of the earthquakes shows that in this zone, the earthquakes have depths less than 200 km or even less than that depth. The shallow earthquakes near the coast are closer to the coastline, which means that the epicenters are closer to inland cities.

The section C-C shows more clearly the typical earthquake distribution of the subduction action of the marine plate, with deeper inland earthquakes. It can be noted that the number of earthquakes is larger in comparison with the northern and central parts of Peru.

Seismic hazard analysis for representative sites of the Andean region of Peru (Cusco city) and the coast of Peru (Lima city) was performed by Cuadra, Tokeshi, and Rojas (2008) and is shown in Fig. 15.3. In this figure, hazard curve for Los Angeles city is also included for comparison. Results of the probabilistic seismic hazard analysis can also be expressed in terms of the return period shown in the right part of Fig. 15.3, where the peak

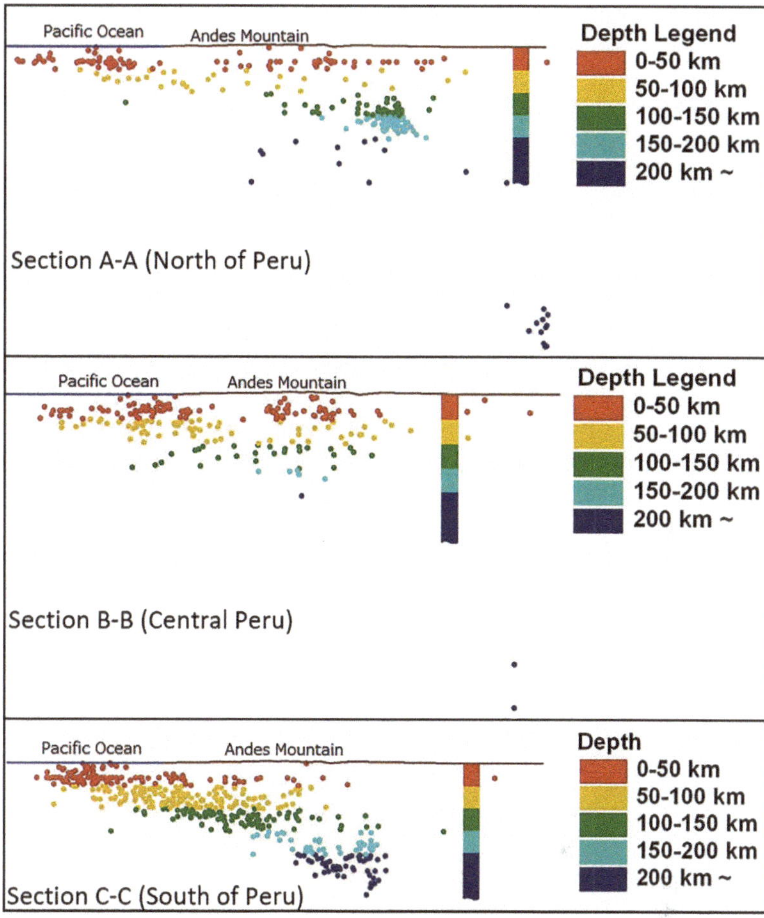

FIG. 15.2 Distribution of earthquakes in elevation.

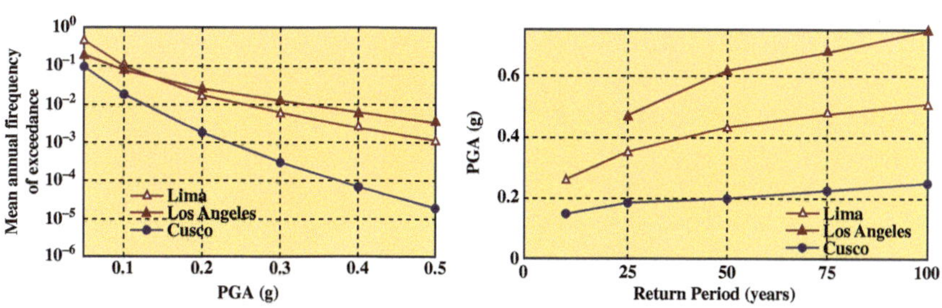

FIG. 15.3 Hazard curve and PGA versus return period for representative sites.

ground acceleration (PGA) values for different return periods with 10% probability of exceedance are plotted. It may be noted that the PGA for 100 years of return period is around 0.25 g for the considered Andean region. For Lima (Coast of Peru), the PGA is 0.5 g. This higher level of peak acceleration may produce failure in buildings in general and the situation becomes critical for weak materials like adobe constructions. In addition, the PGA levels in Fig. 15.3 represents rock or stiff soil conditions, while the actual level of shaking depends on local site conditions at specific heritage sites where the local site amplifications and other effects may be important.

Peru has many sites of cultural and natural heritage reflecting the pre-Inca era, Inca era, as well as the subsequent era, following Spaniard conquistadors in the 17th century. The distribution of these heritages can be observed in Fig. 15.4 and includes the world heritages declared by UNESCO, which are marked by purple marks. Of the 12 Peruvian heritage sites in the UNESCO world heritage list, eight are considered cultural and two are considered natural. The remaining two are considered mixed category, meaning that the sites reflect the cultural as well as the natural heritages. The Andean road system is included in this list; however, it is not marked in Fig. 15.4 because portions of this road are distributed in many

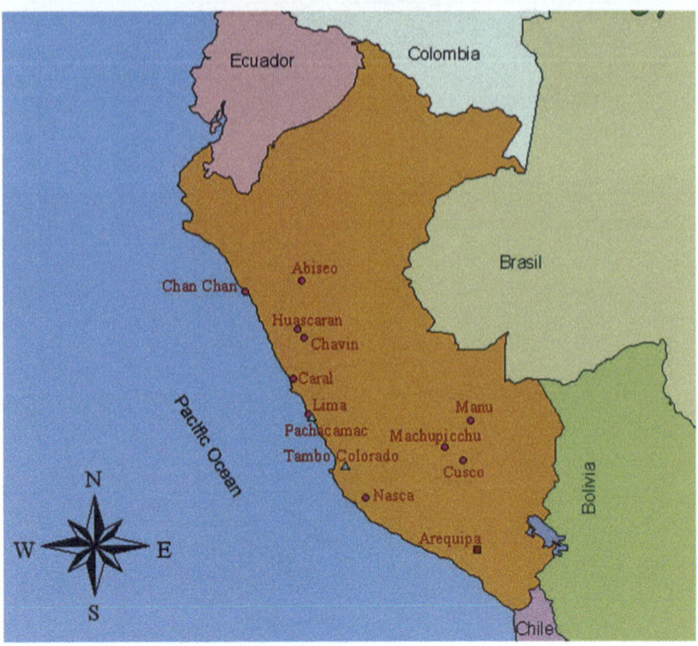

FIG. 15.4 Peruvian world heritages distribution.

places of Peru and neighbor countries. In addition to this UNESCO list, other important architectural heritages like Pachacamac and Tambo Colorado are included in Fig. 15.4 and are marked by triangles.

15.2 Inca stone masonry and equivalent elastic modulus

Three types of stone masonry can be distinguished in the Inca stonework. The first style is a roughly shaped stone fitted with mud mortar, as shown in Fig. 15.5. The second style is a finely shaped masonry with polygonal units fitted without mortar, as shown in Fig. 15.6. The third style is a finely shaped masonry of rectangular blocks forming horizontal layers, as shown in Fig. 15.7.

For these three types of stone masonry, an equivalent elastic modulus is obtained from in situ microtremor measurements and a finite element method for structural analysis.

FIG. 15.5 Roughly shaped stone masonry.

FIG. 15.6 Polygonal stone masonry.

FIG. 15.7 Rectangular stone masonry.

15.2.1 Ambient vibration measurements of Inca stone masonry

15.2.1.1 Roughly shape masonry

Ambient vibration measurements for roughly shaped stone masonry were performed at a representative construction located at Machu Picchu citadel (Cuadra, Karkee, Ogawa, & Rojas, 2004). These constructions are called Colca and Huayrana and are shown in Fig. 15.8. Points of ambient vibration measurements for these buildings are shown in Fig. 15.9 (Ch.7, 8, and 9 on the building and Ch.10 on the ground). Measurements were performed for NS and EW directions separately.

From a Fourier analysis, the natural frequencies of these structures were estimated. Fig. 15.10 shows the transfer functions of measured ambient vibrations. The predominant frequency in the EW direction of the Colca varies from 8 to 12 Hz (Fig. 15.10A), while a clear peak at 5.46 Hz is seen for the NS direction (Fig. 15.10C), which corresponds to the out-of-plane vibration in the triangular gable wall. In the case of Huayrana, the central wall restrains the out-of-plane movement of triangular gable walls. However, a large amplification factor at around 5.52 Hz

FIG. 15.8 Selected roughly shaped stone masonry buildings.

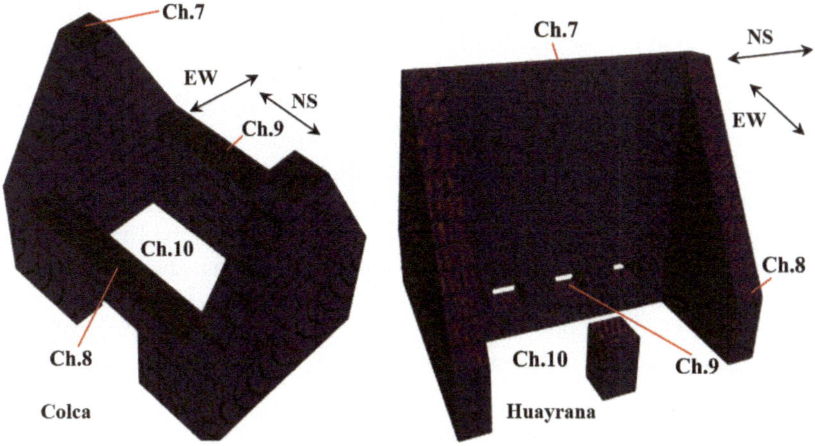

FIG. 15.9 Points of vibration measurements at Colca and Huayrana buildings.

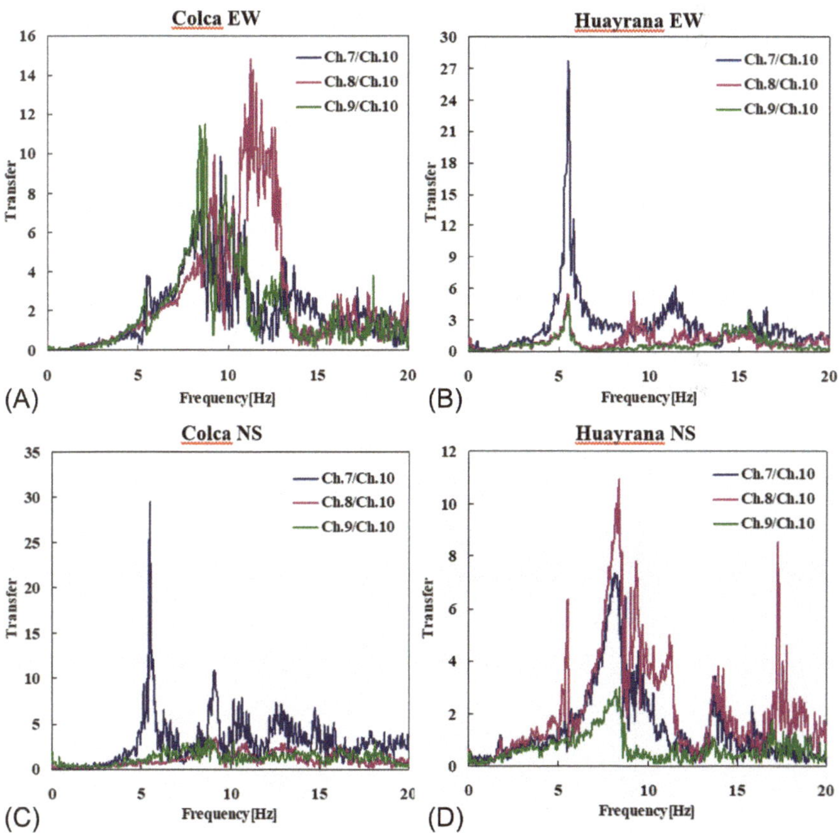

FIG. 15.10 Horizontal transfer functions for Colca and Huayrana.

corresponding to the measurement at the top center of the restraining wall in EW direction is seen, probably due to the large dimensions of the wall (see transfer function for Ch.7/Ch.10 in Fig. 15.10B). In the case of NS direction, Huayrana shows lower amplification factors at around 8.2 Hz for all measured points (Fig. 15.10D). Therefore the first mode of vibration for Colca corresponds to the vibration of the gable wall in the NS direction, with a predominant frequency at 5.46 Hz, and its second mode of vibration corresponds to the vibration of a transversal wall in the EW direction, with a frequency at 10.91 Hz. In the case of Huayrana, the first mode of vibration corresponds to the vibration in the EW direction of the central wall at 5.52 Hz, and the second mode corresponds to the vibration of the structure in the NS direction at 8.34 Hz.

15.2.1.2 Polygonal stone masonry

The temple of the Three Windows, located at the main terrace in Machu Picchu citadel, was selected as a representative construction of polygonal stone masonry (Fig. 15.11). In this case, only horizontal component sensors were set on the structure and on ground. Two horizontal components in NS and EW directions were measured separately. Two sensors were located on the ground surface near the building, one at the external ground (Ch.9) and another at the inner ground (Ch.8). The bottom part of the stone wall acts like a retaining wall, where the ground level inside the building is higher than the external ground. Also, one sensor was located on top of the building, just above the central window (Ch.7). The right part of Fig. 15.11 displays the locations of points of measurements.

To estimate the dynamic characteristics of the Temple of the Three Windows, the transfer function between Ch.9 (external ground) and Ch.7 (wall top) was computed. Then, transfer function between Ch.8 (internal ground) and Ch.7 (wall top) was also calculated. These results along the EW direction are shown in Fig. 15.12. Predominant frequency of the

FIG. 15.11 Temple of the Three Windows and point of vibration measurements.

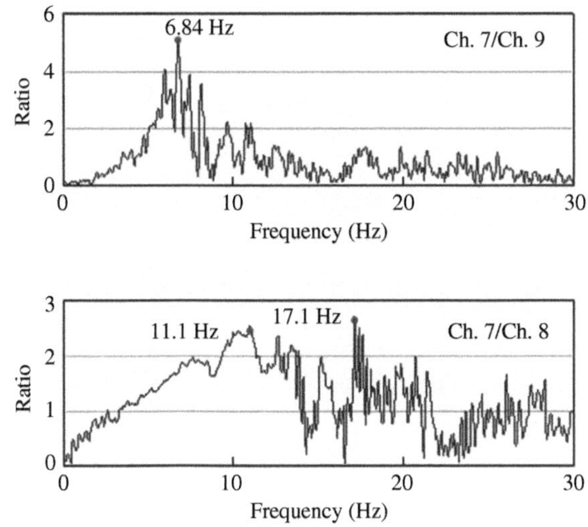

FIG. 15.12 Horizontal transfer functions for the Temple of the Three Windows.

west wall was 6.84 Hz when considering the transfer function from the external ground surface to the top of the wall (Ch.7/Ch.9). When the transfer function from the inner ground surface to the top of the wall is considered, the predominant frequency is 17.1 Hz (Ch.7/Ch.8).

15.2.1.3 Rectangular stone masonry

One building of the archeological park of Pisac, located 33 km from Cusco city on the east, was selected to study the characteristics of a rectangular stone masonry. The selected building is shown in Fig. 15.13. It can be observed that the regular shapes of stone blocks are parallelepipeds, and then the wall surface shows the rectangular shape of block faces. The shape of the building itself is rectangular, with an entrance door and trapezoidal niches and windows. Fig. 15.14 shows the points of measurements that were located on top of the walls (Ch.7, 8, 9, and 10) and on the ground surface (Ch.11).

The vibration characteristics of the selected building correspond mainly to the out-of-plane vibration of the walls. Fig. 15.15 shows the transfer function for EW direction of point (Ch.8 and Ch.10), which are out-of-plane vibrations of longitudinal walls. In this case, the predominant frequency of vibration of the wall with door opening (Ch.10) is 13.28 Hz, while the frequency of the opposite wall is 16.2 Hz.

Fig. 15.16. shows the transfer functions for NS direction, which correspond to the vibration of transversal wall (with smaller length). The predominant frequencies are 25.57 and 14.55 Hz. This difference can be

FIG. 15.13 Target building at Pisac archeological park.

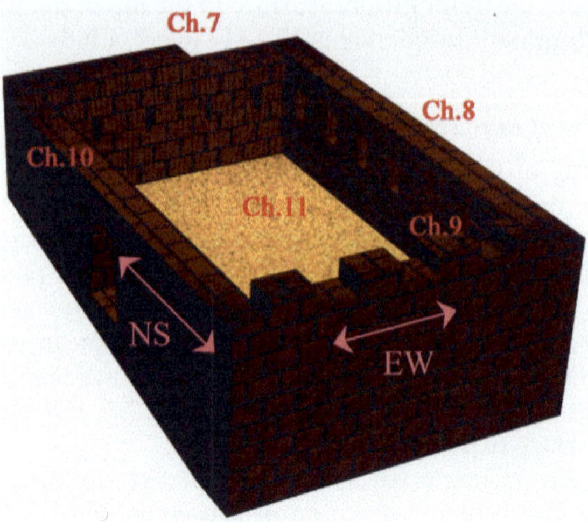

FIG. 15.14 Vibration measurements points for the rectangular stone masonry.

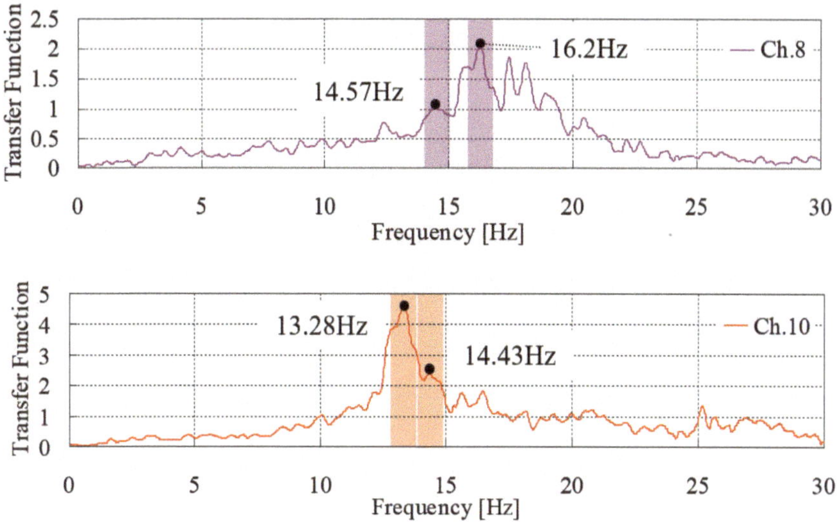

FIG. 15.15 Transfer functions for EW direction (Pisac rectangular stone masonry).

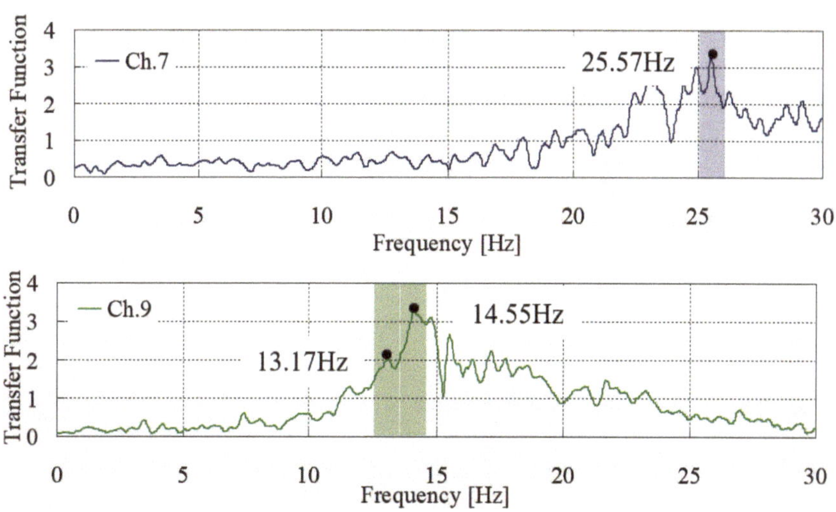

FIG. 15.16 Transfer functions for NS direction (Pisac rectangular stone masonry).

explained by the difference in the height of wall Ch.7 and wall Ch.9. The wall Ch.9 is taller than Ch.7, and therefore it is more flexible and presents a lower frequency.

15.2.2 Finite element modeling and equivalent elastic modulus

An equivalent elastic modulus for each type of masonry was estimated by constructing finite element models and adjusting, by iterations, elastic modulus until an optimal value that reproduces the vibration characteristics comparable with those obtained from ambient vibration measurements is obtained.

Fig. 15.17 shows the finite element model for roughly shaped masonry buildings (Colca and Huayrana). The model was constructed using solid elements. An average elastic modulus of 0.8 kN/mm^2 was used in the analysis. Table 15.1 shows a comparison between measurements results and analysis results, where a reasonable agreement can be observed.

For a polygonal stone masonry, two FEM models were considered, as shown in Fig. 15.18. Because the temple of the Three Windows has the target wall with two heights, the Fig. 15.18A considers the total height from external ground surface until wall top, and Fig. 15.18B considers the bottom part of the wall connected to the inner ground, and therefore the effective height of the wall is considered from the inner ground surface to the wall top. The walls were modeled using solid elements, and the elastic linear analysis was performed to compare the results with those obtained from ambient vibrations. Table 15.2 shows the comparison of measurement results and results from finite element method of analysis. In this case, to obtain these comparable results, an optimum elastic modulus of 1.3 kN/mm^2 was used.

A model for a rectangular stone masonry is shown in Fig. 15.19. In this case, the elastic modulus used for analysis was 1.8. It can be observed that

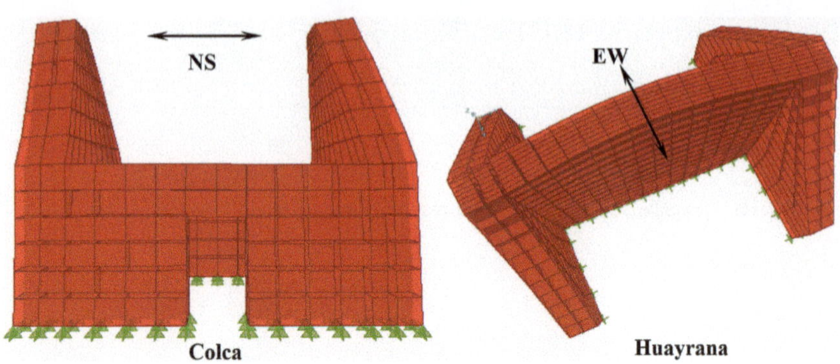

FIG. 15.17 Finite element model for roughly shaped stone masonry.

TABLE 15.1 Measurement results and analysis results for roughly shaped stone masonry.

Building	Direction	Measurements results (Hz)	FEM analysis results (Hz)
Colca	Ch.7 NS	5.46	5.50
	Ch.7 EW	10.91	11.70
	Ch.8 EW	12.03	
	Ch.9 EW	12.62	
Huayrana	Ch.7 EW	5.52	5.49
	Ch.8 EW	5.52	
	Ch.9 EW	5.52	
	Ch.7 NS	8.23	8.07
	Ch.8 NS	8.34	
	Ch.9 NS	8.20	

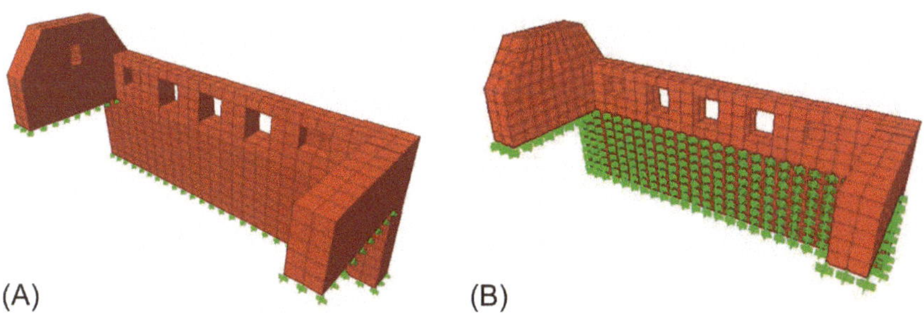

FIG. 15.18 FEM for a polygonal stone masonry.

TABLE 15.2 Measurement results and analysis results for a polygonal stone masonry.

Mode	Measurement results (Hz)	FEM analysis (Hz)
First-mode Model of Fig. 15.18A	6.84	6.93
Second-mode Model of Fig. 15.18B	17.10	17.30

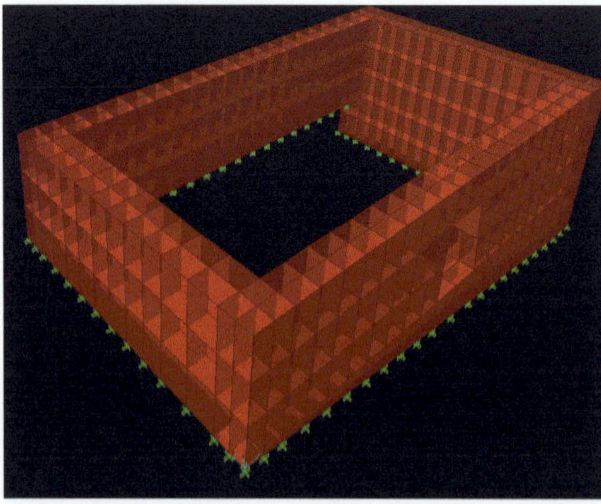

FIG. 15.19 FEM model for a rectangular stone masonry.

walls have different heights due to the difference of the ground levels. All walls are considered as fixed to the ground in their bottom side. A comparison of measurements results and analysis results is shown in Table 15.3. In this case, two modes of vibration are considered for some walls to take into account the multiple peaks observed in the transfer function, which must correspond to multiple modes of vibration.

Table 15.4 summarizes the equivalent elastic modulus for the three types of stone masonry. The roughly shaped stone masonry shows the lowest modulus, which is reasonable, because this kind of masonry uses mud mortar to bond the stone units. The polygonal masonry was expected to have the highest value for equivalent elastic modulus; however, it is the rectangular stone masonry that exhibits the highest value of equivalent elastic modulus. There is an influence of the chosen structure for this

TABLE 15.3 Measurement results and analysis results for the rectangular stone masonry.

Building	Direction	Measurements results (Hz)	FEM analysis results (Hz)
Pisac	Ch.9 NS	13.17	13.28
	Ch.10 EW	13.28	
	Ch.8 EW	14.57	14.56
	Ch.9 NS	14.55	
	Ch.10 EW	14.43	
	Ch.8 EW	16.20	16.04

TABLE 15.4 Comparison of equivalent elastic modulus.

Roughly shaped stone masonry (kN/mm^2)	Polygonal stone masonry (kN/mm^2)	Rectangular stone masonry (kN/mm^2)
0.9	1.3	1.8

result; however, it is believed that the rectangular stone masonry permits a more uniform or homogenous construction in comparison to intricate polygonal shape, resulting in a higher value of an equivalent elastic modulus.

15.3 Influence of the friction coefficient on the seismic behavior of Inca's stone masonry

One notable characteristic of the Inca stone masonry (polygonal and rectangular types) is that finely shaped stone blocks were fitted without any kind of mortar. As mentioned earlier, most of these historical monuments are located on the west side of South America, which is a highly seismic zone. The understanding of the dynamic behavior of this type of masonry is important for its preservation. The shapes of the blocks, as well as the frictional forces generated between these stone blocks without mortar, would generate a particular mechanism to resist lateral forces from earthquakes. This part attempts to explain some of the mechanisms of Inca stone construction through experiments on roughness and on friction coefficient of stone blocks. In situ measurements of the roughness and experimental measurements of the friction coefficients were conducted. A series of shaking table test on a small-scale simple model of stone blocks were performed. Likewise, a mathematical model was developed to estimate the dynamic behavior of simple stone bricks under seismic loads. The theoretical model takes into account the friction forces, as well as the possible impact forces between stone bricks during shaking. According to the results from the mathematical model, a uniform friction would be appropriate to reduce the possible impact forces. However, a high dispersion of the value of roughness (friction coefficient) for the same stone block was obtained from experiments.

To estimate the friction coefficient, in situ measurements were performed in selected building of Machu Picchu citadel and in the Coricancha Temple, in Cusco. To define the characteristic of the model to be used in the shaking table test, surface finishing methods that are used at present (in Japan) were employed to make test models to select the appropriate finishing that reproduces the properties of the Inca stone.

15.3.1 Estimation of the friction coefficient

Two methods were employed in situ to estimate the friction coefficient of Inca stone blocks. First, replicas of the surface were taken by using industrial clay. Then, direct measurements of the surface roughness were performed by using a roughness meter. The selected places for these measurements were the Main Temple and the Temple of the Three Windows in Machu Picchu and the Temple of the Stars in Coricancha Complex in Cusco city.

The selected places in Machu Picchu are shown in Fig. 15.20. Fig. 15.21 shows the Temple of the Stars in Coricancha as well as a detail of the replica taken by using industrial clay.

The conditions of the temples, in Machu Picchu, have permitted to take replicas of the horizontal surfaces between stone layers. In general, replicas of horizontal surfaces and vertical surfaces were obtained from selected locations that were distributed in the respective constructions. To take a replica, the surfaces were cleaned up to take the surface dust, then the industrial clay was warmed to an appropriate temperature (50°C approximately) to obtain adequate plasticity and then was applied on the surface. When the clay gets cold, it takes a solid state and then is

FIG. 15.20 Main Temple and Temple of Three Windows in Machu Picchu.

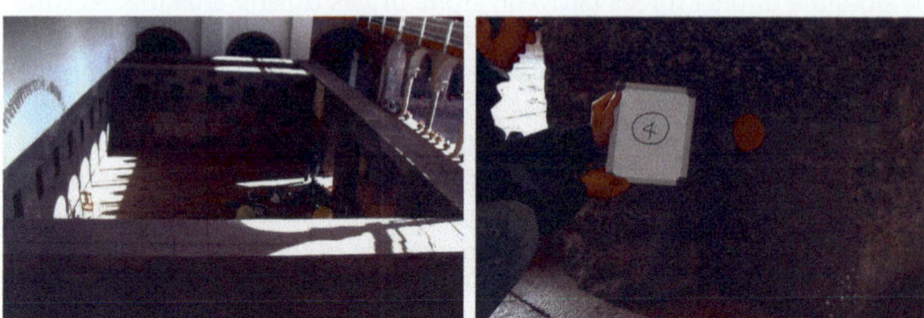

FIG. 15.21 Temple of Stars in Coricancha and detail of replica taking.

retired from the surface and enveloped to be sent to the laboratory. The replicas were analyzed using a tridimensional surface analyzer to obtain the roughness of the replica surface. This roughness is measured as the average of the maximum deformation with respect to the average surface, and it is expressed in micrometers. Results of the estimated roughness are shown in Table 15.5.

Using a portable friction-coefficient-meter, the corresponding values were estimated for 5 points in Machu Picchu and for 4 points in Coricancha. The average results are shown in Table 15.6.

Taking these values as reference, stone surface finishing methods that are used at present in Japan were employed to construct test specimens, and their corresponding roughness and friction coefficients were estimated. The types of finishing used for this research were: bush hammered, dabbed finished, rough grind, rubbing, and polishing. The roughness values and friction coefficients for each type of finishing are shown in Table 15.7.

TABLE 15.5 Stone roughness estimated from replica samples.

Sample	Machu Picchu (μm)	Coricancha (μm)
1	43.26	26.59
2	31.89	51.03
3	32.96	26.70
4	41.83	37.38
5	25.87	27.03
6	36.27	26.45
Average	35.34	32.53

TABLE 15.6 Stone friction coefficient estimated in situ measurements.

Sample	Machu Picchu	Coricancha
1	0.57	0.43
2	0.53	0.39
3	0.41	0.45
4	0.42	0.50
5	0.39	–
Average	0.46	0.44

TABLE 15.7 Roughness and friction coefficient for different types of finishing.

Type of finishing	Roughness (μm)	Friction coefficient
Bush hammered	66.79	0.63
Dabbed finished	30.75	0.48
Rough grind	22.14	0.53
Rubbing	5.68	0.50
Polishing	1.28	0.39

From these results, it can be observed that the dabbed finished type with 30.74 μm of roughness and 0.48 of friction coefficient could be a representative finishing of the Inca stone. Then, this finishing type was selected to construct the model to be used in the shaking table test.

15.3.2 Shaking table test of stone model

To model the friction effect and the forces due to the impact, a simple model that consists of five blocks was selected as specimen for the shaking table test program. As can be observed in Fig. 15.22, the model consists of 3 blocks of 100 × 50 × 100 mm and 2 blocks of 50 × 50 × 100 mm. A base of stone with dabbed finished style was fixed on the shaking table, and the blocks were installed on this base.

As input motion, the signal recorded by the Japan Meteorological Agency during the Kobe earthquake was employed. The input motion was applied five times in successive runs. After each run, the remaining relative displacements between blocks were measured. Fig. 15.23 shows the state of the model after the five runs.

15.3.3 Analytical model for frictional behavior

To develop the numerical model, the blocks are labeled, as is shown in Fig. 15.24. The model considers friction coefficient or frictional forces acting in the horizontal joints between the blocks. In the vertical joints between the adjacent blocks, it is considered that forces due to impact are developed.

For this system, the equation of motion can be expressed as is shown in the following equation:

$$[M]\{a\} + \{f\} = -\{m\}a_g \tag{15.1}$$

FIG. 15.22 Stone model fixed on the shaking table.

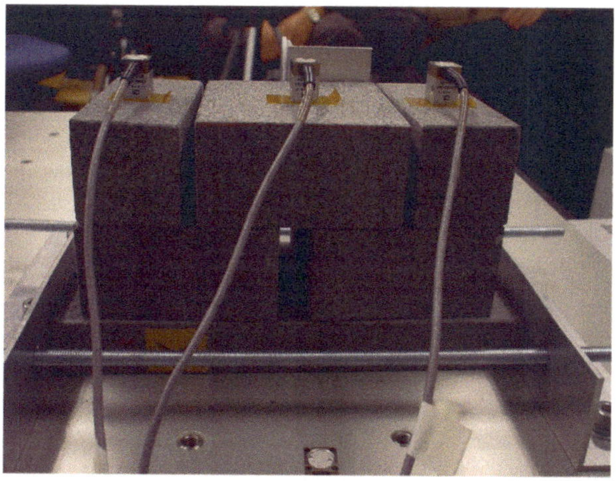

FIG. 15.23 State of the specimen after five runs of Kobe earthquake.

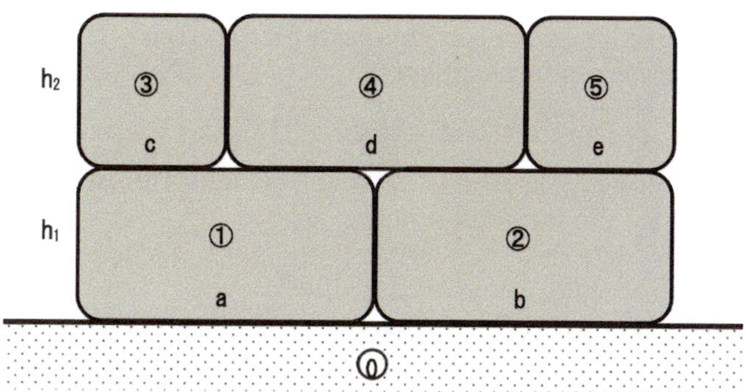

FIG. 15.24 Multiple rigid body model.

where $[M]$ is the mass matrix, $\{a\}$ is the acceleration vector, $\{f\}$ is the force vector due to the friction between blocks, and a_g is the ground or input acceleration. The mass matrix is a diagonal matrix with the masses of each block as elements.

The force vector due to the friction is evaluated considering the equilibrium of the dynamic forces for the multiply rigid body system and is given by the following equation:

$$\{f\} = \begin{Bmatrix} f_1 \\ f_2 \\ f_3 \\ f_4 \\ f_5 \end{Bmatrix} = \begin{Bmatrix} F_{10}\,\mathrm{sgn}v_1 - F_{31}\,\mathrm{sgn}(v_3 - v_1) - F_{41}\,\mathrm{sgn}(v_4 - v_1) \\ F_{20}\,\mathrm{sgn}v_2 - F_{42}\,\mathrm{sgn}(v_4 - v_2) - F_{52}\,\mathrm{sgn}(v_5 - v_2) \\ F_{31}\,\mathrm{sgn}(v_3 - v_1) \\ F_{41}\,\mathrm{sgn}(v_4 - v_1) + F_{42}\,\mathrm{sgn}(v_4 - v_2) \\ F_{52}\,\mathrm{sgn}(v_5 - v_2) \end{Bmatrix} \tag{15.2}$$

here, the F_{ij} represents the friction force between the upper block i and the lower block j. In the analysis, it is considered that the forces become active only in the case that the input acceleration is larger than the friction factor multiplied by the acceleration of gravity. That is, the friction force acts in the dynamic equation only when the input force is larger than the static friction force. Furthermore, the sign of the friction forces depends on the sign of the relative velocities between the blocks. These friction forces F_{ij} are calculated by considering the weight of the blocks and the assumed friction coefficient μ. The weight of each block is calculated from the dimension of blocks and for a given specific weight of the material γ, in this case, stone material. Therefore the following expressions are used to estimate these frictional forces:

$$\begin{aligned} F_{10} &= [at(h_1 + h_2)\gamma]\mu_{10} \\ F_{20} &= [bt(h_1 + h_2)\gamma]\mu_{20} \\ F_{31} &= [cth_2\gamma]\mu_{31} \\ F_{41} &= [(a-c)th_2\gamma]\mu_{41} \\ F_{42} &= [(b-e)th_2\gamma]\mu_{42} \\ F_{52} &= [eth_2\gamma]\mu_{52} \end{aligned} \tag{15.3}$$

The equation of motion (15.1) is solved numerically by means of the Newmark integration scheme. In this case, for the step $i+1$, the expression for the equation of motion is given by:

$$[M]\{a\}_{i+1} + \{f\}_{i+1} = -\{m\}a_{g_{i+1}} \tag{15.4}$$

Then, the following equations for displacement, velocity, and acceleration are used for the iteration scheme:

$$\begin{aligned} \{d\}_{i+1} &= \{d\}_i + \Delta t\{v\}_i + \frac{\Delta t^2}{2}\{a\}_i \\ \{v\}_{i+1} &= \{v\}_i + \frac{\Delta t}{2}(\{a\}_i + \{a\}_{i+1}) \\ \{a\}_{i+1} &= [M]^{-1}\left[-\{m\}a_{g_{i+1}} - \{f\}_{i+1}\right] \end{aligned} \tag{15.5}$$

In addition to the friction forces that are used in the equation of motion, the effect of impact forces between the adjacent blocks are considered. For this initial analysis, elastic impact is assumed, and therefore the coefficient of restitution is equal to 1.

15.3.4 Comparison of analytical and experimental results

Experimental results are shown in Fig. 15.25. Here the residual displacement of each block after the five successive inputs of Kobe earthquake motion is shown. Each set of input motion was called run1, run2, run3, run4, and run5, respectively. All the runs are performed at similar conditions; however, the final displacement differs for each block. In spite of that, the general pattern with large displacements of blocks 3 and 4 and a smaller displacement of block 2 is observed in all runs.

In the case of the analysis, the friction coefficients were first selected randomly around 0.4 with a variation of ±0.02. The results of this analysis are shown in Fig. 15.26. The order of the final displacements around 20 mm agreed with those of the experimental results. However, the general pattern is completely different with respect to the experimental results, with larger displacements in blocks 1 and 2.

The use of almost uniform friction coefficient in the analysis gave as a result that the displacement pattern of each block differs from that obtained experimentally. Therefore a series of sets of friction coefficients with larger variation and randomly selected were used for a new analysis. Approximately, 50 sets were used for analyses, and it was observed that the variation of the friction coefficient has an important influence in the results. From this larger amount of trials, three runs are selected, and their

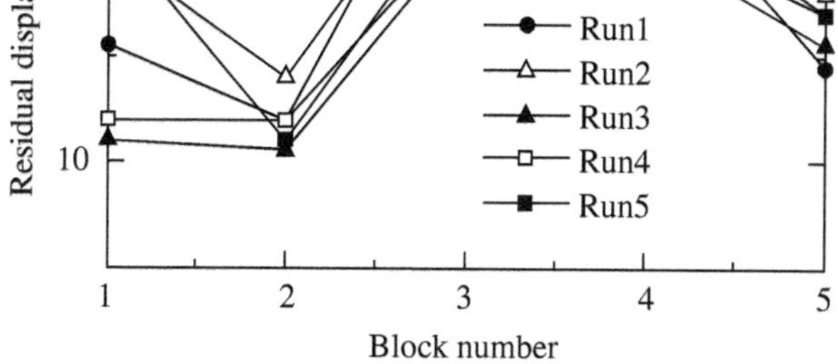

FIG. 15.25 Final displacement after a set of Kobe earthquake input motion.

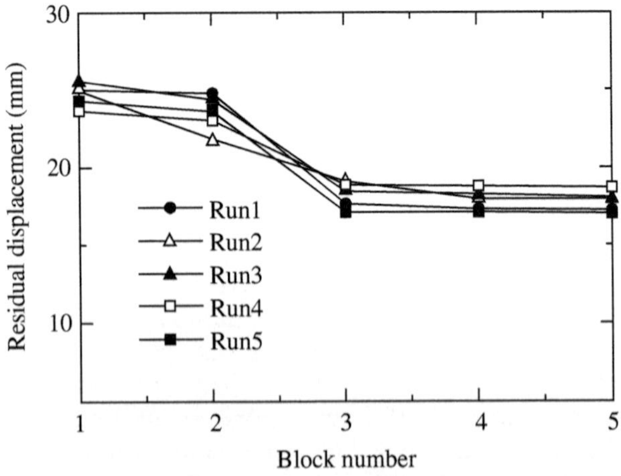

FIG. 15.26 Analytical results with friction coefficient values around 0.4.

results are presented in Fig. 15.27. In Table 15.8, friction coefficients used for these last analyses are shown. As can be observed from Fig. 15.27, it is possible to get a similar pattern of the final state of the blocks compared with experimental results by controlling or changing the values of friction coefficients. It can be noted that larger friction coefficients in the contact of lower blocks and base produce results that agree better with experimental results. Probably, in the case of experiment, the friction at the base is larger than the friction between the blocks because these elements were prepared separately, and owing to the difference in size between the blocks and the base, the conditions for polishing are different.

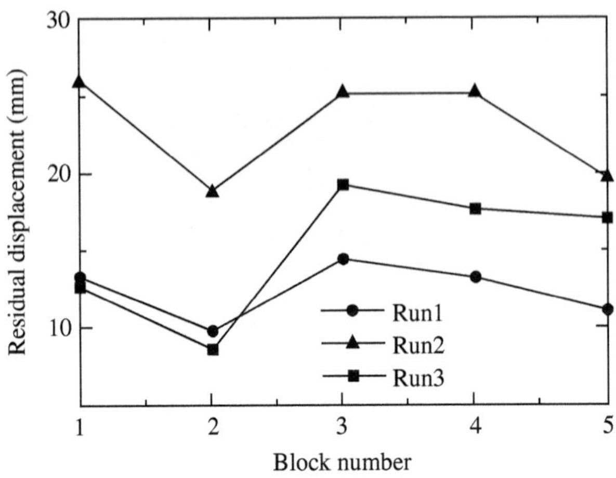

FIG. 15.27 Analytical results with a large variation in the friction coefficient.

TABLE 15.8 Friction coefficients used in analysis with a large variation of the average value.

Set	μ_{10}	μ_{20}	μ_{31}	μ_{41}	μ_{42}	μ_{52}
Run1	0.40	0.50	0.30	0.44	0.44	0.31
Run2	0.47	0.30	0.50	0.39	0.32	0.35
Run3	0.43	0.51	0.52	0.37	0.29	0.45

From the evaluation of the friction coefficient of the Inca stone masonry block, it was found that values ranging from 0.39 to 0.57 were obtained from in situ measurements. These characteristics of friction and roughness agreed well with a present technique used for stone polishing named dabbed finishing.

Experimental small size model was subjected to a series of input motion to simulate its seismic behavior and an analytical model of multiple rigid bodies was employed to compare with experimental results. The analytical model takes into account the friction between the stone blocks and the action of impact forces between the adjacent blocks.

Considering, in the analysis, almost uniform friction among the blocks, the final state of the blocks differs from that obtained experimentally. However, random large variation of these coefficients could reproduce the experimental behavior. Therefore the great influence of the friction coefficient in the response of a simple model of stone masonry was confirmed. Variation of these factors gives, as a result, different patterns of response.

15.4 Case study: Coricancha architectonic complex

Coricancha, probably, was the most important Inca temple complex dedicated primarily to Inti, the Sun God. At present, a small part of the Inca stonework is all that remains of the ancient complex. Most of the complex was demolished by Spaniard conquistadors in the 17th century to make way for the construction of Catholic Church of Santo Domingo on this site, using part of the Inca construction as the foundation. This complex is a fine example of how Inca stonework had been incorporated into the structure of a colonial building. Major earthquakes have severely damaged the church, but the Inca stone walls, built out of huge, finely cut tightly fitting blocks of stone, still stand as a testimony to Inca's architectural skills and sophisticated stone masonry practice.

In this complex, a series of microtremor measurements were carried out to estimate the dynamic characteristics of the structures and the ground. Measurements on buildings include the Inca's masonry walls as well as the tower of the colonial church. Measurements to estimate the ground characteristics were performed in the inner yard and at the outer garden of the complex. In this part, the results of the measurements and corresponding interpretations are discussed. It was found that the predominant period of vibration of the church tower is comparable to the predominant period of the ground, and therefore, the tower appears to be more vulnerable to earthquake occurrence because of the possible resonance phenomenon.

15.4.1 Description of the Coricancha temple complex

It is believed that the complex was dedicated to Inti, the Sun God, and was formed by a group of constructions located around a central square, as can be observed in Fig. 15.28, which is a view of a scale model. Each building was dedicated to one god, respectively. The name of the god corresponding to each temple is indicated in Fig. 15.28. These gods were Inti (Sun), Killa (Moon), Chasca (stars), and Illapa (thunder, rainbow). Also, it is thought that the Incas believed in a creator god, called Wiracocha.

The Coricancha complex was built on a natural hill and between the rivers Shapy-Huatanay and Tullumayo. The construction of the complex is commonly attributed to Pachacutec Inca Yupanqui, around AD 1450. The architectural distribution of the temples was done inside of a space that was circled by high walls. The shapes of the walls are trapezoidal and have a vertical inclination, which is typical of the Inca's architecture. The stones used in this complex were extracted from the quarries Waqoto and Rumicolca that are located 20–30 km away from Cusco city. The stones used by the Incas were plutonic diorite rocks and calcareous rocks. According to the chroniclers, the walls of the temples were covered with gold plates molded with drawings to represent the corresponding god.

FIG. 15.28 Scale model of the Coricancha complex.

In the 16th century, the Christian Catholic Church was built over the Inca stone structure. For the construction of the church, the Spaniards destroyed some part of the temples that formed the complex, and some parts were used as a foundation of the Catholic Church. In this way, the Inca stone structure was incorporated to the colonial construction, resulting in a magnificent monument that shows a mixture of Spanish and Inca cultures. Fig. 15.29 shows the plan view of the current building, where the black part represents the remaining Inca's walls, and the gray part represents the colonial construction.

15.4.2 Ambient vibration measurements

The stars and the thunder temples, the tower of the colonial church, and the inner at outer yards were selected for microtremor measurements. These locations are indicted by encircled numbers in Fig. 15.29. The meaning of the numbering and the purpose of the measurements are described as follows.

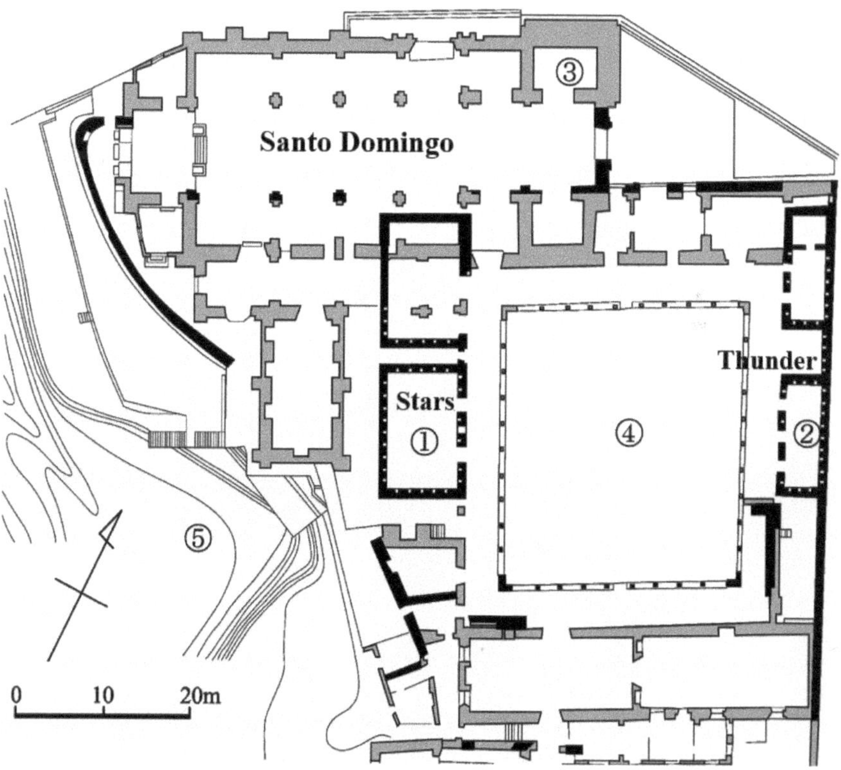

FIG. 15.29 Plan view of the Coricancha complex.

① Star Temple (Chasca): Structure vibration. Measurements of horizontal vibrations in the two principal directions of the building were performed. The sensors were located on the floor level and at walls on all four sides. The sensors at the wall were located on the sill of rectangular offset on the inner wall face closest to the center of the wall. The directions were designated NS and EW, respectively. However, the actual directions are Northeast-Southwest (regarded as NS) and Northwest-Southeast (regarded as EW). The channels for measurement were denominated ch7, ch8, ch9, ch10, and ch11, as shown in Fig. 15.30.

② Temple of Thunder (Illapa): Structure vibration. In a similar manner to the Temple of Stars, measurements of horizontal vibrations in the two principal directions of the building were performed. The sensors were located on the floor level and on walls on four sides. The sensors at the wall were located on the sill of rectangular offset on the inner wall face closest to the center of the wall. Again, the directions were designated NS and EW, respectively.

③ Tower of Santo Domingo Church (Fig. 15.31): Structure vibration. Three components of vibration were measured at the top and bottom of the tower. The directions for measurements are indicated in Fig. 15.31.

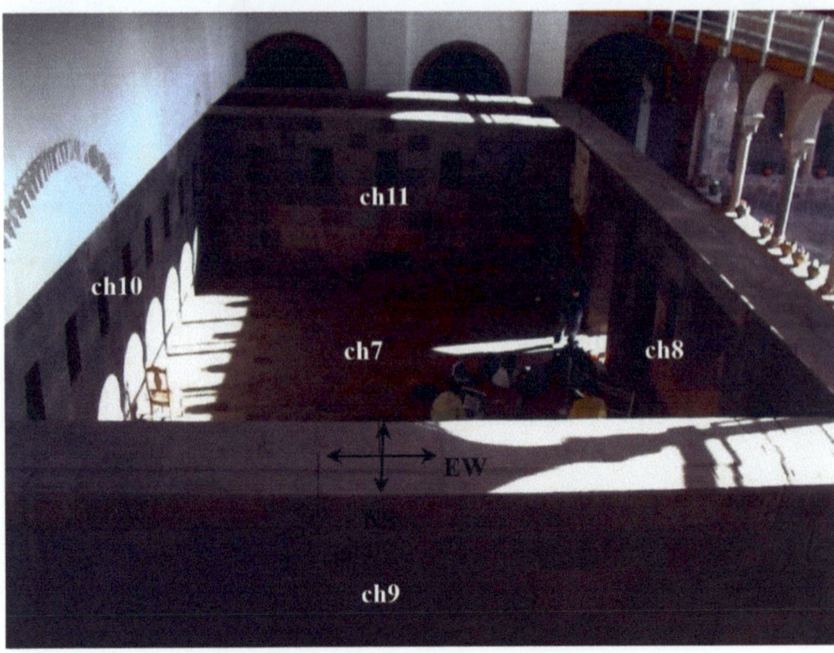

FIG. 15.30 Point of microtremor measurements at the Temple of Stars.

FIG. 15.31 Tower of the Santo Domingo Church.

④ Interior Yard (Ceremonial Fountain): Ground vibration measurement. At this site, array measurements of the vertical components of the ground vibration were carried out. The triangular array configuration of sensors on the yard consisted of approximately 25 m sides triangles. The array measurement is intended for F-K spectral analysis (F: frequency, K: wave number) to obtain the Rayleigh wave dispersion curve, from which the shear wave velocity profile of the ground may be estimated by the inverse analysis.

⑤ Exterior Garden (Sacred Garden of Sun): Ground vibration measurement. In a similar manner to the interior yard, array measurements of the vertical components of the ground vibration were carried out here. The configuration of sensors again consisted of a triangular array with sides of approximately 30 m. As noted earlier, an array measurement is intended for F-K spectral analysis.

15.4.3 Analysis of measurement data

The location of the sensors on the Temple of Stars is indicated in Fig. 15.30. Channels 7–11 were measured simultaneously during 500 s. Then, stable portions of the records were selected to perform the Fourier analysis that permits to estimate the natural period of vibration of the structure. The results of the Fourier analysis are shown in Fig. 15.32. Fig. 15.32A and C show the Fourier spectrum for the EW and NS directions, respectively. The Fourier amplitude spectrum at each measured point of the structure was divided by the one at the ground level, to obtain only the vibration characteristics of the structure. It can be observed from Fig. 15.32A and B that the value of the predominant frequency in the EW direction ranges from 14 to 17 Hz. In the case of the NS direction, these values range from 12 to 16 Hz, as is shown in Fig. 15.32C and D. These differences are due to the presence of a colonial wall on the east wall of the Star Temple. On the other hand, the length of the wall differs for the EW direction and NS direction, and therefore each wall vibrates with different frequency. Therefore the vibration of the structure is intricate, and the predominant frequency is relatively high.

Fig. 15.33 presents the results of the Fourier analysis for the horizontal vibration of the Colonial tower. The directions NS and EW are indicated in Fig. 15.32. The Fourier amplitude spectrum obtained at the top of the tower was divided by the one obtained at the ground level, to obtain only the vibration characteristics of the structure. It can be observed that the value of the predominant frequency in the EW direction is 2.7 Hz and in the NS direction is 3.3 Hz.

Fig. 15.34 presents the dispersion curve obtained from the F-K analysis. This result permits to estimate the predominant frequency of the ground as 2.7 Hz. Then, if the result is compared with that obtained for the tower of the Colonial Church, it can be observed that the frequency of the structure and soil are similar, and therefore the resonance phenomenon is expected to occur. This coincidence of the predominant frequencies could be the explanation of the repeated failure of the tower during the past earthquakes.

The ambient vibration measurements performed in representative structures of Coricancha complex have provided the basis for evaluation of the dynamic characteristics of the Inca stone structure and Colonial stone structure. The measurement at the Inca stone part shows an intricate vibration due to the presence of Colonial walls. It was found that the

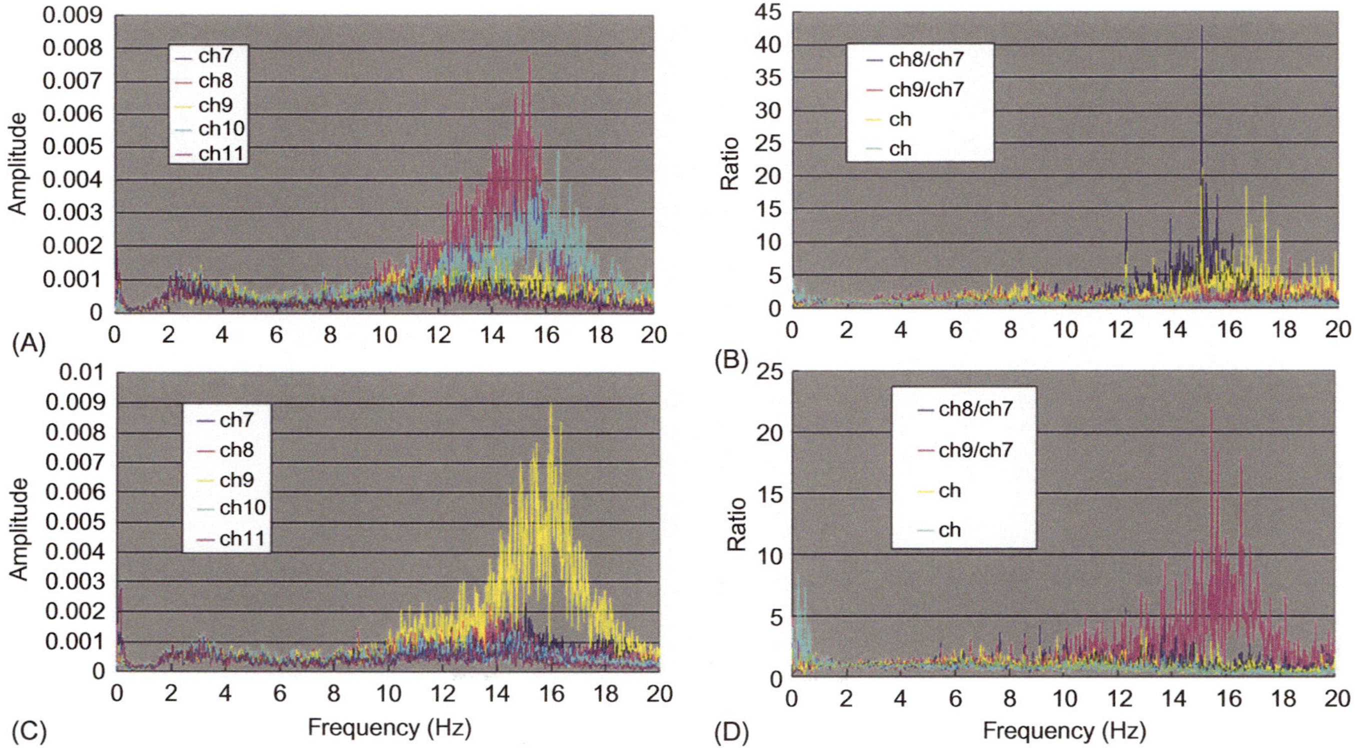

FIG. 15.32 Fourier analysis results for the Temple of Stars.

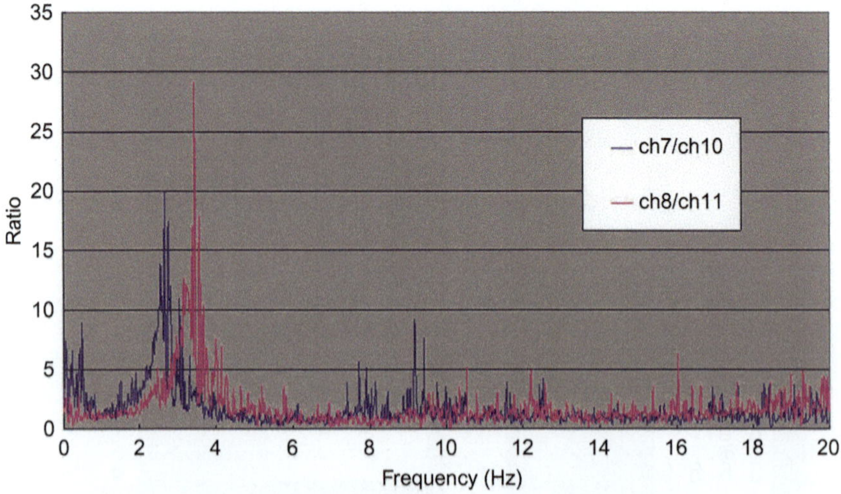

FIG. 15.33 Fourier analysis results for the tower of Santo Domingo Church.

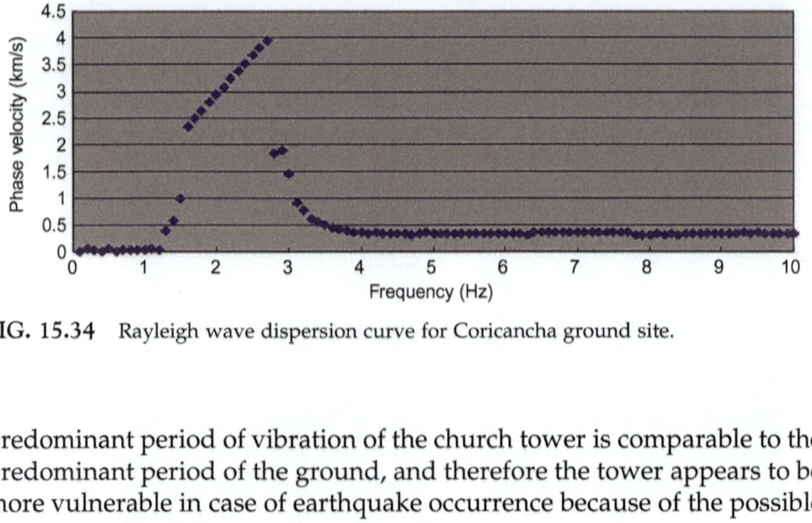

FIG. 15.34 Rayleigh wave dispersion curve for Coricancha ground site.

predominant period of vibration of the church tower is comparable to the predominant period of the ground, and therefore the tower appears to be more vulnerable in case of earthquake occurrence because of the possible resonance phenomenon to be expected.

15.5 Case study of Machu Picchu citadel

In the year 2011, Peru celebrated 100 years since Hiram Bingham discovered the citadel of Machu Picchu. About 500 years ago, the Inca culture reached the peak in its development, which was just before the arrival of Spaniard conquistadors. By the time, the Incas had integrated a vast empire that stretched from the Maule River in Chile to the northern Ecuador along the western side of the Andes Mountain. This territory, as in present days, had been continuously exposed to natural disasters, such as excessive rainfalls, earthquakes, landslides, and floods. In spite of such impending disasters, the Incas were able to develop techniques of construction to withstand such natural forces. The awe-inspiring cities and road networks that remain intact to this day serve as witness to their acumen in construction. The Inca citadel of Machu Picchu is the most famous historical site in Peru, a site that was declared as world cultural heritage by the UNESCO in 1983 and also as one of the New Seven Wonders of the World in 2007. All of its constructions were made of stone masonry. The citadel was constructed by Inca Pachacutec and was abandoned after collapsing the Inca Empire by the year AD 1540. During almost four centuries, the citadel endured and survived under a thick rain forest, until it was discovered by Hiram Bingham in 1911.

The structural system of Inca constructions involves various types, including adobe (sun-dried clay bricks), roughly shaped stones laid down with mud mortar, and finely shaped stones. They also used mud and clay as mortar for surface finishing. Finely shaped stone masonry was used for important buildings like temples, administrative structures, and king's residences. In this type of construction, the adjacent stones are carefully shaped and fit snugly against each other without the use of mortar.

This part attempts to describe mainly the seismic vulnerability of Machu Picchu constructions by observing the structural conditions of historical stone buildings located in the citadel. Although the Andean region of Peru corresponds to a zone of low seismic activity, these low-intensity earthquakes could produce damages in the historical stone constructions. Moreover, the present structural conditions of some buildings lead to think that these constructions have been subjected to the action of past earthquakes or other geodynamic actions. The characteristics of failures are discussed in this part, through field survey, ambient vibration measurements, and dynamic analysis. The field survey on damages consisted of a detailed observation and evaluation of a group of selected buildings. Also, ambient vibration measurements were performed on the ground and on some buildings to estimate their dynamic characteristics. It was observed that for a specific soil condition, the failure in some buildings could be the result of the action of past earthquakes.

15.5.1 Target buildings and evaluation method

Machu Picchu citadel is divided into two zones, namely the Agricultural Zone and the Urban Zone. In this research, a group of stone constructions located in the "Urban sector" of Machu Picchu is analyzed to evaluate the structural condition of buildings. Some buildings are made of stone masonry laid with mud mortar and others are constructions made of finely shaped stone blocks without mortar. For the structural survey, an evaluation sheet was prepared to record the characteristics of each building and describe its structural condition. This survey has permitted to establish that different level of damages can be observed in the stone constructions. In Figs. 15.35–15.39, some data sheets are shown, where main observations of the evaluation are written. It is inferred that some type of damages could be the consequence of the action of previous earthquakes. Damages due to settlement of the underlying ground were also observed. Special attention is focused in two buildings that are located in a terrace called sacred square. These buildings are presented in Figs. 15.38 and 15.39 and are called Temple of the Three Windows and Main Temple, respectively. Both buildings were selected for ambient vibration measurements.

15.5.2 Ambient vibration measurements

The Main Temple and the Temple of the Three Windows are located on the Sacred Plaza sector, which is a terrace. Details of the Temple of the Three Windows and Main Temple were presented in Figs. 15.38 and 15.39. The Main Temple shows damages of displacements between stone blocks of the north wall due to the settlement of the east wall. In the case of the Temple of the Three Windows, relative horizontal displacements of stone blocks can be observed.

Microtremor measurements were performed in the Main Temple and in the Temple of the Three Windows. In the case of the Main Temple, three transducers (two horizontal component sensors and one vertical component sensor) were set at the top of wall and at the ground level, respectively. For the Temple of the Three Windows, only horizontal component sensors were set on the structure and over the ground.

Fig. 15.40 shows the H/V spectral ratio obtained from microtremor measurements at the ground of Sacred Plaza. A predominant frequency of 23.7 Hz was estimated, which would mean that the depth of the bedrock is shallow. Usually, the natural frequency of the ground ranges between 1 Hz for soft soil deposits and 10 Hz for hard soil deposits. Therefore the estimated predominant frequency of 23.7 Hz indicates that the thickness for soft soil layer would be very thin (around 1 m), and that the bedrock would be close to the ground surface. In consequence, these buildings would be founded directly on the bedrock. Therefore the apparent soil settlement failure observed in Fig. 15.39 would correspond to a failure or rupture of the underlying rock.

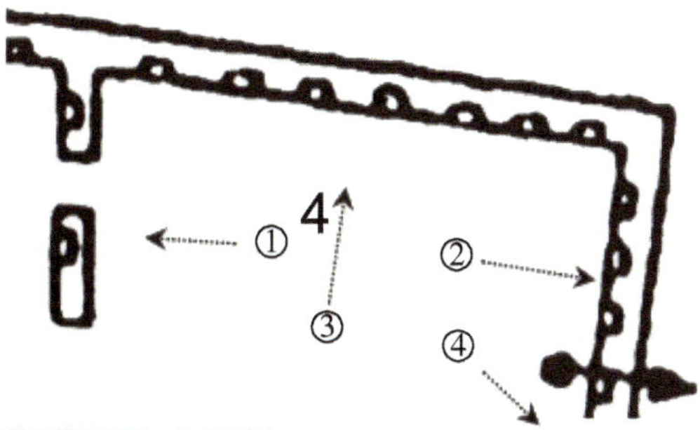

Photo ①

Photo ②

Photo ③

Photo ④

Type of structure: Fine shaped stone
Location: Inclined ground
Wall inclination: Slightly
Damages on the upper part: Yes
Collapse of walls: Partial
Wall damages: Stone slip in 25 sites
- Gaps of 5~10[mm]: 4 places
- Gaps of 10~15[mm]: 1 place
- Gaps of 15~20[mm]: 6 places
- Gaps of 20~25[mm]: 5 places
- Gaps of more than 25[mm] : 9

FIG. 15.35 Damage evaluation at Building No. 2-4.

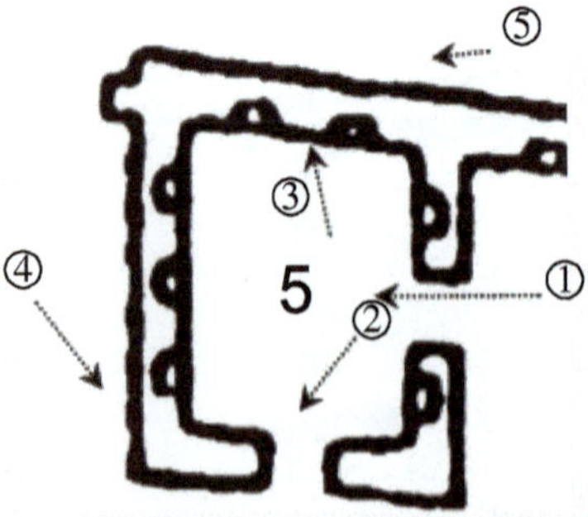

Photo ①

Photo ②

Photo ③

Photo ④

Photo ⑤

Type of structure: Fine shaped stone combined with stone laid with mud mortar
Location: Inclined ground
Wall inclination: Slightly
Damages on the upper part: Yes
Collapse of walls: Gable wall collapse
Wall damages: Stone slip in 35 sites
  Gaps of 10~15[mm]: 6 place
  Gaps of 15~20[mm]: 3 places
  Gaps of 15~20[mm]: 6 places
  Gaps of more than 25[mm] : 25 places

The construction apparently had triangular gable walls to form the roof of the buldings. These gable walls has collapsed and it is believed that the overturning was due to the action of horizontal force acting perpendicularly to the wall plane.

FIG. 15.36 Damage evaluation at Building No. 2-5.

Structure: Fine shaped stone and stone laid with mud mortar (6.44×4.66[m])
Inclined ground, plane terrace
No wall inclination but damages on the upper part (Gable wall collapse)
Wall damages: Stone slip in 16 sites
Gaps of 5~10[mm]: 6 places
Gaps of 10~15[mm]: 8 place
Gaps of 15~20[mm]: 1 places
Gaps of 20~25[mm]: 1 places

FIG. 15.37 Damage evaluation at Building No. 3-3.

Type of structure: Fine shaped stone
Dimensions: 4.53 × 10.4 [m]
Location: Hill terrace
Wall inclination: Slightly
Damages on the upper part: Yes
Collapse of walls: Partial
Wall damages: Stone slip in 12 sites
Horizontal slips of big stone blocks could be due to the action of pase earthquakes

FIG. 15.38 Damage evaluation at Building No. 4-2 (Temple of the Three Windows).

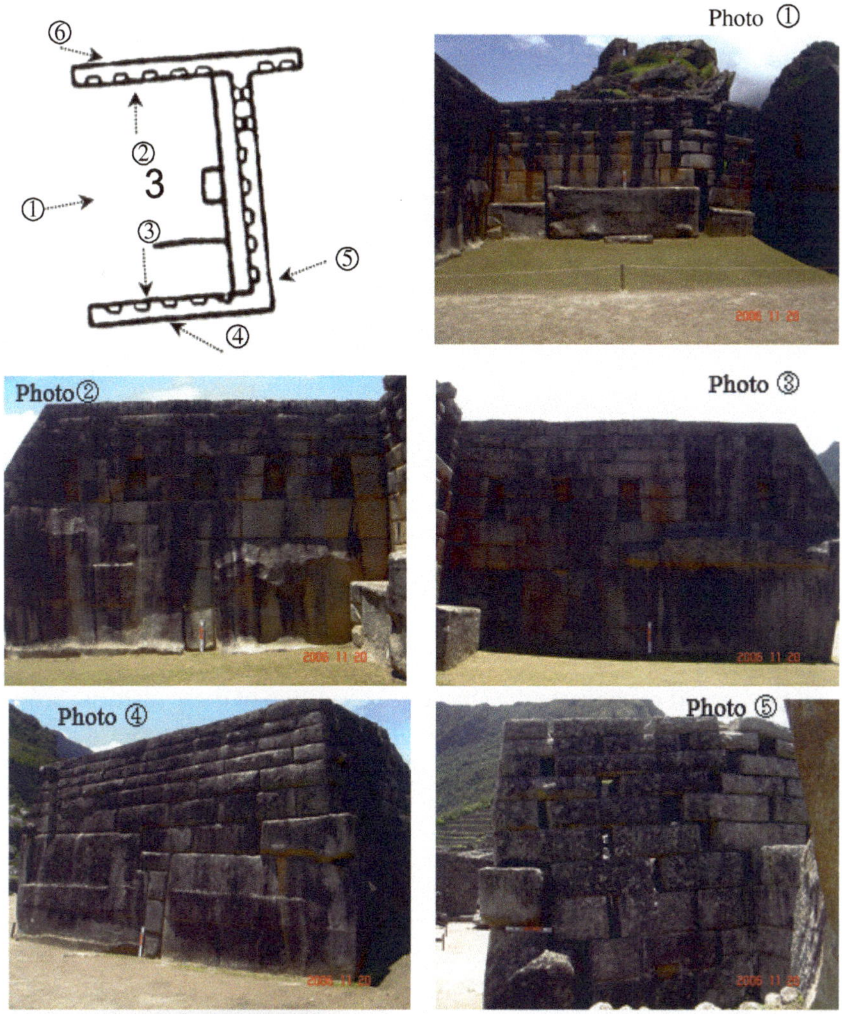

Type of structure: Fine shaped stone
Dimensions: 6.40 × 7.95 [m]
Wall inclination: Slightly
Damages on the upper part: Yes
Collapse of walls: Partial
Wall damages: Stone slip in 36 sites
- Gaps of 25~50[mm]: 11 places
- Gaps of 50~100[mm]: 7 places
- Gaps of 100~200[mm]: 13 places
- Gaps of 200~300[mm]: 5 places

FIG. 15.39 Damage evaluation at Building No. 4-3 (Main Temple).

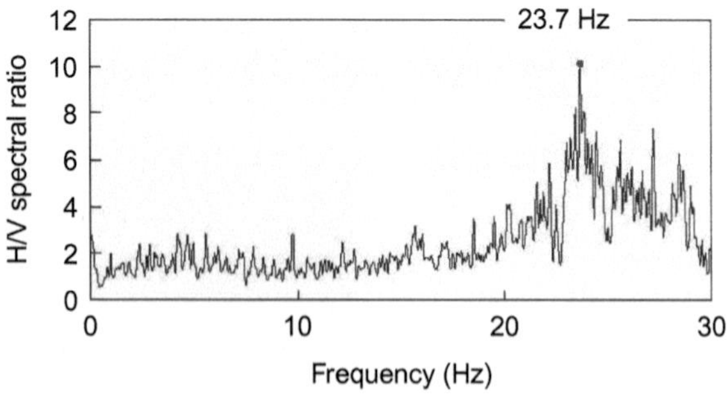

FIG. 15.40 H/V spectral ratio at the ground of Sacred Plaza.

In the case of the Temple of the Three Windows, two horizontal components in NS direction and EW direction were measured separately. The results of these measurements were used to estimate the equivalent elastic modulus for polygonal stone masonry, and the results can be found in Part 2 of this chapter. It is important to remark that the horizontal sliding between stone blocks in the Temple of the Three Windows could be a consequence of strong seismic forces of past earthquakes.

The field survey performed on selected buildings of Machu Picchu citadel has served to evaluate the structural condition of these historical constructions. Damages were detected in many buildings and, an evaluation to establish the causes of these damages was done in this research. As examples of this evaluation, the results of the selected buildings are shown.

Slips of stone blocks are observed as gaps between the adjacent blocks in stone masonry walls. In some cases, the movement of the blocks could be due to the settlement of the underlying ground. However, in some cases, there is no ground settlement, and therefore it is supposed that the movement of the blocks could be the result of the action of past earthquakes.

Probable action of past earthquakes on the failure of these buildings or portion of these historical buildings can be inferred from the observation of the structural condition of gable walls. Most of them have collapsed, and from a detailed observation of apparently undamaged walls, it can be observed that most of them have been reconstructed.

The evaluation of the settlement observed in the Main Temple shown in Fig. 15.39 (Building No. 4-3) conduces to think that it is probably due to the fracture of the subjacent bedrock in which the building is founded. Apparently, the bedrock is very shallow, and the fracture of this rock can be explained only by the action of large geodynamic forces.

The particular horizontal sliding between the stone blocks at one wall corner of the Temple of the Three Windows, shown in Fig. 15.38 (Building No. 4-2, Photograph ⑥), could be a consequence of strong seismic forces of past earthquakes.

Ambient vibration measurements performed, in this case, in two temples of the Sacred Plaza of Machu Picchu provided valuable information for evaluating the dynamic characteristics of Inca stone heritage structures.

References

Cuadra, C., Karkee, M., Ogawa, J., & Rojas, J. (2004). An evaluation of earthquake risk to Inca's historical constructions. In: *13th World conference on earthquake engineering, Vancouver, B.C., Canada, August 1–6, 2004, CD ROM*. Paper No. 150.

Cuadra, C., Tokeshi, K., & Rojas, J. (2008). Seismic damage to stone structures in Machu Picchu. In: *Proceedings of the 14th World conference on earthquake engineering, October 12–17, 2008, Beijing, China*. Paper No. S11-001.

Further reading

Bingham, H. (1948). *Lost city of the Incas: The story of Machu Picchu and its builders*. Westport, CT: Greenwood Press (reprint 1981) a division of congressional information service, Inc..

Cuadra, C. (2011). Vulnerability of Machu Picchu citadel 100 years after its scientific discovery. In: *12th International conference on structural studies, repairs and maintenance of heritage architecture, STREMAH XII, Chianciano Terme, Italy, September 2011*.

Cuadra, C. (2015). Structural condition of Peruvian heritage architecture to resist earthquake actions. In *The 2015 World congress on advances in structural engineering and mechanics (ASEM15). Paper ID: ES366-1879, 25–29 August 2015 at Songdo Convensia in Incheon, Korea*.

Cuadra, C., & Furui, Y. (2018). Preliminary study for retrofitting of a historical wooden structure using base isolation system. *International Journal of Heritage Architecture, 2*(1), 81–89. https://doi.org/10.2495/HA-V2-N1-81-89.

Cuadra, C., Karkee, M. B., & Tokeshi, K. (2008). Earthquake risk to Inca's historical constructions in Machupicchu. *Advances in Engineering Software, 39*, 336–345. Computational Modelling of Historical Constructions, Elsevier. Published online 14 February 2007 (2008). www.sciencedirect.com.

Cuadra, C., Saito, T., & Zavala, C. (2013). Diagnosis for seismic vulnerability evaluation of historical buildings in Lima, Peru. *Journal of Disaster Research, 8*(2), 320–327. Special Issue on Enhancement of Earthquake and Tsunami Disaster Mitigation Technology in Peru.

Cuadra, C., Saito, T., Zavala, C., & Diaz, M. (2010). The challenges of protect historical adobe constructions in Peru. In: *14th European conference on earthquake engineering (14ECEE), Ohrid, Macedonia, August 30–September 3, 2010*. Paper Number: 392.

Cuadra, C., Sato, Y., Tokeshi, J., Kanno, H., Ogawa, J., Karkee, M., & Rojas, J. (2005). Preliminary evaluation of the seismic vulnerability of the Inca's Coricancha temple complex in Cusco. In *Ninth international conference on structural studies, repairs and maintenance of heritage architecture, STREMAH IX, Malta, June 2005* (pp. 245–253).

Cuadra, C., Yamada, Y., & Narumi, R. (2013). Evaluation of dynamic properties of traditional wooden shrines in Yurihonjo, Japan. *WIT Transactions on The Built Environment*. Vol. 131. WIT Presshttps://doi.org/10.2495/STR130391.

Dimova, S. (1999). Modeling of collision in sliding systems subjected to dynamic excitations. In: *Proceedings of the fourth European conference on structural dynamics EURODYN99, Prague, Czech Republic* (Vol. I, pp. 175–179).

Karkee, M., Cuadra, C., & Sunuwar, L. (2005). The challenges of protecting heritage architecture in developing countries from earthquake disasters. In: *Ninth international conference on structural studies, repairs and maintenance of heritage architecture, STREMAH IX, Malta, June 2005*, pp. 407–419.

Ogawa, J., Cuadra, C., Karkee, M., Tokeshi, K., Kanno, H., & Sato, Y. (2006). Influence of the friction coefficient on the seismic behavior of Inca stone masonry. In: *First European conference on earthquake engineering and seismology (a joint event of the 13th ECEE & 30th General Assembly of the ESC), Geneva, Switzerland, September 3–8, 2006*. Paper Number: 709.

Taguchi, T., & Cuadra, C. (2014). Mechanical properties of brick masonry with different types of bonds. In *15th European conference on earthquake engineering, Istanbul, Turkey, Paper No.400*.

Wright, K. R., & Valencia, A. (2000). *Machu Picchu: A civil engineering marvel*. Reston, VA: American Society of Civil Engineers ASCE PRESS.

Index

Note: Page numbers followed by *f* indicate figures, and *t* indicate tables.

MIX
Papier | Fördert
gute Waldnutzung
FSC® C083411

Kolibri 360 GmbH
Ferdinand-Jühlke-Straße 7
99095 Erfurt, Deutschland
produktsicherheit@kolibri360.de